Plant Ecology: Concepts and Applications

Plant Ecology: Concepts and Applications

Edited by Jude Boucher

SYRAWOOD
PUBLISHING HOUSE

New York

Published by Syrawood Publishing House,
750 Third Avenue, 9th Floor,
New York, NY 10017, USA
www.syrawoodpublishinghouse.com

Plant Ecology: Concepts and Applications
Edited by Jude Boucher

International Standard Book Number: 978-1-68286-571-2 (Hardback)

Cataloging-in-Publication Data

Plant ecology : concepts and applications / edited by Jude Boucher.
    p. cm.
Includes bibliographical references and index.
ISBN 978-1-68286-571-2
1. Plant ecology. 2. Ecology. 3. Botany. I. Boucher, Jude.
QK901 .P53 2018
581.7--dc23

TABLE OF CONTENTS

PREFACE

I am honored to present to you this unique book which encompasses the most up-to-date data in the field. I was extremely pleased to get this opportunity of editing the work of experts from across the globe. I have also written papers in this field and researched the various aspects revolving around the progress of the discipline. I have tried to unify my knowledge along with that of stalwarts from every corner of the world, to produce a text which not only benefits the readers but also facilitates the growth of the field.

Plant ecology is a sub-discipline of ecology which studies the distribution and interaction between plants and organisms. Plant ecology can be divided into plant ecophysiology, community ecology, landscape ecology and biosphere ecology. Releasing oxygen into the Earth's atmosphere is a primary process that falls into the discipline of plant ecology. While understanding the long-term perspectives of the topics, the book makes an effort in highlighting their impact as a modern tool for the growth of the discipline. A number of latest researches have been included to keep the readers up-to-date with the global concepts in this area of study.

Finally, I would like to thank all the contributing authors for their valuable time and contributions. This book would not have been possible without their efforts. I would also like to thank my friends and family for their constant support.

Editor

1-MCP Releasing Complex for Open-Field Application

Majher I. Sarker[1], Peggy Tomasula[1] & LinShu Liu[1]

[1] Dairy and Functional Foods Research Unit, Eastern Regional Research Center, Agricultural Research Service, US Department of Agriculture, 600 E. Mermaid Lane, Wyndmoor, PA 19038, USA

Correspondence: LinShu Liu, Eastern Regional Research Center, Agricultural Research Service, US Department of Agriculture, 600 E. Mermaid Lane, Wyndmoor, PA 19038, USA.
E-mail: LinShu.Liu@ARS.USDA.GOV

Abstract

1-Methylcyclopropene (1-MCP) is a gas at room temperature which makes it difficult to handle and limits its application to use in closed environments. Open field application of 1-MCP can be a solution to protect crops from environmental stresses like drought or water logging. Our previous studies showed that Boronized-MCP is stable at ambient conditions and can gradually release 1-MCP when in contact with water. In this study the new complexes, suitable for being used directly in open crop fields have been generated and analyzed and also the effectiveness of the previously reported complexes in an open environment has been investigated. This new generation of boron complexes releases 1-MCP in a controlled way upon water contact without emitting any other volatile substances and thus expands their field of application.

Keywords: Boron derivatives of methylene cyclopropane, 1-MCP, open field application, plant growth regulator

1. Introduction

Ethylene is a plant hormone which plays a role at different stages of plant physiological changes, such as in the opening of flowers, the ripening of fruits or the yellowing of leaves. There are three groups of compounds that bind to the ethylene receptors of plant tissue (Sisler & Serek, 2003). 1-MCP belongs to one of the three groups in which a single exposure of plant tissue to these compounds is enough to prevent ethylene from binding even at very high levels of ethylene concentration, although this action disappears after a certain time either due to diffusion of the compounds from the binding sites or the development of new receptors (Bayer 1976a, 1976b, 1978; Sisler & Blankenship, 1993; Sisler et al., 1996a, 1996b, 1999; Sisler & Serek, 1997, 1999; Veen, 1983). The tight binding characteristic of 1-MCP to the ethylene receptor in plants blocks the activities of ethylene (Serek et al., 1995a, 1995b; Sister & Serek, 2003; Blankenship & Dole, 2003). Previous studies showed the affinity of 1-MCP for the receptors is about 10 times greater than that for ethylene (Watkins, 2006). Under environmental stresses like drought or waterlogging the level of endogenous ethylene is further increased in the plant (Kapuya & Hall, 1984; El-beltagy & Hall, 1974) due partly to decreased diffusion and partly to increased synthesis, which overall hampers agricultural yields around the world.

1-MCP is a bio-pesticide approved by the EPA for use on fruits and vegetables because of its nontoxic mode of activity (EPA 2002), negligible residues and effectiveness at low concentrations (Serek et al., 1995; Sisler & Blankenship, 1996; Feng et al., 2004). It is used commercially to keep fruit, flowers or vegetables fresh by preventing or delaying the natural ripening process. It is also used to prevent premature wilting, leaf yellowing, premature opening of flowers as well as premature death (Chow et al., 2006; De Paepe & Van der Straeten, 2005) and can also reduce the effect of water stress level in plants by inhibiting the ethylene response (Kawakami et al., 2010). However, because 1-MCP is a gas at ambient temperature it is difficult to handle and ship and to control its release. Due to these difficulties, 1-MCP cannot be used in an open crop field which could have a positive impact on increasing overall crop production.

In 2010, about $161.6 billion worth of food was discarded in the USA alone (Buzby et al., 2014), most of which was rotting fruits and vegetables. So the ability to keep fruits and vegetables fresher for a longer period of time could reduce food waste. Among all other factors, drought is the main environmental constraint to crop productivity worldwide. About 45% of the total agricultural lands worldwide (Bot et al., 2000) are subjected to continuous or frequent drought conditions which reduce the overall crop yields substantially. It has been reported that (Kawakami & Oosterhuis, 2006; Wang & Asiimwe, 2010) 1-MCP has good impact on crop yield during the dry seasons. For this purpose, investigations have been conducted for last two decades on 1-MCP particularly to address handling the volatile 1-MCP so that it could be used directly in open crop fields. A number of encapsulation agents have been developed to incorporate 1-MCP like molecular sieves (Daly & Kourelis, 2001), saw dust (Sisler & Blankenship, 1996), Cucurbit[6]uril (Zhang et al., 2011), modified starch (Gao, 2005) or α-cyclodextrin (α-CD) (Basel & Kostansek, 2009; Chong et al., 2002; Jacobson & Wehemyer, 2005). Among them α-CD is the most preferred one as it can easily release 1-MCP as a gas when the 1-MCP capsulated complex (1-MCP/α-CD) is dissolved in water. The 1-MCP/α-CD complex powder is commercially available under the trade name SmartFresh. However, α-CD would be expensive for widespread use and encapsulation does not allow for a controlled release of 1-MCP. Therefore, it can only be used in enclosed sites, such as coolers, truck trailers, greenhouses, storage facilities and shipping containers. Another challenge of using uncontrolled 1-MCP is the treatment dose as it relates to the maturity stage of the agricultural product. For example, if immature fruit is treated with 1-MCP it might not mature or ripen. On the other hand 1-MCP may not have any influence on matured fruit and in this case there is no justification for the expense in relation to shelf-life extension. To overcome this limitation, a non-volatile 1-MCP derivative named N,N-dipropyl (1-cycloproperrylmethyl) amine (DPCA) has been developed (Sisler et al., 2009). Less effective results have been reported even after 24 hours treatment and physical contact of the liquid chemical with food is required which is undesirable.

The objective of our research is to develop a complex which can release 1-MCP gradually so that it can be used in an open environment to protect crops from drought conditions or any other environmental stresses. To meet this goal we have realized that encapsulation of 1-MCP is not enough but a chemical reaction should be involved in this phenomenon which will liberate 1-MCP as the reaction moves forward. In this case the releasing rate of 1-MCP will depend on the kinetics of the particular chemical reaction. To design such a complex, a number of characteristic properties should be taken under consideration. (a) Sensitivity: The complex should be sensitive to environmentally friendly elements such as air or water to initiate the reaction. (b) Stability and volatility: The complex should be stable under ambient conditions and have a high boiling point. (c) Reaction properties: The 1-MCP releasing reaction should be carried out at ambient conditions without generating any volatiles or low boiling point products except for 1-MCP itself. (d) Toxicity: All products or byproducts should be non-toxic including the complex itself.

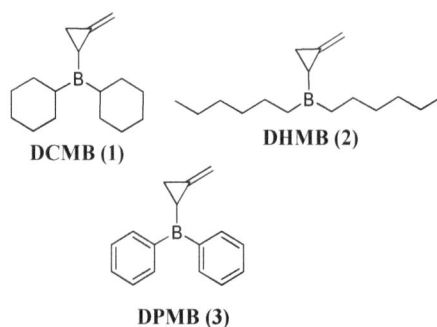

Figure 1. Structure of Boron complexes 1-3

In our previous study (Sarker et al., 2015), the reported synthesized Boron complexes (Figure 1, 1-3) have shown encouraging results toward the gradual release of 1-MCP when reacting with water. The releasing rate of 1-MCP depends on the kinetics of the particular hydrolysis reaction which gives the advantage of selective application. The most encouraging fact about these complexes is that they have the potential of releasing 1-MCP over a long period. The effectiveness of the complexes (1 and 3) has been proved by applying on green tomatoes in a closed environment. However, those compounds also released other compounds such as cyclohexane and cyclohexanol from complex 1 or hexane and hexanol from complex 2 and benzene from complex 3 due to the breakage of other Boron-Carbon bonds. Although these released compounds have much higher boiling points than that of 1-MCP (~ 12 °C), but they are volatile and some are toxic. Boric acid is also produced as the final product which is practically nontoxic to birds, fish and aquatic invertebrates. Boric acid and its sodium borate salts are active ingredients in pesticide products used as insecticides, acaricides, algaecides, herbicides, fungicides and as wood preservatives (Boric acid, EPA report 2006). In addition to pesticidal uses, boric acid and borate salts may be used as soil amendments in boron-deficient soils (OMRI Generic Material List 2013).

As our reported complexes show a strong potential for use in an open environment as a source of 1-MCP in a controlled way, we are encouraged to analyze their effectiveness in an open environment at the first place and secondly, to address the associated limitations of release of volatile chemicals generated by the hydrolysis reactions. This goal is achieved by synthetic structural modification of the complexes. A modification with poly aromatic hydrocarbons (PAH) (Figure 2, 4-10) for the side chains can result in complexes that can be used in dual purposes, gradually releasing 1-MCP and also acting as a pesticide. In this article, we have reported the synthesis and analysis of the complex modified by biphenyl group, **BPMB (11)** because, biphenyl is solid and prevents the growth of molds and fungus, and is therefore used as a preservative. It can be degraded biologically by conversion into nontoxic compounds (Linden and Sun 2014; Lee et al. 2011; Sakai et al. 2005), some bacteria are able to hydroxylate biphenyl and its polychlorinated biphenyls. Modification with other poly aromatic hydrocarbons will generate the options for particular applications.

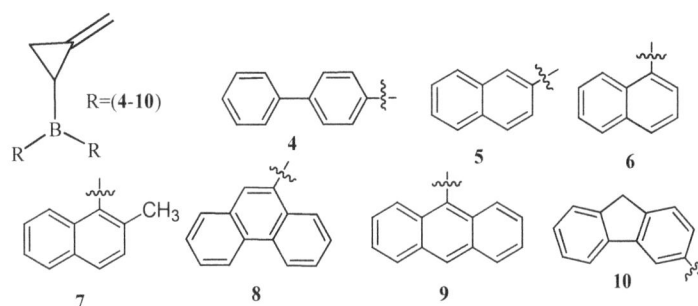

Figure 2. A list of polycyclic aromatic hydrocarbons (PAH)

2. Materials and Methods

All the chemicals and solvents used for the synthetic purposes were purchased from Sigma-Aldrich unless otherwise stated. All solvents used were of HPLC grade and moisture dry.

2.1 Synthesis of Complex BPMB (11) (Figure 3)

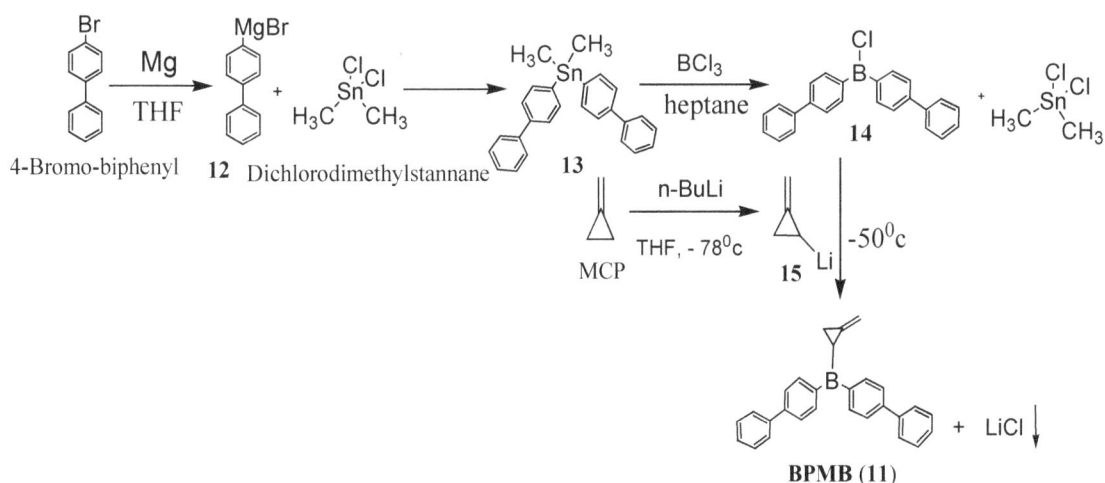

Figure 3. Synthetic scheme of complex BPMB (11)

BPMB (11) has been prepared by the reaction of Bis-biphenyl-4-yl-chloro-borane (14) with lithiated methylenecyclopropane (15), whereas, bis-biphenyl-4-yl-chloro-borane (14) was previously made from the reaction between Bis-biphenyl-4-yl-dimethyl-stannane (13) and boron trichloride in heptane. Bis-biphenyl-4-yl-dimethyl-stannane (13) was synthesized by the reaction of dichlorodimethylstannane with Grignard reagent of 4-bromo-biphenyl (12). Dichlorodimethylstannane and 4-Bromo-biphenyl were purchased from Sigma Aldrich. Methylene cyclopropane (MCP) was prepared from the reaction between potassium-[bis(trimethylsilyl)]amide and methallyl chloride (Binger et al., 2002).

2.1.1 Synthetic Procedure of Bis-biphenyl-4-yl-dimethyl-stannane (13)

An oven-dried 200 mL Schlenk flask cooled under argon was charged with 2.4 g Mg turnings (100 mmol) and 40 mL of anhydrous THF. To the mixture, 16.6 g (71.2 mmol) of 4-bromo-biphenyl dissolved in 30 mL of anhydrous THF was slowly added via a cannula. The reaction mixture was refluxed for 2 h and cooled at room temperature. The freshly prepared Grignard reagent was transferred via cannula in to a 200 mL 3-neck flask equipped with a condenser and cooled at 0 ^{0}C. A solution of dichlorodimethylstannane (5 g, 22.8 mmol) in 10 mL of dry THF was added in to the 3-neck flask via a cannula. The mixture was stirred for 30 minutes at room temperature before it was refluxed for 3 h. The reaction mixture was stirred at room temperature for 16 h. After the reaction the solution was cooled at 0 ^{0}C, treated little by little with total 10 mL of saturated NH_4Cl solution and extracted with dichloromethane in a separating funnel. The organic layer was washed three times with total of 150 mL water, concentrated in vacuo to give crude product mixed with biphenyl. The crude product was mixed with 25 mL of hexane and filtered under vacuum to obtain 9.6 g (89 % yield) of 13 as white powder. ^{1}H-NMR (400 MHz, $CDCl_3$): δ 0.56 (3H, s), 7.33 (1H, t, J = 7.5 Hz), 7.42 (2H, t, J = 7.5 Hz), 7.54-7.64 (6H, m); ^{13}C NMR (100 MHz, $CDCl_3$) δ -9.98, 126.9, 127.1, 127.3, 128.7, 136.6, 139.3, 141.1, 141.4.

2.1.2 Synthetic Procedure of Bis-biphenyl-4-yl-chloro-borane (14)

An oven dried 100 mL thick-walled flask with Teflon screw cap cooled under argon was charged with 5 g (11 mmol) of bis-biphenyl-4-yl-dimethyl-stannane (13), 11 mL of 1M borontrichloride solution (11 mmol) in heptane and 50 mL of anhydrous heptane in a nitrogen saturated glove box. The mixture was stirred for 30 minutes at room temperature and then heated at 110 ^{0}C for 48 h. After the reaction, the solution was cooled and subjected to vacuum filtration under nitrogen in the glove box. The solid was washed three times with a total 15 mL of dry dichloromethane to remove dichlorodimethylstannane. The rest of dichlorodimethylstannane was removed by sublimation technique to obtain 1.6 g of 14 (41.3 % yield) as a white powder. ^{11}B NMR ($CDCl_3$): δ 63.2; ^{1}H-NMR (400 MHz, $CDCl_3$): δ 7.40 (1H, t, J = 7.6 Hz), 7.48 (1H, t, J = 7.6 Hz), 7.68 (1H, d, J = 7.6 Hz), 7.74 (1H, d, J = 7.6 Hz), 8.12 (1H, d, J = 8.4 Hz); ^{13}C NMR (100 MHz, $CDCl_3$) δ 126.5, 127.3, 128.0, 128.9, 136.4, 137.6, 140.3, 145.5.

2.1.3 Synthetic Procedure of bis-biphenyl-4-yl-(2-methylene-cyclopropyl)-borane, BPMB (11)

An oven-dried 100 mL Schlenk flask cooled at -78 ^{0}C was charged with MCP (0.44 g, 8.2 mmol) in 15 mL of anhydrous THF. To the solution, 2.5 mL of 2.5 M n-BuLi in hexane (6.2 mmol) was added slowly and stirred at room temperature for 3 h. The mixture was cooled at -50 ^{0}C and 2 g (5.7 mmol) of bis-biphenyl-4-yl-chloro-borane (14) dissolved in 20 mL of dry THF was added slowly over 15 min via a cannula. The reaction mixture was stirred at room temperature for 24 h. The solution was filtered to remove salt and concentrated in vacuo. The solid is dissolved in 15 mL of dichloromethane and filtered with a syringe filter to get a clear solution. The solution was concentrated in vacuo and dried under vacuum to obtain 1.6 g (76% yield) of BPMB (11) as a solid. ^{11}B NMR ($CDCl_3$): δ 33.0.

2.2 Sample Preparation of BPMB (11) for the GC Analysis

GC analysis was conducted to study the controlled release capability of 1-MCP. For BPMB (11), 248 mg of sample was mixed with 0.5 mL of H_2O in a 1.5 mL air tight vial. The vapor collected from the head space of the solution was injected in a GC.

For quantitative GC analysis, a Hewlett-Packard 5890 GC with capillary column (30 m × 0.25 mm i.d.) coated with a 0.25 μm film of 5% phenyl methyl silicon and a flame ionization detector was used. The temperature of GC was programmed at 30 °C isothermal with an injection point temperature of 50 °C. The detector was operated at 230 °C and sample was injected under split less condition. Helium was used as carrier gas with a 1.5 mL/min column flow.

2.3 Comparative Study of Release of 1-MCP from DCMB (1) and BPMB (11) in Closed Environment at Different Time Segments

Two air tight 1 liter flasks were separately charged with equivalent amount (1.6 mmol) of DCMB (378 mg) and BPMB (595 mg) respectively and 2.5 mL of water in each flask. The mixtures in both flasks were kept under vigorous stirring. The vapor collected from the head space of the flasks was injected in a GC to quantify the accumulated 1-MCP for three different time segments, 0-24 h, 25-48 h and 49-94 h. After each segment the flasks were kept under vacuum for 20 minutes and flashed with nitrogen.

2.4 Treatment of Tomatoes for Quality Analysis in an Open Environment.

Mature green tomatoes were purchased through a major fruit company (Gargiulo, Inc., Naples, FL.) and stored at 10 °C before analysis. For the experiment three 1-gal glass jars without lids were used to treat six green tomatoes each. The tomatoes in two jars were treated with 1-MCP released from 1.14 g (4.9 mmol) of DCMB (1) and 1.13 g (3.0 mmol) of BPMB (11) respectively. The hydrolysis reaction was initiated with 4.0 mL of water and carried out in a 14 mL vial with continuous stirring for both cases. The vial was placed at the bottom of the jars. The third jar was kept in an identical condition without placing any boron complex inside. The treatment was carried out for 7 days at 22 °C. Theoretically, the total of 4.9 mmol and 3.0 mmol of 1-MCP gas are released in DCMB (1) and BPMB (11) jars respectively as one mole 1-MCP is liberated from every equivalent mole of these boron derivatives during hydrolysis reaction. Quality analysis (color and firmness) was performed at 22 ± 2 °C. Color was measured with a Hunter UltraScan® VIS colorimeter (Hunter Associates Lab, Reston, VA) and firmness was evaluated with a TA-XT2i Texture Analyzer (Texture Technologies Corp., Scarsdale, NY). Four measurements were taken for each tomato for color and firmness. The color of the tomatoes was measured six times during the monitoring period and the firmness was determined in the beginning and at the end of investigation.

3. Results and Discussion

The capability of gradual release of 1-MCP from boron complexes encouraged us to perform further research on these compounds. The major limitation associated with the reported complexes (1-3) of our previous study (Sarker et al., 2015) was the generation of volatile solvents as byproducts due to the breakage of additional two B-C bonds other than the desired B-MCP bond in hydrolysis reaction. This problem mainly generates cyclohexane, hexane and benzene as by products from 1, 2 and 3 complexes respectively after reacting with water along with boric acid in common. Although the boiling point of these compounds are much higher (68-81 °C) than that of released 1-MCP (12 °C) making these complexes useful in practical application but the volatility of the released compounds from these complexes limits the extension of their application especially in hot environment. To deal with this challenge we have modified the structure of the complex. Our first effort was to introduce biphenyl, a polycyclic aromatic hydrocarbon (PAH) as it is a solid with pesticidal properties and is also bio-degradable. We have also suggested a series (Figure 2, 4-10) of PAH which have potential to be used as a replacement to generate more complexes suitable for open field application. In synthetic aspect, BPMB (11) includes a four steps synthetic scheme (Figure 3) with a good yield at every step. An important part of this synthetic route is the recovery of dichlorodimethylstannane in the following step that can be reused reducing the synthetic cost of BPMB (11).

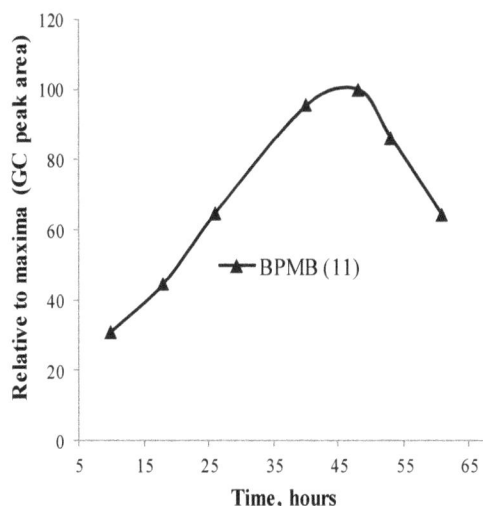

Figure 4. Controlled release of 1-MCP from BPMB (11)

3.1 Control Release of 1-MCP from BPMB (11)

From Figures 4 it is clear that BPMB (11) is capable of gradual release of 1-MCP when in contact with water. BPMB (11) requires 48 h to reach its highest point of release which is long enough to treat the agricultural products either in closed or open environments. After 48 hours, the descending curve indicates the finish line of releasing 1-MCP from the complex. But as we took more 1-MCP out for continued analysis the concentration of accumulated 1-MCP was reduced. It shows better resistance in releasing 1-MCP than DCMB, where 21 hours were needed (Sarker et al., 2015) to reach the highest point. Capability of sustained release of 1-MCP makes BPMB (11) a better choice for the application where longer treatment is necessary.

3.2 Comparative Study of Releasing 1-MCP from DCMB (1) and BPMB (11)

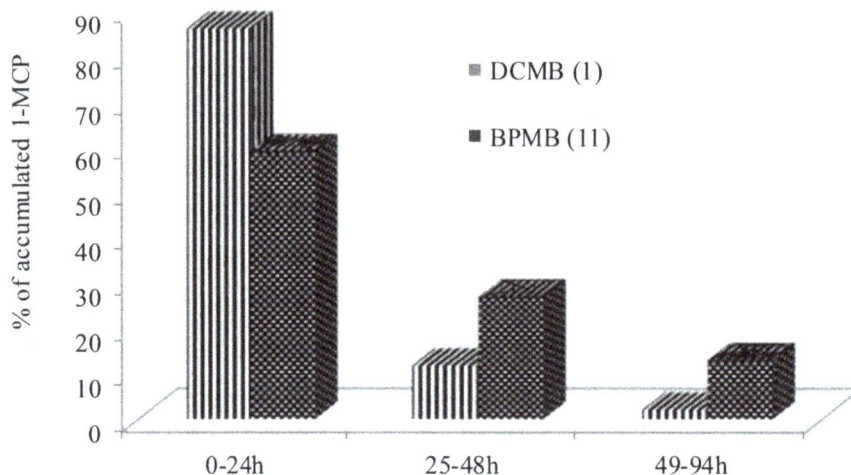

Figure 5. Accumulation of 1-MCP from DCMB (1) and BPMB (11) at different time segments at 22±2 °C

Comparative study at individual time segment shows that (Figure 5) the releasing pattern of 1-MCP from both complexes is same as they released the major portion in first 24 hours. However, there is a significant difference in releasing rate because DCMB (1) accumulated 86% of released 1-MCP in 0-24 hours whereas, BPMB (11) accumulated only 59%. In between 25-48 hours, DCMB (1) and BPMB (11) accumulated 12% and 27% of released 1-MCP respectively, whereas, the rest of 2% and 13% of released 1-MCP were accumulated by DCMB (1) and BPMB (11) respectively in next 49-94 h. This unique characteristic of releasing 1-MCP over a long period from these complexes would be useful for selective application and also for multi batch treatment in a closed environment, which will substantially reduce the cost of treatment.

3.3 Quality Analysis of Tomatoes Treated with DCMB (1) and BPMB (11) in an Open Environment

Figure 6. Comparative color changes of treated (DCMB & BPMB) and non-treated (NT) tomatoes in an open environment

Before claiming efficacy of any of the complexes for use in an open field, it is essential to demonstrate the idea in a practical setting. The result shows that both DCMB (1) and BPMB (11) have the capability to retard the ethylene response on tomatoes even in an open environment. Distinctive color differences were observed after 4 days; where 1-MCP (released from BPMB and DCMB) treated tomatoes remained green but some of the untreated tomatoes (NT) turned yellow or red (Figure 6). At the end of 7th day of investigation, all of the non-treated tomatoes (NT) became red or yellow whereas, 66% of the BPMB treated tomatoes and all of the DCMB treated tomatoes were still green. No injuries were observed on the surfaces of the treated tomatoes.

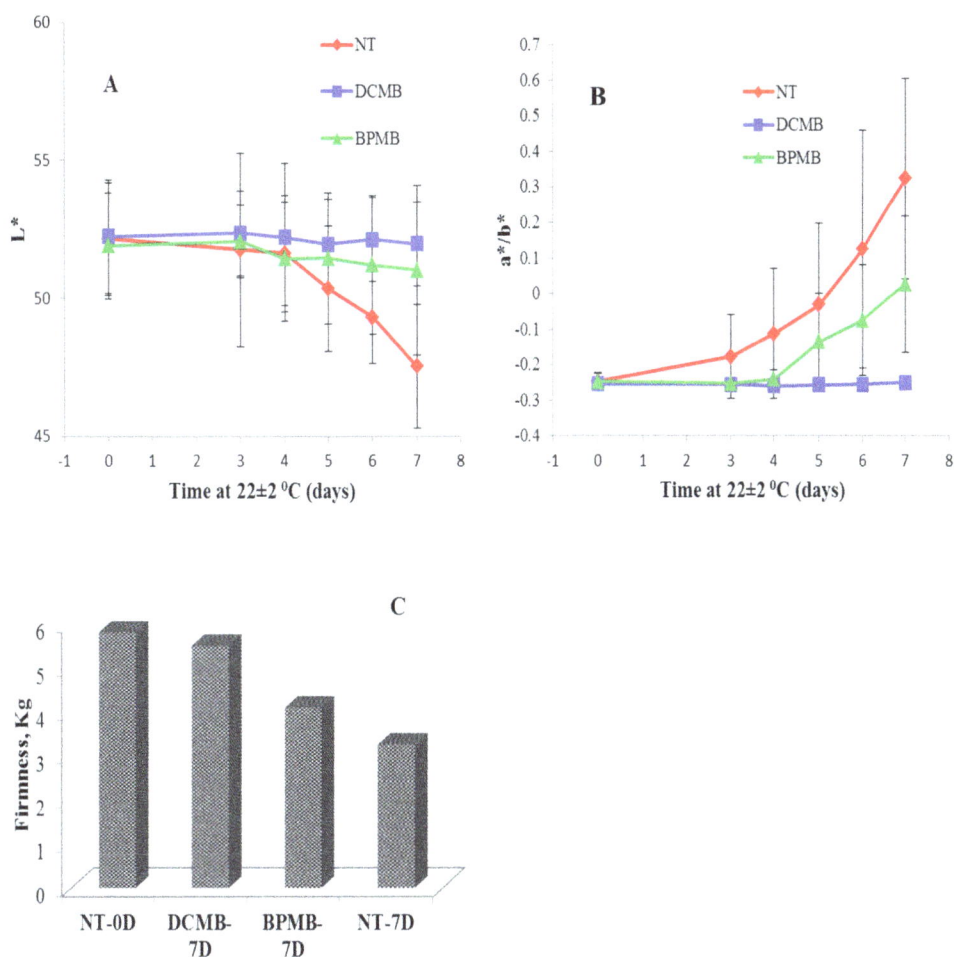

Figure 7. Comparative changes in L* (A), a*/b* (B) and firmness (C) of treated (DCMB and BPMB) and non-treated (NT) tomatoes during storage time in an open environment

For quality analysis, L* values indicate the darkness of the tomato surface color (Figure 7A). L* values of the tomatoes which were treated with Boron complex, DCMB (1) and BPMB (11) in open jars were not significantly changed at the end of the 5th day, but the L* value of the non-treated tomatoes (NT) began decrease significantly from day 5 as some of them became dark red in color. At the end of the 7th day the differences in L* values were observed by 4 and 5 points between NT-BPMB and NT-DCMB treated tomatoes respectively.

On the other hand, the a* and b* values represent the redness and yellowness of tomatoes respectively. The higher the a* values, the redder the tomatoes were. Over 7 days of the experiment, the a*/b* value of the treated tomatoes (BPMB and DCMB) increased slowly or remained constant, where, the a*/b* value for the non-treated (NT) tomatoes increased rapidly (Figure 7B). These results suggested that the treated tomatoes were less red in comparison to the non-treated tomatoes (NT) even after 7 days of investigation which is apparent as shown in Figure 6.

The tomatoes which were treated with boron complexes BPMB (11) and DCMB (1) were firmer than the non-treated tomatoes (NT) (Figure 7C) at the end of 7 days. The values of firmness were 5.5, 4.1 and 3.2 kg for the DCMB (DCMB-7D), BPMB (BPMB-7D) and non-treated (NT-7D) tomatoes respectively, which is a decrease of 5%, 29% and 45% respectively from the initial value of 5.8 Kg (NT-0D) at day 0. From overall quality analysis it is evident that both complexes are able to retard the ripening process of tomatoes in an open environment. However, DCMB (1) shows better protection over BPMB (11) which is simply because more equivalent amount of DCMB (4.9 mmol) was used than the equivalent amount of BPMB (3.0 mmol) in this experiment. The different equivalent amounts were taken intentionally to have an idea which level of concentration of released 1-MCP is required to block the ethylene receptor successfully.

4. Conclusion

In this study we have shown that previously reported boron complex DCMB (**1**) and newly synthesized BPMB (**11**) can work on tomato fruits even in open environment to retard their ripening process, although a certain concentration level of released 1-MCP is required for better performance. BPMB (**11**) has the potential to be used directly in open crop field as it gradually releases 1-MCP over a long period so the vicinity of the crop field will have 1-MCP continuously released by boron derivative. It is also important not to produce volatile byproducts due to hydrolysis reaction of the complex like BPMB (**11**), which generates solid biphenyl as byproduct. These complexes can also be used in close environment for multi batch operation or selective application. Replacement with other Poly Aromatic Hydrocarbons (PAH) can results better complex to be used in open crop field.

Acknowledgment

We thank Ran Li for her help with acquiring and processing the data for quality analysis of tomatoes.

References

Bayer, E. M. (1976a). Silver ion: a potent anti-ethylene agent in cucumber and tomato. *HortScience, 11*, 195-196.

Bayer, E. M. (1976b). A potent inhibitor of ethylene action in plants. *Plant Physiology, 58*, 268-271. http://dx.doi. org/10.1104/pp.58.3.268

Bayer, E. M. (1978). Method for overcoming the anti-ethylene effects of Ag^{+1}. *Plant Physiology, 62*, 616-617. http://dx.doi.org/10.1104/pp.62.4.616

Basel, R. M., & Kostansek, E. C. (2009). Compositions with cyclopropenes and adjuvants. *U.S. Patent: 20090088323A1.*

Binger, P., Brinkmann, A., & Wedemann, P. (2002). Highly Efficient Synthesis of Methylene cyclopropane. *Synthesis, 10*, 1344-1346. http://dx.doi.org/10.1055/s-2002-33122

Blankenship, S. M., & Dole, J. M. (2003). 1-Methylcyclopropene: a review. *Postharvest Biol Technol, 28*, 1-25. http://dx.doi.org/10.1016/S0925-5214(02)00246-6

Boric Acid/Sodium Borate Salts. (2006). HED Chapter of the Tolerance Reassessment. Eligibility Decision Document (TRED); *U.S. Environmental Protection Agency, Office of Prevention, Pesticides, and Toxic Substances, Health Effects Division, U.S Government Printing Offices: Washington, DC.*

Bot, A. J., Nachtergaele, F. O., & Young. (2000). A Land resource potential and constraints at regional and country levels. World Soil Resources Reports 90. *Land and Water Development Division, FAO, Rome.*

Buzby, J., Wells, H., & Hyman, J. (2014). The Estimated Amount, Value and Calories of Postharvest Food Losses at the Retail and Consumer Levels in the United States. *Economic Information Bulletin Number, 121*, USDA, February 2014.

Chong, J. A., Farozic, V. J., Jacobson, R. M., Synder, B. A., Steephens, R. W., & Mosley, D. W. (2002). Continuous process for the preparation of encapsulated cyclopropenes. *U.S. Patent: 20020043730A1.*

Chow, B., & McCourt, P. (2006). Plant hormone receptors; perception is everything. *Genes Dev, 20*(*15*), 1998-2008.

Daly, J., & Kourelis, B. (2001). Synthesis methods, complexes and delivery methods for the safe and convenient storage, transport and application of compounds for inhibiting the ethylene response in plants. *U.S. Patent: 6017849.*

De Paepe, A., & Van der Straeten, D. (2005). Ethylene biosynthesis and signaling; an overview. *Vitam Horm, 72*, 399-430. http://dx.doi.org/10.1016/S0083-6729(05)72011-2

El-beltagy, A. S., & Hall, M. A. (1974). Effect of water stress upon endogenous ethylene levels in Vicia Faba. *New Phytol, 73*, 47-60. http://dx.doi.org/10.1111/j.1469-8137.1974.tb04605.x

EPA (Environmental Protection Agency). (2002). *Fed Regist, 67*, 796-800.

Feng, X. Q., Apelbaum, A., Sisler, E. C., & Goren, R. (2004). Control of ethylene activity in various plant systems by structural analogues of 1-methylcyclopropene. *Plant Growth Regul, 42*, 29-38. http://dx.doi.org/10.1023/B:GROW.0000014900.12351.4e

Gao, R. T. (2005). Microcapsuled fruit, vegetable and flower antistaling agent and preparation thereof. *Chin.*

Patent: 10089065.

Jacobson, R. M., & Wehemyer, F. L. (2005). Humidity activated delicery systems for cyclopropenes. *Eur. Patent: 1593306A2.*

Kapuya, J. A., & Hall, M. A. (1984). Plant Sensitivity to Endogenous Ethylene in Relation to Species Characteristics. *Zeitchrift fur Pflanzenphysiologie, 113*, 461-464. http://dx.doi.org/10.1016/S0044-328X(84)80102-6

Kawakami, E. M., Oosterhuis, D. M., & Snider, J. L. (2010). Physiological effects of 1-Methylcyclopropene on well-watered and water-stressed cotton plants. *J. Plant Growth Regul, 29*, 280-288. http://dx.doi.org/10.1007/s00344-009-9134-3

Kawakami, E. M., & Oosterhuis, D. M. (2006). Effect of 1-MCP on the Physiology and Growth of Drought-Stressed Cotton Plants. *Summaries of Arkansas Cotton Research,* 62-66.

Lee, T. K., Lee, J., Sui, W. J., Iwai, S., Chai, B., Tiedje, J. M., & Park, J. (2011). Novel Biphenyl-Oxidizing Bacteria and Dioxygenase Genes from a Korean Tidal Mudflat. *Appl Environ Microbiol, 77*, 3888-3891. http://dx.doi.org/10.1128/AEM.00023-11

Linden, D., & Sun, Z. (2014). Biphenyl Pathway Map. Retrieved June 30, 2014, from http://eawag-bd.ethz.ch/bph/bph_map.html

OMRI Generic Materials List. (2013). Organic Materials Review Institute. Retrieved from http://www.omri.org/sites/default/files/app_materials/OMRI-GML-Stan-2013small_0.pdf

Sakai, M., Ezaki, S., Suzuki, N., & Kurane, R. (2005). Isolation and characterization of a novel polychlorinated biphenyl-degrading bacterium, Paenibacillus sp. KBC101. *Appl. Microbiol Biotechnol, 68*, 111-116.

Sarker, M., Xuetong, F., & Liu, L. (2015). Boron derivatives: As a source of 1-MCP with gradual release. *Scientia Horticulturae, 188*, 36-43.

Serek, M., Sisler, E. C., & Reid, M. S. (1995a). Effects of 1-MCP on the vase life and ethylene response of cut flowers. *Plant Growth Reg, 16*, 93-97. http://dx.doi.org/10.1007/BF00040512

Serek, M., Tamari, G., Sister, E. C., & Boro, A. (1995b). Inhibition of ethylene-induced cellular senescence symptoms by 1-methylcyclopropene, a new inhibitor of ethylene action. *Physiol Plant, 94*(2), 229-232.

Sisler, E. C., & Blankenship, S. M. (1993). Diazocyclopentadiene (DACP) a light sensitive reagent for the ethylene receptor in plants. *Plant Growth Reg, 12*, 125-132. http://dx.doi.org/10.1007/BF00144593

Sisler, E. C., & Blankenship, S. M. (1996). Method on counteracting an ethylene response in plants. *U.S. Patent: 5518988.*

Sisler, E. C., Dupille, E., & Serek, M. (1996a). Effect of 1-methylcyclopropene and methylenecyclopropane on ethylene binding and ethylene action on cut carnations. *Plant Growth Reg, 18*, 79-86. http://dx.doi.org/10.1007/BF00028491

Sisler, E. C., Serek, M., & Dupille, E. (1996b). Comparison of cyclopropene, 1-methylcyclopropene and 3,3-dimethylcyclopropene as ethylene antagonists in plants. *Plant Growth Reg, 18*, 169-174. http://dx.doi.org/10.1007/BF00024378

Sisler, E. C., Goren, R., Apelbaum, A., Serek, M. (2009). The effect of dialkylamine compounds and related derivatives of 1-methylcyclopropene in counteracting ethylene responses in banana fruit. *Postharvest Biology and Technology, 51*, 43-48. http://dx.doi.org/10.1016/j.postharvbio.2008.06.009

Sisler, E. C., & Serek, M. (1997). Inhibition of ethylene responses in plants at the receptor level: Recent developments. *Physiol Plant, 100*, 577-582. http://dx.doi.org/10.1034/j.1399-3054.1997.1000320.x

Sisler, E. C., Serek, M., Dupille, E., & Goren, R. (1999). Inhibition of ethylene responses by 1-methylcyclopropene and 3-methylcyclopropene. *Plant Growth Reg, 27*, 105-111. http://dx.doi.org/10.1023/A:1006153016409

Sisler, E. C., & Serek, M. (1999). Compounds controlling the ethylene receptor. *Bot. Bull. Acad. Sin, 40*, 1-7.

Sister, E. C., & Serek, M. (2003). Compounds interacting with the ethylene receptor in plants. *Plant Biol., 5*(5), 473-480. http://dx.doi.org/10.1055/s-2003-44782

Veen, H. (1983). Silver thiosulphate: an experimental tool in plant science. *Scientia Horticulturae, 20*, 211-224. http://dx.doi.org/10.1016/0304-4238(83)90001-8

Wang, G., & Asiimwe, R. K. (2010). Effects of 1-MCP and Quadris on cotton growth and yield. *Arizona Cotton Report (P-159)*, 28-33.

Watkins, C. B. (2006). The use of 1-methylcyclopropene (1-MCP) on fruits and vegetables. *Biotechnology Advances, 24*, 389-409. http://dx.doi.org/10.1016/j.biotechadv.2006.01.005

Zhang, Q., Zhen, Z., Jiang, H., Li, X. G., & Liu, J. (2011). Encaplulation of the Ethylene Inhibitor 1-Methylcyclopropene by Cucurbit[6]uril. *J. Agric Food Chem, 59*, 10539-10545. http://dx.doi.org/10.1021/jf2019566

A Pancreatic Lipase Inhibitory Activity by Mango (*Mangifera indica*) Leaf Methanolic Extract

Kimihisa Itoh[1], Kazuya Murata[2], Yuta Nakagaki[2], Ayaka Shimizu[2], Yusuke Takata[3], Kohsuke Shimizu[1], Tetsuya Matsukawa[1,3], Shin'ichiro Kajiyama[3], Masahiko Fumuro[1], Morio Iijima[1,4] & Hideaki Matsuda[1,2]

[1] The Experimental Farm, Kindai University, Wakayama, Japan

[2] Faculty of Pharmacy, Kindai University, Osaka, Japan

[3] Faculty of Biology-Oriented Science and Technology, Kindai University, Wakayama, Japan

[4] Faculty of Agriculture, Kindai University, Nara, Japan

Correspondence: Kimihisa Itoh, The Experimental Farm, Kindai University, 2355-2 Yuasa, Yuasa-cho, Arida-gun, Wakayama 643-0004, Japan. E-mail: itoh_k@nara.kindai.ac.jp

Abstract

The objective of this study was to identify pancreatic lipase inhibitory active ingredients of mango leaves, and to examine a relationship between leaves maturation and pancreatic lipase inhibitory activity. A methanolic extract of old dark green mango leaves (OML-ext) showed a porcine pancreatic lipase inhibitory activity. The pancreatic lipase inhibitory activity of OML-ext was attributable to 3-C-β-D-glucosyl-2,4,4',6-tetrahydroxybenzophenone (**2**) and mangiferin (**1**). The pancreatic lipase inhibitory activity of young mango leaf extract was superior to that of old leaf extract. It was suggested that the activity is correlated with the content of **2** in these extract. Considering the amounts of leaves obtained from pruning, old dark green leaves may be a reasonable natural resource for the preparation of ingredients with lipase inhibitory activity.

Keywords: 3-C-β-D-glucosyl-2,4,4',6-tetrahydroxybenzophenone, lipase inhibitory activity, *Mangifera indica*, mangiferin, obesity

1. Introduction

Obesity is recognized as a major life style disorder especially in developing countries and it is prevailing at an alarming speed in the world due to fast food intake, industrialization, and reduction of physical activity (Cairns, 2005).

Pancreatic lipase is a key enzyme for lipid absorption by hydrolysis of total dietary fats (Seyedan et al., 2015). Two pancreatic lipase inhibitors, namely orlistat (Xenical®) (Ballinger & Peikin, 2002) in U.S.A. and cetilistat (Oblean®) (Gras, 2013) in Japan, have been approved for the treatment of obesity syndrome. In order to find new pancreatic lipase inhibitors from natural resources, screening of plant extracts has been considered as one of strategies. Hitherto, several extracts of the plants, such as chokeberry fruit (Sosnowska et al., 2015), *Nelumbo nucifera* leaves (Liu et al., 2013), *Coffea arabica* seeds (Patui et al., 2014) and *Panax japonicus* rhizomes (Han et al., 2005), have been reported to have lipase inhibitory activities. As a folk tradition in India and Thailand when the high fat diet was taken, young mango (*Mangifera indica* Linne) leaves have been eaten together for health (Iwasa, 1984). Lipid metabolic enzyme inhibitory (Moreno et al., 2006) and cholesterol esterase inhibitory activities (Gururaja et al., 2015) of mango leaves have been reported, however pancreatic lipase inhibitory activity of young mango leaves has never been examined.

Several types of C-glucosyl-polyphenols, such as **1** and **2**, were isolated from mango leaves (Severi et al., 2009). Moreno et al. (2006) reported that 95% ethanolic extracts of mango leaves and bark inhibited human pancreatic lipase inhibitory activity at a high concentration of 1 mg/ml, and they suggested that these extracts may provide botanical therapeutics useful in weight control. However, they did not identify any active constituents. Severi et al. (2009) reported that mango leaf extract showed antiulcerogenic activity and the activity was attributable to phenolic compounds such as **2**. Zhang et al. (2011, 2013) isolated several benzophenone C-glycosides namely foliamangiferosides, in addition to **1** and **2** from 70% ethanolic extract of mango leaves as active inhibitor of triglyceride accumulation in 3T3-L1 cells. Recently, Gururaja et al. (2015) reported that the extracts of mango leaves exhibited cholesterol esterase inhibitory activity, and fractions containing 3-β-taraxanol or **2** showed a potent activity. Compound **1** has been reported as active constituent of lipolytic effect in rat epididymal fat-delivered cultured adipocytes, but **1** did not inhibit pancreatic lipase in the literature (Yoshikawa et al., 2002).

These reports prompt us to reexamine the pancreatic lipase inhibitory activity of mango leaf extracts and identify active constituents. Considering with the folk tradition described above, we also examined pancreatic lipase inhibitory effect of young mango leaf extract.

2. Materials and Methods

2.1 Plant Materials

Leaves of *M. indica* (cv. Irwin) were collected in the Experimental Farm, Kindai University (34° 2′ N, 135° 11′ E, 17 m ASL), located in Wakayama Prefecture, Japan. The *M. indica* trees planted in the ground are commercially grown in a plastic greenhouse {temperature: winter, min. 5°C (room) and 13°C (soil); summer, max. 39°C (room) and 30°C (soil)}. 'Irwin' mango is a representative cultivar especially in Japan and Taiwan because of its early ripening and relative good cold resistance. The leaves were collected from 250 mango trees which were propagated by grafting (the height of trees; 2 m, the age of trees; 18-26 years old, the life span of trees; 40-50 years). The collection date of the materials were as follows, old leaves; July 2013 and August 2015, young dark brown and yellow leaves; August 2015. These young and old leaves were collectable at the same time of pruning in late summer due to mango is an evergreen plant. The collected leaves were visually classified by the color of leaf into three groups, namely dark brown, yellow and dark green (Photo 1). The samples were identified by the Experimental Farm at Kindai University, air-dried at 50°C for 72 h in an automatic air-drying apparatus (Vianove Inc., Tokyo, Japan), and powdered. Voucher specimens of leaves (Mango old Leaf: OML201307DG-S and OML201508DG-S, Mango young dark brown Leaf: YML201508DB-S, Mango young yellow leaf: YML201508Y-S) are deposited in the Experimental Farm, Kindai University.

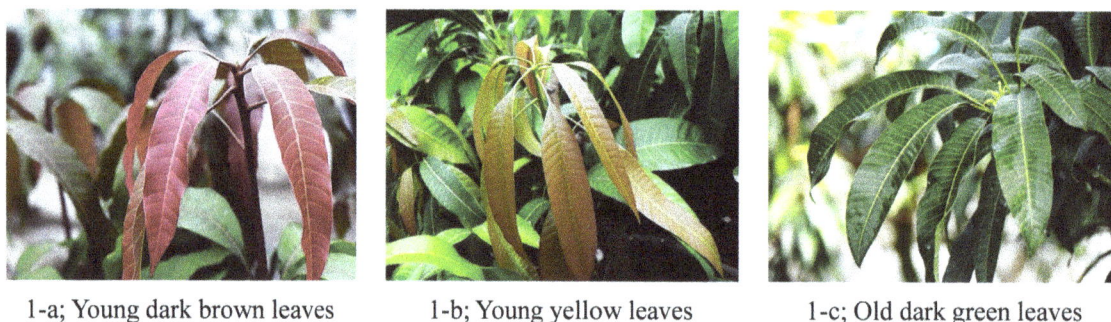

1-a; Young dark brown leaves   1-b; Young yellow leaves   1-c; Old dark green leaves

Photo 1. Photographs of typical mango leaves at various stages of development

2.2 Extraction

The leaf powder (10 g) was extracted with methanol (MeOH, 200 ml) for 72 h at room temperature. Each extract solution was evaporated under reduced pressure to give each MeOH extract. The yields of MeOH extract of young dark brown leaves, young yellow leaves, and old dark green leaves (OML-ext) were 28%, 25%, and 23%, respectively.

2.3 Reagents

4-Metylumbeliiferyl oleate, lipase (type II, from porcine pancreas, Lot #: SLBN3801V), authentic **1** were purchased from Sigma-Aldrich (St. Louis, MO, U.S.A.). Orlistat was purchased from Tokyo Chemical Industry (Tokyo, Japan). Other chemical and biochemical reagents were of reagent grade and were purchased from Wako Pure Chemical Industries, Ltd. (Osaka, Japan) and/or Nacalai Tesque, Inc. (Kyoto, Japan) unless otherwise noted.

2.4 In Vitro Pancreatic Lipase Inhibition Assay

Porcine pancreatic lipase (type II, from porcine pancreas) activity was measured according to the method of Nakai et al. (2005) with minor modification. The test sample was dissolved with dimethyl sulfoxide (DMSO) and diluted with 13 mM Tris-HCl buffer containing 150 mM NaCl, 1.3 mM CaCl$_2$ (pH 8.0) to a final DMSO concentration of 2.5% v/v. 4-Methylumbelliferyl oleate (4-MU) was used as a substrate. The substrate and enzyme were both diluted in above mentioned buffer immediately before use. An aliquot of 25 μl of the test solution and 50 μl of 0.1 mM 4-MU solution were mixed in back colored microtiter plates, 25 μl of 0.2 mg/ml enzyme solution was then added to each well to start the reaction. After incubation for 30 min at 37°C, 100 μl of 0.1 M citrate buffer (pH 4.2) was added to stop the reaction. The fluorescence associated with enzymatically released 4-methylumbelliferone product was monitored at an excitation wavelength of 355 nm and emission of 460 nm using a multi-label counter (PerkinElmer 2030 ARVO X4，PerkinElmer Life and Analytical Sciences). Orlistat, a known inhibitor of pancreatic lipase, was used as reference compound. The activity of negative control was also evaluated by adding stop solution before enzymatic reaction. The inhibition activity was calculated using the following formula:

$$\% \text{ inhibition} = [(A-B)-(C-D)]/(A-B) \times 100$$

where A is the fluorescence with enzyme and substrate, but without test substance (adding stop solution after enzymatic reaction); B the fluorescence with enzyme and substrate, but without test substance (adding stop solution before enzymatic reaction); C the fluorescence with enzyme, substrate and test substance (adding stop solution after enzymatic reaction); and D the fluorescence with enzyme, substrate and test substance (adding stop solution before enzymatic reaction).

Each concentration of samples was confirmed in triplicate (P value < 0.01). IC_{50} value represents the concentration required to inhibit 50% of pancreatic lipase activity.

2.5 Spectroscopy

ESI mass spectra were recorded on a Triple TOF 5600+ mass spectrometer (SICEX) combined with an LC-20A HPLC system (Shimadzu, Kyoto, Japan). The samples were eluted with 0.1% formic acid aqueous solution and acetonitrile at a constant flow rate of 0.2 ml/min.

^{1}H NMR (800 MHz), ^{13}C NMR (200 MHz), and 2D NMR (COSY, HSQC, and HMBC) spectra were obtained with JNM-ECA 800 MHz NMR spectrometer (JOEL, Tokyo, Japan) using a 5 mm probe, DMSO-d_6 as solvents.

2.6 Isolation of 3-C-β-D-glucosyl-2,4,4',6-tetrahydroxybenzophenone (2) and Mangiferin (1)

The OML-ext (5.44 g) was submitted to column chromatography over 125 g of silica gel (Wakogel C-100, Wako, Osaka, Japan, 36 mm × 205 mm). Elution with stepwise gradient of 1.2 l of CHCl$_3$/MeOH (10:0 v/v), (30:1 v/v), (10:1 v/v), (5:1 v/v) and (0:10 v/v) to give 30 fractions. The lipase inhibitory activity at 25 μg/ml of each fraction was assayed. Fractions (1-6) eluted with CHCl$_3$/MeOH 10:0 v/v were almost inactive. Active fractions (7-25) eluted with CHCl$_3$/MeOH 30:1 v/v, 5:1 v/v and 0:10 v/v were further purified by a preparative HPLC (column: Inertsil ODS-3 column, 20 mm × 50 mm, GL Sciences, Tokyo, Japan. elution; 15% acetonitrile aqueous solution, flow rate; 9.0 ml/min, detection; UV 280 nm) gave compound **A** and compound **B**. Identification of the isolated compound **A** as **1** were done by co-chromatography and comparison of spectrum analysis with authentic **1** under the same HPLC condition as described in the section of HPLC Analysis conditions. Compound **B** was identified as **2** by comparison of ^{1}H, ^{13}C-NMR and mass spectra data (Severi et al., 2009) as described in the section of Identification of 3-C-β-D-glucosyl-2,4,4',6-tetrahydroxybenzophenone (**2**).

2.7 Identification of 3-C-β-D-glucosyl-2,4,4',6-tetrahydroxybenzophenone (2)

^{1}H-NMR δ: 3.17 (1H, dd, J=2.3, 2.5 Hz, H-5"), 3.20 (2H, m, H-3", -4"), 3.49 (1H, m, H-6"a), 3.57 (1H, dd, J=8.9, 9.5 Hz, H-2"), 3.61 (1H, m, H-6"b), 4.58 (1H, d, J =9.5 Hz, H-1"), 5.93 (1H, s, H-5), 6.76 (2H, d, J =8.7 Hz, H-3', 5'), 7.55 (2H, d, J =8.7 Hz, H-2', 6'). ^{13}C-NMRδ: 60.7 (C-6"), 69.8 (C-4"), 72.1 (C-2"), 74.9 (C-1"), 78.5 (C-3"), 81.2 (C-5"), 103.9 (C-5), 107.1 (C-1), 114.8 (C-3', -5'), 131.0 (C-1'), 131.6 (C-2', -6'), 157.0 (C-2), 161.0 (C-3), 161.3 (C-4'), 194.8 (C-7). ESI-MS: [M+H]$^+$ = m/z 409.1.

2.8 HPLC Analysis Conditions

An HPLC system consisted of LC-20A pump and SPD-M20A photodiode array detector (Shimadzu, Kyoto, Japan), which were used for detection and UV spectrometry. The samples were analyzed by using an Inertsil ODS-3 reverse phase column (4.6 × 150 mm, GL Sciences, Tokyo, Japan) and gradient elution with acetonitrile aqueous solution at a constant flow rate of 0.8 ml/min. The elution was carried out using linear gradient condition as follows; initial condition was set at 5% acetonitrile and maintained for 5 min., followed by a linear gradient from 5% to 100% acetonitrile for 35 min. The column temperature was set at 40°C, and eluted compounds were detected at a range of 200 to 700 nm. Under this condition, **1** and **2** were eluted at the retention time of 20 min and 17 min, respectively. Concentration of **1** and **2** were calculated from linear calibration curves made from external standards, authentic **1** and the isolated **2**. Linear calculation curves in the range of 0.5 to 5 μg were made from the peak areas analyzed at 280 nm, and the correlation coefficients of **1** and **2** were 0.967 and

0.996, respectively. Dried extracts were dissolved in MeOH, and an appropriate amount of sample solution was applied to HPLC analysis in triplicate. Values represent the mean ± standard deviation.

2.9 Statistical Analysis

The experimental data were evaluated for statistical significance using Bonferroni/Dunn's multiple-range test with GraphPad Prism for Windows, Ver. 5 (GraphPad Software Inc., 2007).

3. Results and Discussion

3.1 Identification of Pancreatic Lipase Inhibitory Active Ingredients of OML-ext

In the preliminary evaluation of mango leaf extract on porcine pancreatic lipase inhibitory activity using 4-metylumbeliiferyl oleate as a substrate, the methanolic old mango leaf extract (OML-ext) inhibited a pancreatic lipase activity with the IC_{50} value of 13.9 μg/ml. To identify the active constituents, we carried out activity-guided fractionation of OML-ext using lipase inhibitory assay. Silica gel column chromatographic fractionation of OML-ext gave active fractions eluted with $CHCl_3$/MeOH 30:1 v/v, 5:1 v/v and 0:10 v/v, and inactive fractions eluted with $CHCl_3$/MeOH 10:0 v/v. Further purification of active fractions using preparative HPLC led to isolation of **1** and **2** as active constituents. The IC_{50} values (Table 1) of **1** and **2** were 273 and 226 μM, respectively. As shown in Table 1, the IC_{50} value of orlistat as a reference compound was 0.1 μM (= 0.0495 μg/ml) in accordance with the reported IC_{50} value (0.05 μg/ml) (Ado et al., 2013). Thus, a part of the pancreatic lipase inhibitory activity of OML-ext is attributable to these two compounds. To the best of our knowledge, this is the first report on lipase inhibitory activity of **2**.

Table 1. Inhibitory activities of 3-C-β-D-glucosyl-2,4,4',6-tetrahydroxybenzophenone (**2**) and mangiferin (**1**) on pancreatic lipase

Samples	IC_{50} values [a] (μM)
3-C-β-D-glucosyl-2,4,4',6-tetrahydroxybenzophenone (**2**)	226 μM
Mangiferin (**1**)	273 μM
Orlistat	0.1 μM

Orlistat was used as reference compound. a); IC_{50} value represents the concentration required to inhibit 50% of pancreatic lipase activity.

As described above, **1** is a xanthone found in various parts of mango and has several biological activities, *e.g.* antitumor (Yoshimi et al., 2001), antiviral (Zheng & Lu, 1990), antidiabetic (Miura et al., 2001), anti-inflammatory (Garrido et al., 2004) and potent antioxidant (Sato et al., 1992) activities.

Yoshikawa et al. (2002) described that **1** showed lipolytic effect in rat epididymal ·fat-delivered cultured adipocytes, but **1** did not inhibit pancreatic lipase. In addition, Koga et al. (2013) reported IC_{50} value of **1** as lipase inhibitor was above 2962 μM. In our experiments, **1** inhibited pancreatic lipase (IC_{50} value; 273 μM, Table 1) in contrast to the reports by Yoshikawa et al. (2002) and Koga et al. (2013). Because they did not use a reference compound such as orlistat, it may be difficult to discuss the reasons anymore. We assumed that a part of the discrepancy of the activity of **1** might be due to the some differences in experimental conditions, such as substrate and evaluation method.

3.2 A Relationship Between Leaves Maturation and Pancreatic Lipase Inhibitory Activity

Considering with the use of mango leaves and also the folk tradition in India and Thailand described above, we examined the inhibitory activity of young and old leaves on pancreatic lipase. The color of mango leaves turns from dark brown to dark green with an increase in area after unfolding. Since the mango is an evergreen plant, young dark brown, young yellow and old dark green leaves were collectable at the same time of pruning in late summer after fruit harvest. The collected leaves were visually classified by the color of leaf into three groups, namely dark brown, yellow and dark green.

As shown in Table 2, lipase inhibitory activity of young dark brown leaf extract showed most potent activity with the IC_{50} value of 2.9 μg/ml. In accordance with the leaves mature from yellow to dark green, activities of these extracts were slightly decreased. The IC_{50} values of young yellow leaves and old dark green leaves were 4.5 and 13.9 μg/ml, respectively (Table 2).

Table 2. Inhibitory activities of MeOH extracts of young dark brown and young yellow leaves and old dark green mango leaves on pancreatic lipase

Samples	IC$_{50}$ values [a] (µg/ml or µM)
Young dark brown leaf extract	2.9 µg/ml
Young yellow leaf extract	4.5 µg/ml
Old dark green leaf extract (= OML-ext)	13.9 µg/ml
Orlistat	0.1 µM

Orlistat was used as reference compound. a); IC$_{50}$ value represents the concentration required to inhibit 50% of pancreatic lipase activity.

The contents (mg/g extract) of **2** and **1** in these leaf extracts were determined by HPLC analysis. As a result, young dark brown leaf extract contained 400.0 ± 11.0 mg/g of **2** and 78.6 ± 1.2 mg/g of **1**. The corresponding content data for other two leaf extracts were as follows; young yellow leaf extract, 278.6 ± 15.5 and 79.4 ± 1.0 mg/g, and OML-ext, 205.9 ± 7.6 and 85.1 ± 0.5 mg/g. The pancreatic lipase inhibitory activity of young dark brown and young yellow leaf extracts was superior to that of OML-ext. These results suggested that the inhibitory activity of extract was in accordance with the content of **2** in extracts.

High contents of **2** and **1** in the extracts indicated that these two compounds would be partly responsible to their inhibitory activities. On the other hand, the potent inhibitory activities of leaf extracts can't exclude a hypothesis that other ingredients may also contribute to the activity. To identify other active ingredients, further studies are required, and now undergoing.

In the cultivation process of mango fruit, pruning in summer after fruit harvest is important so as to obtain excellent flower buds in next spring and a rich harvest of superior fruit in next summer, and pruned leaves are usually discarded. As shown in Photo 1, anthocyanin content in young dark brown leaves is high (Photo 1-a), but during their enlargement, the anthocyanin content is decreased (Photo 1-b), while the chlorophylls content is increased. Concurrently, the leaves become thick and rigid (Photo 1-c) (Ali et al., 1999). The period of young dark brown and young yellow leaves were short, leaves rapidly turned to dark green with developing due to rapid decrease of anthocyanin content. Therefore pruned amounts of young dark brown and young yellow leaves are much less than that of old dark green leaves. Considering the amount of leaves obtained from pruning, old dark green leaves may be a reasonable natural resource for the preparation of ingredients with lipase inhibitory activity.

4. Conclusion

Young dark brown and young yellow mango leaf extracts and OML-ext exhibited pancreatic lipase inhibitory activities. The activities of these young mango leaf extracts was superior to that of old mango leaves, this is the first report to reveal a relationship between leaves maturation and pancreatic lipase inhibitory activity. These findings may support the folk tradition in India and Thailand. It was revealed that a part of the pancreatic lipase activity of leaf extract was attributable to **2** and **1**. This is the first report on lipase inhibitory activity of **2**. Hitherto, pruned mango leaves were unworthy and discarded during the cultivation process of mango fruit, this finding suggested that pruned mango leaves may be a useful resource for the preparation of ingredients for the treatment of obesity. However, further investigations are required to examine of administration safety and the mechanisms involved and to reveal other active constituents.

Acknowledgments

We are grateful to all technical staffs of Yuasa Experimental Farm, Kindai University for the collection of mango leaves. I am deeply grateful to Dr. Shunsuke Naruto for his invaluable guidance and advice.

References

Ado, M. A., Abas, F., Mohammed, A. S., & Ghazali, H. M. (2013). Anti- and pro-lipase activity of selected medicinal, herbal and aquatic plants, and structure elucidation of an anti-lipase compound. *Molecules, 18*, 14651-14669. http://dx.doi.org/10.3390/molecules181214651

Ali, K., Koeda, K., & Nii, N. (1999). Changes in anatomical features, pigment content and photosynthetic activity related to age of 'Irwin' mango leaves. *Journal of the Japanese Society for Horticultural Science, 68*(6), 1090-1098.

Ballinger, A., & Peikin, S. R. (2002). Orlistat: its current status as an anti-obesity drug. *European Journal of Pharmacology, 440*, 109-117.

Cairns, E. (2005). Obesity: the fat lady sings? *Drug Discovery Today, 10*(5), 305-307.

Garrido, G., González, D., Lemus, Y., García, D., Lodeiro, L., Quintero, G., ... Delgado, R. (2004). In vivo and in vitro anti-inflammatory activity of *Mangifera indica* L. extract (VIMANG®). *Pharmacological Research, 50*, 143-149. http://dx.doi.org/10.1016/j.phrs.2003.12.003

Gras, J. (2013). Cetilistat for the treatment of obesity. *Drugs Today, 49*(12), 755-759. http://dx.doi.org/10.1358/dot.2013.49.12.2099318

Gururaja, G. M., Mundkinajeddu, D., Dethe, S. M., Sangli, G. K., Abhilash, K., & Agarwal, A. (2015). Cholesterol esterase inhibitory activity of bioactives from leaves of *Mangifera indica* L. *Pharmacognosy Research, 7*(4), 355-362. http://dx.doi.org/10.4103/0974-8490.159578

Han, L. K., Zheng, Y. N., Yoshikawa, M., Okuda, H., & Kimura, Y. (2005). Anti-obesity effects of chikusetsusaponins isolated from *Panax japonicus* rhizomes. *BMC Complementary and Alternative Medicine, open access*, 1-10. http://dx.doi.org/10.1186/1472-6882-5-9

Iwasa, S. (1984). *Tropical fruit magazine* (2nd ed.) (In Japanese). Tokyo, Japan: Kokinsyoin.

Koga, K., Hisamura, M., Kanetaka, T., Yoshino, K., Matsuo, Y., & Tanaka, T. (2013). Proanthocyanidin oligomers isolated from *Salacia reticulata* leaves potently inhibit pancreatic lipase activity. *Journal of Food Science, 78*(1), 105-111. http://dx.doi.org/10.1111/1750-3841.12001

Liu, S., Li, D., Huang, B., Chen, Y., Lu, X., & Wang, Y. (2013). Inhibition of pancreatic lipase, α-glucosidase, α-amylase, and hypolipidemic effects of the total flavonoids from *Nelumbo nucifera* leaves. *Journal of Ethnopharmacology, 149*, 263-269. http://dx.doi.org/10.1016/j.jep.2013.06.034

Miura, T., Ichiki, H., Hashimoto, I., Iwamoto, N., Kato, M., Kubo, M., ... Tanigawa, K. (2001). Antidiabetic activity of a xanthone compound, mangiferin. *Phytomedicine, 8*(2), 85-87.

Moreno, D. A., Ripoll, C., Ilic, N., Poulev, A., Aubin, C., & Raskin, I. (2006). Inhibition of lipid metabolic enzymes using *Mangifera indica* extracts. *Journal of Food Agriculture & Environment, 4*(1), 21-26.

Nakai, M., Fukui, Y., Asami, S., Toyoda-Ono, Y., Iwashita, T., Shibata, H., ... Kiso, Y. (2005). Inhibitory effects of oolong tea polyphenols on pancreatic lipase in vitro. *Journal of Agricultural and Food Chemistry, 53*, 4593-4598. http://dx.doi.org/10.1021/jf047814+

Patui, S., Clincon, L., Peresson, C., Zancani, M., Conte, L., Terra, L. D., ... Braidot, E. (2014). Lipase activity and antioxidant capacity in coffee (*Coffea arabica* L.) seeds during germination. *Plant Science, 219-220*, 19-25. http://dx.doi.org/10.1016/j.plantsci.2013.12.014

Sato, T., Kawamoto, A., Tamura, A., Tatsumi, Y., & Fujii, T. (1992). Mechanism of antioxidant action of pueraria glycoside (PG)-1 (an isoflavonoid) and mangiferin (a xanthonoid). *Chemical and Pharmaceutical Bulletin, 40*(3), 721-724.

Severi, J. A., Lima, Z. P., Kushima, H., Brito, A. R., Santos, L. C., Vilegas, W., & Hiruma-Lima, C. A. (2009). Polyphenols with antiulcerogenic action from aqueous decoction of mango leaves (*Mangifera indica* L.). *Molecules, 14*, 1098-1110. http://dx.doi.org/10.3390/molecules14031098

Seyedan, A., Alshawsh, M. A., Alshagga, M. A., Koosha, S., & Mohamed, Z. (2015). Medicinal plants and their inhibitory activities against pancreatic lipase: a review. *Evidence-Based Complementary and Alternative Medicine, 2015*, 1-13. http://dx.doi.org/10.1155/2015/973143

Sosnowska, D., Podsędek, A., Redzynia, M., & Żyżelewicz, D. (2015). Effects of fruit extracts on pancreatic lipase activity in lipid emulsions. *Plant Foods for Human Nutrition, 70*, 344-350. http://dx.doi.org/10.1007/s11130-015-0501-x

Yoshikawa, M., Shimoda, H., Nishida, N., Takada, M., & Matsuda, H. (2002). *Salacia reticulata* and its polyphenolic constituents with lipase inhibitory and lipolytic activities have mild antiobesity effects in rats. *Biochemical and Molecular Actions of Nutrients, 132*, 1819-1824.

Yoshimi, N., Matsunaga, K., Katayama, M., Yamada, Y., Kuno, T., Qiao, Z., ... Mori, H. (2001). The inhibitory effects of mangiferin, a naturally occurring glucosylxanthone, in bowel carcinogenesis of male F344 rats. *Cancer Letters, 163*, 163-170.

Zhang, Y., Han, L., Ge, D., Liu, X., Liu, E., Wu, C., ... Wang, T. (2013). Isolation, structural elucidation, MS profiling, and evaluation of triglyceride accumulation inhibitory effects of benzophenone C-glucosides from leaves of *Mangifera indica* L. *Journal of Agricultural and Food Chemistry, 61,* 1884-1895. http://dx.doi.org/ 10.1021/jf305256w

Zhang, Y., Qian, Q., Gc, D., Li, Y., Wang, X., Chen, Q., ... Wang, T. (2011). Identification of benzophenone *C*-glucosides from mango tree leaves and their inhibitory effect on triglyceride accumulation in 3T3-L1 adipocytes. *Journal of Agricultural and Food Chemistry, 59,* 11526-11533. http://dx.doi.org/10.1021/jf2028494

Zheng, M. S., & Lu, Z. Y. (1990). Antiviral effect of mangiferin and isomangiferin on herpes simplex virus. *Chinese Medical Journal, 103*(2), 160-165.

3

Spring Wheat Breeding: Evaluation of Selected Adapted Spring Wheat Germplasm in Eastern Canada

Shahrokh Khanizadeh[1], Harvey Voldeng[1], Xuelian Wang[1], Allen Xue[1], Mirko Tabori[1], Hana Moidu[1], Richard Martin[2], Allan Cummiskey[2], Mark Etienne[3], Ellen Sparry[4] & Jean Goulet[5]

[1] Ottawa Research and Development Centre, Agriculture and Agri-Food Canada, K.W. Neatby Building, 960 Carling Ave, Ottawa, Ontario, K1A 0C6, Canada

[2] Crops and Livestock Research Centre, Agriculture and Agri-Food Canada, 440 University Ave, Charlottetown, Prince Edward Island, C1A 4N6, Canada

[3] Dow AgroSciences rental field in Kincardine, Ontario, N0M 1A0, Canada

[4] C&M Seeds, 6180 5th Line, Palmerston, Ontario, N0G 2P0, Canada

[5] Centre de Recherche Semican, 1290 Route 116 Ouest, Princeville, Quebec, G6L 4K7, Canada

Correspondence: Shahrokh Khanizadeh, Ottawa Research and Development Centre, Agriculture and Agri-Food Canada, K.W. Neatby Building, 960 Carling Ave, Ottawa, Ontario, K1A 0C6, Canada. E-mail: Shahrokh.Khanizadeh@AGR.GC.CA, http://khanizadeh.info

Abstract

Twenty-five hard red spring wheat (*Triticum aestivum*) lines, including three known cultivars used as checks, were grown in seven locations across Eastern Canada. The objective of this multi-location experiment was to evaluate selected Eastern Cereal and Oilseed Research Centre advanced lines (ECAD lines) from the Spring Wheat Breeding Program in order to identify the best lines for performance and grower trials. The lines from this trial performed very well compared to the check varieties, especially at the Ontario locations. Overall, the ECAD lines were on a par with or superior to the checks in terms of several attributes, including yield, protein content, and Fusarium head blight resistance.

Keywords: Triticum aestivum, breeding, cultivar description, AC Carberry, AC Scotia, AC Sable

1. Introduction

Wheat (*Triticum aestivum*) breeding programs across Canada aim to produce high-yielding varieties (McCaig and DePauw 1995) with improved resistance to diseases, especially *Fusarium* head blight (FHB) (Gilbert & Tekauz 2000; McCaig & DePauw, 1995), early maturity (Thomas & Graf, 2014), and high protein content (Wang et al., 2002). Through these programs, crop losses have been greatly reduced and grain quality has been increased (Thomas and Graf 2014). Of particular interest to Canadian wheat growers are cultivars that have a short growing season, produce high yields, and are resistant to FHB (Preston et al. 1991; McCaig & DePauw 1995). Common attributes of interest to commercial mills are grain protein, falling number, test weight, and deoxynivalenol (DON) values (Darby, 2015).

Fusarium head blight resistance is currently of great importance for wheat production in Canada. Epidemics of FHB in 1980 in Eastern Canada and in 1993 in Manitoba sparked interest in developing FHB-resistant cultivars and FHB management strategies (Gilbert & Tekauz, 2000). There is currently no single control method for FHB; the best approach involves the use of a cultivar of intermediate resistance in combination with fungicide applications, and suitable farming practices (Gilbert & Tekauz, 2000). More than 0.25% *Fusarium*-damaged kernels by weight can downgrade wheat to the point of significant economic loss (Fernandez et al., 2005). In a spring wheat analysis conducted by Darby (2015), 13 out of 19 tested varieties had FHB symptoms, indicating that the disease is a severe problem. The risk of FHB is increased by warm and humid weather, short plant height, conservation tillage, and wheat–maize rotations (Gilbert & Tekauz, 2000). Practices that reduce FHB risk consist of shredding maize residues to decrease DON values, performing conventional tillage, and using fungicides such as tebuconazole (Gilbert & Tekauz, 2000). Despite such practices, FHB is difficult to manage, because shorter plants are more susceptible, and more than half of the world's wheat cultivars contain two major dwarfing genes that are gibberellic-acid-insensitive (Gilbert & Tekauz, 2000). Furthermore, FHB thrives in warm, humid climates, and climate change may create environmental conditions that are favorable for the disease (Gilbert & Tekauz, 2000).

Grain yield is very important to wheat growers, and thus many studies on how to increase yield have been published. The optimal test weight is considered to be 56 to 60 lb/bushel (Darby, 2015), and newly registered wheat cultivars often have increased yields relative to check varieties with which they are compared. In previous studies, two examples of this are 'Snowbird', which gave higher yields than all check varieties except 'McKenzie' (Humphreys et al., 2007) and 'Carberry', which yielded 9.6% more than all checks (DePauw et al., 2011). Wheat breeders attempting to produce high-yielding cultivars can adopt an approach that involves using hybrids of spring wheat and winter wheat: the mean yields of spring/winter hybrids were found to surpass those of all controls (McKenzie & Grant, 1974). A study conducted by Hucl and Baker (1987) found that crop grain yield was correlated with biological yield and that cultivars with a later heading date produced more spikelets and more kernels per spike and also had increased yield. The same study found that some cultivars produced significantly fewer spikes per square meter, but the kernels were heavier, resulting in a net 25% increase in crop yield (Hucl & Baker 1987). Another study on yield reported that grain yield was not associated with tillering capacity or mortality but was instead associated with kernel weight (Hucl & Baker, 1988). That study concluded that grain yield showed no association with tiller density or tiller mortality (Hucl & Baker, 1988). Biomass production can be modeled by the new DNDC-CSW crop model, which was found to be acceptable for describing crop growth processes (Kröbel et al., 2011).

In addition to FHB resistance and high yield, grain protein is an important attribute of newly registered spring wheat cultivars. Industry standards for protein content are 12% to 14% (Darby 2015). In a study that used mouse bioassays to evaluate the protein quality and digestibility of wheat, it was found that barley protein is superior to wheat protein and that lysine is the first-limiting amino acid in wheat (Bell & Anderson 1984). That study also found certain wheat cultivars with superior performance: 'Neepawa' had the highest protein digestibility and 'Twin' had a superior protein rating (Bell & Anderson, 1984). Nevertheless, in another study, grain yield and grain protein concentration were found to be negatively correlated (Löffler & Busch, 1982).

The objective of this multi-location experiment was to evaluate selected Eastern Cereal and Oilseed Research Centre advanced lines (ECAD lines) from the Spring Wheat Breeding Program at seven different locations across Eastern Canada in order to identify the best lines for performance and grower trials.

2. Materials and Methods

A total of 25 selected hard red spring wheat germplasm lines, including three known cultivars, were chosen from three populations, namely, eight lines from Ontario (ECO406.1-8, ECO427.1-19, ECO439.1-20, ECO441.1-32, ECO446.1-29, ECO448.1-38, EC0330-9, and AW775), eleven lines from Quebec (11NQW-28, 11NQW-112, 11NQW-161, 11NQW-294, 11NQW-372, 11NQW-624, 11NQW-697, 11NQW-842, 11NQW-956, 10NQW-228, and FL62R1), and three lines were selected from the western breeding program (11BS2210, 11BS2288, and 11BW0292), along with three known cultivars to be used as "checks" for comparison purposes ('AC Scotia', 'AC Carberry', and 'Sable'), in seven locations in Eastern Canada (Table 1). Two of the field experiment sites were located in Ottawa, Ontario, at the Eastern Cereal and Oilseed Research Centre: the Central Experimental Farm Advisory Council field (CEF9C) (45° 23′ N, 75° 43′ W) and the Central Experimental Farm no-till field (CEFNT) (45° 22′ N, 75° 43′ W). With regard to the rest of the fields, one was located in Kincardine, Ontario (44° 10′ N, 81° 38′ W), one in Palmerston, Ontario (43° 50′ N, 80° 50′ W), one in St. Isidore, Ontario (45° 23′ N, 74° 54′ W), one in Princeville, Quebec (46° 10′ N, 71° 52′ W), and one in Harrington, Prince Edward Island (46° 21′ N, 63° 10′ W). The 25 wheat lines were arranged or planted in plots (5 × 1.5 m) of six rows spaced 20 cm apart. Depending on the lines, most of the plots were seeded by the beginning of May and harvested in late August or early September. A completely randomized block design with three replicates was used in all locations. Yield, days to head, test weight, thousand-kernel weight, height, protein content (%), lodging, mildew, FHB, DON, and Fusarium-damaged kernels were evaluated at each site.

Table 1. Pedigree of selected lines

Variety	Pedigree
ECO406.1-8	00H04*J3/3/ECO159.13.5.B(BW307/2*HOFFMAN HRF)
ECO427.1-19	BD57-4 (3BS,Lr21)/3/ECO159.13.5.B(BW307/2*HOFFMAN HRF)
ECO439.1-20	NORWELL/AC-06FL-1
ECO441.1-32	AC06FL-75-B/NORWELL
ECO446.1-29	W984-8767(AC BRIO/AC BARRIE)/2/AC06FL-87
ECO448.1-38	BD57-4 (3BS,Lr21)/BAICHUN
11NQW-28	6N-564 / FL62.R1
11NQW-112	F4 PL162.A1 F5 / McKenzie
11NQW-161	FL62R1 / BC21B-83-18 // F5 PL223.C2C F6 / BA83-EC8
11NQW-294	BA83-EC8 / FL62R1 // 03TAB86A1 (4W149.1C) / AC Barrie
11NQW-372	BW297a / NyuBay
11NQW-624	F4 PL259.B1 F5 / FL62R1
11NQW-697	05SFV-106.A2 / 06FL-75.A // GS-1-EM0168 / 06FL-1
11NQW-842	AC Cadillac / F5 PL223.C2C F6
11NQW-956	06FL62A / 06FL-1 // Cadillac / FL62 R1
11BS2210	BG51A-47-7-4
11BS2288	BG51B-40-9-10
11BW0292	BF31A-5-8
10NQW-228	AC Cadillac/FL62R1
AW775	AW622/BD57-4
EC0330-9	AC03W-104(QG22.24/Alsen/2/Blomidon/Alsen)/3/AC Barrie
AC Scotia	AC Helena//Quantum/AC Walton
AC Carberry	Alsen/Superb
FL62R1	QG22.24/Alsen/2/Blomidon/Alsen
Sable	TG3S/B58664HCH

3. Statistical Analysis

Analyses of data obtained from the 25 wheat lines as well as all data obtained from the seven locations in Eastern Canada were analyzed using the MIXED procedure of SAS 9.3 (SAS Institute, Cary, NC, USA) after the homogeneity of the experimental error was examined, and the means were compared using least significant differences (0.05) when the differences were significant.

4. Results

No interaction between line and location, indicating that the relative performance of the lines was similar.

In general, all the advanced lines performed as well as or better than the check varieties (Table 2). The check line FL62R1 had the second lowest mean FHB index. In terms of yield, all varieties except 11NQW-372 (2707 kg/ha) performed better than 'AC Carberry' (2972 kg/ha), and 'AC Scotia' had the best yield (4096 kg/ha). For days to head, line EC0330-9 had the best performance (54 d), equaling that of 'AC Carberry'. For test weight, all lines performed better than 'AC Scotia' (75.3 kg/hl). For thousand-kernel weight, ECO446.1-29, ECO446.1-38, and ECO446.1-19 (40.0, 39.8, and 39.6 g, respectively) had the highest weights, even better than 'Sable' and 'AC Carberry' (34.8 and 34.1 g, respectively). For plant height, all the lines performed better than 'AC Carberry' and 'Sable' (74.5 and 83.7 cm, respectively), with ECO446.1-29 being the tallest (105.4 cm). For protein content, all lines performed better than 'AC Scotia' (12.4%), with ECO446.1-32 having the highest protein content (15.1%). For lodging, all lines performed better than 'AC Scotia' (4.9), with ECO446.1-32 being the line with the best performance (0.5). Of all the lines, ECO439.1-20 had the lowest mildew value (0.1), which was better than the values for 'AC Scotia', 'Sable', and 'AC Carberry' (1.6, 2.1, and 3.3, respectively). For mean FHB index, line 11NQW-294 showed exceptional FHB resistance (6.8%), and lines 11NQW-112 and 11NQW-956 (7.6% and 8.4%, respectively) performed much better than 'AC Scotia', 'Sable', and 'AC Carberry' (10.3%, 21.4%, and 28.4%) did. For DON values, the lines that performed best were 11NQW-956 and 11NQW-624 (6.5 and 7.5 ppm, respectively), achieving lower values than 'Sable', 'AC Carberry', and 'AC Scotia' (7.8, 14.0, and 15.3 ppm, respectively). For Fusarium-damaged kernels, line 11NQW-294 (4.4) performed the best, better than 'Sable', 'AC Scotia' and 'AC Carberry' (7.5, 8.0, and 8.5, respectively) did (Table 2).

Table 2. Average attributes of 25 spring wheat lines tested at seven locations in Eastern Canada in 2014

Variety	Yield (kg/ha)	Rank	Days to head	TSTWT (kg/hl)	TKW (g)	Height (cm)	PROT (%)	LODG (0–9)	Mildew (0–9)	Natural FHB (%)	FHB Index (%)	DON (ppm)	FDK (0-9)
ECO406.1-8	3134	16	56	77.0	37.5	98.0	14.3	1.5	3.4	17.6	21.9	19.8	8.2
ECO427.1-19	3382	5	57	76.6	39.6	98.1	13.6	2.0	4.0	16.9	19.3	18.4	8.0
ECO439.1-20	3199	12	59	76.5	37.4	92.8	14.5	0.8	0.1	8.4	22.0	18.3	7.3
ECO441.1-32	3162	14	59	77.3	37.3	86.5	15.1	0.5	2.1	10.6	15.9	15.8	7.5
ECO446.1-29	3191	13	59	76.7	40.0	105.4	14.2	3.2	1.3	14.5	11.1	9.6	7.5
ECO448.1-38	3322	8	58	77.5	39.8	99.4	14.1	2.2	0.3	23.1	16.9	12.2	7.5
11NQW-28	3145	15	59	77.1	35.4	100.2	13.7	2.9	1.1	13.3	9.5	9.7	6.4
11NQW-112	3378	6	59	78.2	33.7	101.2	14.2	2.4	1.8	3.8	7.6	8.7	6.3
11NQW-161	3105	18	55	78.1	33.3	87.1	14.7	1.7	4.1	7.9	18.8	8.3	7.8
11NQW-294	3323	7	61	78.0	31.4	96.0	14.9	1.0	2.3	4.7	6.8	10.3	4.4
11NQW-372	2707	25	55	77.3	30.0	92.4	14.4	3.0	3.4	12.2	12.0	13.1	8.8
11NQW-624	3013	22	63	76.4	33.8	101.0	14.5	2.7	1.4	6.1	4.2	7.5	5.8
11NQW-697	3212	10	62	76.0	34.9	102.1	13.8	1.5	0.8	5.3	11.3	11.4	8.8
11NQW-842	3095	19	58	77.9	36.1	95.7	14.2	2.7	1.9	4.8	9.5	9.1	7.3
11NQW-956	3050	20	60	78.8	37.3	103.2	14.2	2.7	2.9	7.6	8.4	6.5	7.4
11BS2210	3045	21	57	76.7	35.0	94.0	14.9	1.3	4.1	9.1	14.0	13.4	8.0
11BS2288	3217	9	55	78.5	37.1	91.5	14.1	0.9	3.3	11.9	22.0	23.7	8.8
11BW0292	2992	23	58	76.8	33.3	92.5	13.6	2.5	3.6	9.3	14.1	15.4	8.5
10NQW-228	3212	11	59	76.6	35.7	101.7	14.3	4.7	0.9	8.1	12.8	10.1	6.8
AW775	3873	2	55	76.5	38.7	96.4	12.7	1.7	1.7	18.6	23.4	13.4	8.3
EC0330-9	3126	17	54	76.0	35.5	90.2	14.3	1.8	0.3	13.7	29.3	19.4	7.8
AC Scotia	4096	1	59	75.3	42.7	106.8	12.4	4.9	1.6	9.4	10.3	15.3	8.0
AC Carberry	2972	24	54	77.3	34.1	74.5	14.5	0.3	3.3	7.9	28.4	14.0	8.5
FL62R1	3625	4	61	78.1	33.6	100.0	14.1	4.0	2.0	1.8	4.5	10.5	4.0
Sable	3631	3	57	78.1	34.8	83.7	14.0	1.1	2.1	33.6	21.4	7.8	7.5
LSD ($p = 0.05$)	1431.8		22.6	2.8	3.5	16.6	2.0	1.9	3.2	2.0	3.8	2.6	<0.001

Legend: TSTWT – test weight; TKW – thousand-kernel weight; PROT – protein content; LODG – lodging; FHB – *Fusarium* head blight; DON – deoxynivalenol; FDK – *Fusarium*-damaged kernels; LSD – least significant difference at the 0.05 level.

When the yield parameter was compared among the seven locations (Table 3), the CEFNT location was found to have the highest mean value (4222 kg/ha), which was significantly higher than the mean values recorded for the Kincardine, Princeville, St. Isidore, and Harrington locations (3833, 3427, 1892, and 1020 kg/ha, respectively) but did not differ significantly from the mean values for the Palmerston and CEF9C locations (4206 and 4128 kg/ha, respectively). Comparing the days to head parameter, the CEF9C location had the lowest mean value (47), which was significantly lower than the mean values for the Harrington locations (60) but did not differ significantly from the mean values for the Kincardine, CEFNT, St. Isidore, and Palmerston locations (48, 49, 49, and 49, respectively). In terms of the test weight parameter, the Harrington location had the highest mean value

(94.0 kg/hl), which was significantly higher than the mean values at the Princeville, CEFNT, CEF9C, St. Isidore, and Palmerston locations (78.9, 77.9, 77.5, 77.5, and 76.9 kg/hl, respectively). When the thousand-kernel weight parameter was compared among the seven locations, the Palmerston location had the highest mean value (40.0 g), which was significantly higher than the mean values for the CEF9C, CEFNT, Harrington, St. Isidore, and Princeville locations (37.6, 36.9, 34.5, 34.0, and 33.0 g, respectively). With regard to the plant height parameter, the Palmerston location had the highest mean value (117.9 cm), which was significantly higher than the mean values recorded for the Princeville, CEF9C, CEFNT, Kincardine, St. Isidore, and Harrington locations (105.4, 101.1, 99.2, 89.2, 84.8, and 73.3 cm, respectively). Comparing the protein content parameter among the seven locations, the Princeville location had the highest mean value (14.9%), which was significantly higher than the mean values for the CEF9C, St. Isidore, and CEFNT locations (14.6%, 14.1%, and 12.3%, respectively) but did not differ significantly from the mean value at the Palmerston location (14.7%).

Table 3. Average attributes per location of 25 spring wheat lines tested at seven locations in Eastern Canada in 2014

Locations	Yield (kg/ha)	Rank	Days to head	TSTWT (kg/hl)	TKW (g)	Height (cm)	PROT (%)	LODG (0–9)	Mildew (0–9)	Natural FHB (%)
CEF9C	4128a	2	47c	77.5ab	37.6b	101.1c	14.6b	1.72	-	-
CEFNT	4222a	1	49c	77.9ab	36.9b	99.2c	12.3d	-	-	-
St. Isidore	1892d	6	49c	77.5ab	34.0cd	84.8d	14.1c	-	-	-
Harrington	1020e	7	60b	74.1c	34.5c	73.3e	-	-	0.7	-
Palmerston	4206a	3	49c	76.9b	40.0a	117.9a	14.7ab	3.3	2.0	17.8
Kincardine	3833b	4	48c	-	-	89.2d	-	2.3	-	10.5
Princeville	3427c	5	-	78.9a	33.0d	105.4b	14.9a	1.1	3.8	5.4
LSD ($p = 0.05$)	159		2.6	1.8	1.0	2.6	0.6	-	-	-

Legend: TSTWT – test weight; TKW – thousand-kernel weight; PROT – protein content; LODG – lodging; FHB – *Fusarium* head blight; DON – deoxynivalenol; FDK – *Fusarium*-damaged kernels; CEF9C – Central Experimental Farm field, CEFNT – Central Experimental Farm field with no-till, both located at the Government of Canada research center in Ottawa, Ontario; St. Isidore – Government of Canada Rental field in St. Isidore, Ontario; Harrington – field located at the Government of Canada research center in Harrington, Prince Edward Island; Palmerston – field located at the C&M Seeds company's research facility in Palmerston, Ontario; Kincardine – field located at the Dow AgroSciences research facility in Kincardine, Ontario; Princeville – field located at the Semican company's research facility in Princeville, Quebec; " - " – attribute not measured at that location.

LSD – least significant difference at the 0.05 level.

5. Discussion

A In the current study, the ECAD lines generally performed best among all the tested lines based on the data analysis results for the different attributes. This is probably due to the pedigree of the lines, considering, for example, that 'Norwell' has performed well in terms of most of the attributes described (Canadian Food Inspection Agency, 2015). The lines at the research station in Quebec also have a resistant pedigree thanks to varieties such as 'Cadillac', which has been shown to be resistant to mildew disease and which performs better in drier and cooler environments. As well, FL62R1, which is in the pedigree of some of the varieties, has been shown to be resistant to FHB, which prefers wet and humid environments; the best performance for FL62R1 is obtained in cooler and drier environments.

The results obtained for the average attributes per location for the 25 spring wheat lines (Table 3) clearly demonstrate that some sites contain extreme outliers for certain attributes. These extreme highs and lows can alter the values given in Table 2, which represent the average values for each ECAD line across the seven locations tested. For example, the very low yield and protein content values for the Harrington location could explain the lower per-line yield and protein content estimates shown in Table 2. If these outliers are taken into account, the yield and protein content values for the ECAD lines would likely be much higher than the respective estimates in Table 2, whereas the values for days to head, test weight, and height are likely lower than their respective estimates in Table 2 are. Considering this, the ECAD varieties performed very well when compared to the check varieties.

Given the range of conditions to which the ECAD lines were exposed at the seven different locations, they performed quite well, especially at the CEF9C, CEFNT, Palmerston, and Kincardine locations.

The ECAD lines trial performed very well compared to the checks in this experiment, especially at the locations in Ontario. Further experimentation is required to evaluate whether the extreme outliers obtained for the St. Isidore, Harrington, and Princeville locations are the result of seasonal variations or whether those locations are not suitable growing sites for the ECAD lines.

References

Bell, J. M., & Anderson, D. M. (1984). Protein quality evaluations of selected cultivars of wheat, measured by chemical scores and mouse bioassays. *Can. J. Anim. Sci., 64*(1), 127-137. http://dx.doi.org/10.4141/cjas84-016

Canadian Food Inspection Agency. (2015). "Norwell" variety description. Online. Retrieved from http://www.inspection.gc.ca/english/plaveg/pbrpov/cropreport/whe/app00005506e.shtml

Darby, H. (2015). Heirloom Spring Wheat Seeding Rate Trial, University of Vermont Extension.

DePauw, R. M., Knox, R. E., McCaig, T. N., Clarke, F. R., & Clarke, J. M. (2011). Carberry hard red spring wheat. Can. *J. Plant Sci., 91*, 529-534. http://dx.doi.org/10.4141/cjps10187

Fernandez, M. R., Selles, F., Gehl, D., DePauw, R. M., & Zentner, R. P. (2005). Crop production factors associated with Fusarium head blight in spring wheat in eastern Saskatchewan. *Crop Science, 45*(5), 1908-1916. http://dx.doi.org/10.2135/cropsci2004.0197

Gilbert, J., & Tekauz, A. (2000). Review: Recent developments in research on fusarium head blight of wheat in Canada. *Can. J. Plant Pathol., 22*(1), 1-8. http://dx.doi.org/10.1080/07060660009501155

Hucl, P., & Baker, R. J. (1987). A study of ancestral and modern Canadian spring wheats. *Can. J. Plant Sci., 67*(1), 87-97. http://dx.doi.org/10.4141/cjps87-010

Hucl, P., & Baker, R. J. (1988). An evaluation of common spring wheat germplasm for tillering. *Can J. Plant Sci., 68*(4), 1119-1123. http://dx.doi.org/10.4141/cjps88-133

Humphreys, D. G., Townley-Smith, T. F., Czarnecki, E., Lukow, O. M., McCallum, B., Fetch, T., ... Menzies, J. (2007). Snowbird hard white spring wheat. *Can J. Plant Sci., 87*(2), 301-305. http://dx.doi.org/10.4141/P06-139

Kröbel, R., Smith, W., Grant, B., Desjardins, R., Campbell, C., Tremblay, N., ... McConkey, B. (2011). Development and evaluation of a new Canadian spring wheat sub-model for DNDC. *Can. J. Soil Sci., 91*(4), 503-520. http://dx.doi.org/10.4141/cjss2010-059

Löffler, C. M., & Busch, R. H. (1982). Selection for grain protein, grain yield, and nitrogen partitioning efficiency in hard red spring wheat. *Crop Science, 22*(3), 591-595. http://dx.doi.org/10.2135/cropsci1982.0011183X002200030038x

McCaig, T. N., & DePauw, R. M. (1995). Breeding hard red spring wheat in western Canada: Historical trends in yield and related variables. *Can. J. Plant Sci., 75*(2), 387-393. http://dx.doi.org/10.4141/cjps95-065

McKenzie, H., & Grant, M. N. (1974). Evaluation for yield potential of spring wheat strains from four spring × winter crosses. *Can. J. Plant Sci., 54*(1), 45-46. http://dx.doi.org/10.4141/cjps74-008

Preston, K. R., Kilborn, R. H., Morgan, B. C., & Babb, J. C. (1991). Effects of frost and immaturity on the quality of Canadian hard red spring wheat. *Cereal Chemistry, 68*(2), 133-138.

SAS Institute Inc. (2012). SAS/AF User's Guide, Release 6. SAS Institute Inc., Cary, NC, USA.

Thomas, J. B., & Graf, R. J. (2014). Rates of yield gain of hard red spring wheat in western Canada. *Can. J. Plant Sci., 94*(1), 1-13. http://dx.doi.org/10.4141/cjps2013-160

Wang, H., McCaig, T. N., DePauw, R. M., Clarke, F. R., & Clarke, J. M. (2002). Physiological characteristics of recent Canada Western Red Spring wheat cultivars: Yield components and dry matter production. *Can. J. Plant Sci., 82*(2), 299-306. http://dx.doi.org/10.4141/p01-107

4

Effects of Root-trimming and Cutting-heights on Growth Performance of Potted Native Warm-season Grasses

Vitalis W. Temu[1], David Johnson[1] & Maru K. Kering[1]

[1] Agricultural Research Station, Virginia State University, Petersburg, Virginia, USA

Correspondence: Vitalis W. Temu, Agricultural Research Station, Virginia State University, 238 M.T. Carter Bldg, P.O. Box 9061 Petersburg, Virginia, USA. E-mail: vtemu@vsu.edu

Abstract

Mechanized transplanting of native warm-season grass (NWSG) seedlings raised in biodegradable strip-cups may require trimming outgrown and entwined roots to facilitate individual placement and complete root covering. During establishment, mowing is often used to reduce weed competition and promote tillering. In two randomized complete block split-split-plot design experiments, effects of root-trimming and cutting-height on growth and biomass production of potted NWSGs [big bluestem (BB, *Andropogon gerardii* Vitman), eastern gamagrass (GG, *Tripsacum dactyloides* L.), indiangrass (IG, *Sorghastrum nutans* L.), and switchgrass (SG, *Panicum virgatum* L.)] were assessed. Six-week old seedlings were transplanted, with or without root-trimming, and four of each type and species, assigned to 10-, 15-, or 20-cm cutting-height. All plants were fertilized uniformly and watered sufficiently. After a 7-d adjustment period, plants were clipped to 10 cm which promoted tillering. A three-week regrowth was then allowed before the first of three forage harvests, at assigned cutting-heights. Plant heights were recorded every two weeks after transplanting and on each harvest date. Data were analyzed for effects of root-trimming, cutting-height, and species. Root-trimming had no effect on the parameters. Cutting-height had no effect on plant heights except for second GG and SG regrowths, and/or the third BB and SG. Cutting-height affected only SG forage biomass significantly ($P < 0.05$) during year1 and every species during year2 with 100%+ greater values at the 20- than the 10-cm. All 20-cm average growth rates and belowground biomass in year2 were greater ($P < 0.001$) than the 10-cm by > 100%, but with similar root:total biomass ratios. Overall, species yield increased in the order; IG<BB<GG<SG. With adequate soil moisture and fertility, results indicate that root-trimming may not affect growth or forage biomass of NWSGs during establishment. Mowing NWSGs, during establishment, for up to three 20-cm cuts at ≥ 3-week intervals, may not impact recovery growth or belowground biomass, negatively. Results from field studies are required ahead of practical establishment management recommendations.

Keywords: native grass, crown, transplant, seedling, root-trimming cutting-heights, yield, growth rate

1. Introduction

1.1 Establishment Methods

Establishing NWSG stands from seeds is difficult due to a number of factors including poor germination, improper seeding depths and weed competition (Springer, 2005). Low seedling vigor makes them unable to compete with weeds for resources with low moisture and droughty conditions being frequently responsible for establishment failure (Blake, 1935). Following successful germination and seedling emergence, it usually takes at least two years for NWSG stands to be harvest-ready, and even longer if reseeding of failed patches is involved (Miller & Dickerson, 1999; Temu et al., 2016). Slow development of extensive root systems during early establishment phase is known to be responsible for delayed vegetative growth of NWSGs (Miller & Dickerson, 1999). It is reported that, rapid establishment of adventitious roots is essential for seedling survival (Hyde et al., 1971). Even with effective weed control, the window of favorable growth that ensures availability of sufficient root biomass towards fall is significantly reduced. During this phase, resources are preferentially channeled to initiation of dormant tiller buds and energy reserves for the next spring rather than current vegetative growth. As a result, significant vegetative growth of new NWSG stands can only occur in the next growing season after germination.

As an alternative establishment approach, NWSG seedlings can be raised in a modified environment and be ready for transplanting as soon as field temperatures become favorable for growth. Use of transplants give the NWSGs a growing advantage over weeds at stand establishment, which enables them to finish their first growing season with energy-rich crowns and many dormant tillers for a robust spring growth (Temu, et al., 2016). However, the effectiveness of transplanting in NWSG establishment may depend on factors like seedling age, root biomass, root-covering at planting, soil moisture availability, weed challenges, and early defoliation management as well as their interactions. Therefore, decisions on the appropriate timing and frequency of specific management practices, including harvesting, should take into account possible impacts of the individual factors and/or their interaction effects.

1.2 Transplants and Defoliation Management

When planting is delayed in anticipation of favorable growing conditions, seedlings in biodegradable strip-cups may have their roots outgrow and entwine as to impact mechanized planting operations. Such outgrown and entwined roots may have to be trimmed in order to facilitate individual placement and achieve complete root covering. Information on how root-trimming affects seedling growth and establishment success is scarce and not many studies have been done on it. In the case of GG, for example, root-trimming has been found to result in reduced shoot growth (Roden et al., 2002). However, information about root-trimmed seedlings of other NWSGs remains scarce or non-existent.

Similarly, while strategic mowing during establishment is intended to control weeds and increase the NWSG tiller densities (Meyer et al., 1999), negative impacts on growth and yield can occur if not done appropriately. To ensure a quick recovery, cutting-heights should be high enough to leave sufficient leaf area and minimize loss of growing points. Furthermore, defoliation should not be too intense during the establishment year when the plants are short of energy reserves for recovery growth. Therefore, to avoid undesirable consequences, decisions on defoliation management are usually based on anticipated changes in re-growth rates and subsequent biomass production. However, species differences in leaf morphology and tiller orientation may influence their response to similar defoliation treatments. Information on how the growth of NWSG transplants could be affected by combined root-trimming and the intensity of early defoliation is an important tool for designing appropriate establishment management strategies. Therefore, this study assessed the effects of root-trimming at planting and three cutting-heights on growth rates and forage yields of four potted NWSG species.

2. Materials and Methods

2.1 Study Location and Experimental Layout

The study was conducted in a well ventilated (open-walls) high tunnel at Virginia State University's research farm (Randolph Farm) located in Chesterfield county, Virginia at 37° 13" 43' N; 77° 26" 22' W, and 45 m above sea level. The area has a 20-year average June, July, and August day temperatures of 30.2, 32.1, and 31.2 °C, respectively (Satellite N.O.A.A., 2013). In two successive years, the experiments were ran in a randomized complete block design with a split-split-plot treatment allocation for effects of root-trimming (main-plot factor), species (sub-plot factor), and cutting-height (sub-sub-plot factor).

Degradable strip-cups, 2×2 cm top and 2 cm deep, filled with germination pot media (Premier ProMix Germinating Mix 3.8CF PGX, Griffins Greenhouse supplies, Richmond, VA) were arranged on perforated flat trays and seeded with BB, GG, IG, and SG. Seeded trays were placed on greenhouse tables covered with a plastic sheet and kept moist by bottom-up watering. After six weeks of growth in the high tunnel, 24 seedlings of each species were transplanted into plastic greenhouse pots (28 cm-top, 20 cm-bottom, and 30 cm-deep) filled with the same pot medium. At transplanting, 12 seedlings had their outgrown roots trimmed with a matching set of 12 left intact. All potted plants received the same fertilizer treatment and were watered as needed to ensure sufficient moisture throughout the experimental periods. All potted plants were allowed a 7-d long initial growth period, so they could adjust to the new environmental conditions and recover from stresses associated with transplanting. After the adjustment period, the 12 root-trimmed and 12 intact plants of each species were randomly assigned to 10-, 15-, or 20-cm cutting-heights (treatments) and thus replicating each treatment four times. All plants were then clipped at 10-cm height, to promote tillering, and allowed a 30-d long regrowth before the first forage harvest. The pre-treatment clipped biomass was insignificant and, therefore, discarded. For each species, the set of 24 pots was arranged on a separate table (2.4-m L × 1.2-m W). Similarly, a second year trial was set, but with longer respective regrowth periods (23-, 25-, and 40-d) to the first, second, and third harvests, respectively. At the same cutting-heights, increasing the regrowth period was considered necessary to allow significant recovery and substantial biomass production.

2.2 Measurements and Data Collection

2.2.1 Aboveground Measurements

During the 30-d long first re-growth in the first year, two bi-weekly plant leaf height (LH) measurements as vertical distance from the exposed pot media (synonymous with ground surface) to the top bending point of most leaves were recorded on four random tillers per pot between 9:00 and 10:00 am. After the second bi-weekly height measurement, plants were clipped as assigned and first re-growth forage biomass determined. For each potted plant, the clipped material was weighed before and after oven-drying (at 65 °C) to constant weight. Similarly, subsequent LH measurements as well as the second (15-d) and third (17-d) re-growth shoot biomass were obtained to assess yield response to root-trimming and cutting-height treatments. During the second year, growth and yield data were collected and processed similarly. However, due to unexplained lack of treatment differences in the recorded year1 plant heights, LH measurements were considered potentially unreliable and excluded from the second year data set.

To assess how the treatments might affect plant response to defoliation, the average growth rate (AGR) on the basis of mean shoot biomass production was calculated as [AGR = DW/number of days between successive cuts]. To correct for differences in reserve carbohydrates immediately before the preceding harvest on the rate of recovery growth, the calculated AGR values were expressed as proportions of their preceding harvested DWs to get the respective relative growth rates (RGR). In doing so, the first and second harvest yields were regarded as indicators of the plant sizes that produced the respective second and third harvest weights. Thus [RGR2 = (AGR2/DW1)*100] and [RGR3 = (AGR3/DW2)*100].

2.2.2 Belowground Biomass

After the final harvest, during the second year, each pot was repeatedly watered with a garden shower head then emptied into a 40-L plastic container in which the contents were gently moved up and down to free the loosely held potting media. Then the semi-cleaned root-crown mass was repeatedly flushed with the shower head over a wire mesh that trapped any roots breaking loose. The roots were trimmed off the crown and separately oven-dried to constant weight after which the dry weights were recorded. For each pot, the root and crown dry weights were added to the cumulative forage weight to get total (above- and belowground) biomass. The root weight was then divided by the total biomass to get the respective root:total biomass (RTB) ratio and used for establishing changes in allometric relationships.

2.2.3 Statistical Analyses

The data were analyzed using a computer-based statistical software, the proc GLM, SAS 9.4, Copyright (c) 2002-2014 by SAS Institute Inc., Cary, NC, USA. During the statistical analyses, individual harvest weights for each treatment, within species, were combined as total forage biomass. Respective treatment means were compared within and between species. Means separation was according to Fishers LSD at α = 0.05.

3. Results and Discussion

For convenience, the results on measured and derived forage biomass responses to treatment from year1&2 data sets are discussed in separate subsections. Root-trimming had no effect on all parameters determined (Table 1) and was, therefore, omitted from subsequent analyses. Also, due to significant species*cutting-height interactions, results are reported separately for cutting-heights within species and the *vice versa*.

3.1 Effects of Root-Trimming on Growth

The observed lack of root-trimming effect on plant heights and forage DM yield suggests that the root loss imposed was probably not severe enough to impact shoot growth. In fact, in the case of GG, a < 50% loss of root mass is known to have no impact on subsequent shoot growth (Roden et al., 2000). It is likely that the uniform watering and fertilizer application adopted in the current study helped to mask differences, if any, associated with the root biomass at planting. Because root-trimming had no effect on all parameters determined, as shown in the summary of ANOVA for total forage biomass (Table 1), only effects of cutting-height and species are discussed.

3.1.1 Plant Regrowth Heights

As summarized in Table 2, and for each species, the first year plant heights recorded during the first 30-d regrowth showed no treatment difference. However, following the first forage harvest, regrowths in GG and SG were taller for the 20-cm than 10-cm (Table 2). For the second regrowth, SG and BB cut at 20-cm had taller plants than those cut at 10-cm height. For each harvest, regrowth heights in IG were similar for all cutting-heights.

Table 1. Summary of analysis of variance for effects of root-trimming (Root loss), Cutting-height (Cut-height), and species on cumulative forage biomass of potted big bluestem, gamagrass, indiangrass, and switchgrass recorded during the first (Year1) and second (Year2) experiments

Source of variation		First year			Second year		
	DF	SS	F Value	Pr > F	SS	F Value	Pr > F
Model	41	7238.5	4.97	<.0001	41566.55	32.99	<.0001
Rep	3	11.99	0.11	0.9524	18.26	0.2	0.8972
Root loss	1	24.29	0.68	0.4121	3.88	0.13	0.7237
Rep*Root loss	3	90.37	0.85	0.474	46	0.5	0.6845
Cut height	2	174.39	6.3	0.0135	11692.64	190.26	<.0001
Root loss*Cut-height	2	38.6	0.54	0.5842	78.52	1.28	0.287
Rep*Root loss*Cut-height	12	166.14	0.39	0.9617	175.8	0.48	0.9197
Species	3	5117.36	47.99	<.0001	27238.75	295.49	<.0001
Species*Root loss	3	65.3	0.61	0.6099	27.73	0.3	0.8246
Species*Cut-height	6	312.27	1.46	0.208	2188.58	11.87	<.0001
Species*Root loss*Cut-height	6	98.39	0.46	0.8338	96.39	0.52	0.7885
Error	54	1919.48	-	-	1659.28	-	-
Corrected Total	95	9157.99	-	-	43225.84	-	-

The first year data set was from potted plants harvested at 30-, 15-, and 17-d intervals while the corresponding second year data set was of plants harvested at 23-, 25-, and 40-d intervals, respectively, but at same heights.

Table 2. Growth response of potted big bluestem, gamagrass, indiangrass, and switchgrass to cutting-heights (cm) in an open-sided high tunnel based on bi-weekly mean plant height[†] measurements between August and September, inclusively

Cut-height (cm)	Species plant-heights			
	Initial 30-d growth[‡]		Bi-weekly regrowth	
	1st 15-d	2nd 15-d	First	Second
-------cm-------				
-------Big bluestem-------				
10	11.3	15.4	20.5	11.4b[‡‡]
15	10.9	14.3	19.8	13.0b
20	10.9	16.0	25.3	15.7a
Pr > α[§]	0.77	0.15	0.17	<0.001
-------Gamagrass-------				
10	12.8	20.9	16.0b	17.1
15	11.8	19.5	16.9b	17.9
20	13.2	21.6	19.9a	19.9
Pr > α	0.33	0.50	0.02	0.06
-------Indiangrass-------				
10	13.6	15.1	20.0	11.0
15	12.6	14.3	17.9	11.5
20	13.5	15.3	19.0	12.5
Pr > α	0.49	0.44	0.67	0.31
-------Switchgrass-------				
10	10.2	14.9	16.8b	16.4b
15	11.4	16.1	26.9a	17.4ab
20	11.0	16.3	29.1a	19.1a
Pr > α	0.22	0.07	<.001	0.035

[†]Plant heights recorded as vertical distance from the port surface to the bending of topmost leaf blades (three pot[-1]). [‡]Unlike all other, there was no harvest event preceding the second bi-weekly height measurements; [‡‡]Means of the same species within a column followed by the same letter are not significantly different at α = .05. [§]The probability of difference between means of the same species within column.

3.2 Effects of Cutting-Heights on Subsequent Growth Performance

3.2.1 Forage Biomass

From the year1 data set, species-wise results of forage biomass pot^{-1} are summarized in Table 3. At the 30-d long first harvest, the GG and SG forage biomass showed no treatment difference but the matching values for BB and IG were consistently greater for the 10-cm than the 15- and 20-cm cut heights (Table 3). At the second regrowth harvest, only SG showed treatment differences with 20-cm cutting-height producing significantly ($P=0.05$) greater biomass (14.4 g DM pot^{-1}) than the other cutting-heights. At the third regrowth harvest, SG and GG cut at 20-cm height produced significantly greater biomass than that cut at 10-cm for both species. While total forage biomass for all species portrayed an upward trend as the cutting-heights increased, it is only in SG that the 20-cm cutting-height produced a significantly greater biomass (29 g pot^{-1}) compared to the 10-cm treatment. The observed SG total forage biomass values were 9-units greater for the 20-cm (29 g) than the 10-cm, but similar to that for the 15-cm (25 g) harvest-height.

Table 3. Effects of cutting-heights (cm) on initial 30-d and subsequent bi-weekly regrowth forage yields and growth rates of potted big bluestem, gamagrass, indiangrass, and switchgrass in an open-sided high tunnel between August and September, inclusively, during the first year

Height (cm)	Regrowth dry matter yield				Average growth rate[†]			Relative growth rate[‡]	
	Cut1[‡‡]	Cut2	Cut3	Total	Cut1	Cut2	Cut3	Cut2	Cut3
	----------------g pot^{-1}----------------				-----------mg d^{-1} pot^{-1}-----------			----- mg g^{-1} d^{-1} -----	
					------Big bluestem------				
10	6.1a^{AB§}	6.5^{BC}	5.6^{B}	18^{B}	204a^{AB}	433^{BC}	331^{B}	75c^{B}	58^{AB}
15	3.1b^{BC}	7.0^{C}	5.6^{B}	16^{C}	104b^{BC}	467^{C}	331^{B}	176b^{AB}	49^{A}
20	3.0b^{CB}	7.0^{B}	6.7^{B}	17^{C}	100b^{BC}	467^{B}	397^{B}	194a^{B}	60^{A}
Pr > α[¶]	<0.001	0.79	0.51	0.36	<0.001	0.79	0.51	0.05	0.66
					------Gamagrass------				
10	8.5^{A}	10.5^{A}	9.6b^{A}	29^{A}	283^{A}	700^{A}	566b^{A}	87b^{B}	63^{A}
15	7.6^{A}	13.1^{A}	10.4ab^{A}	31^{A}	254^{A}	875^{A}	610ab^{A}	124a^{B}	47^{A}
20	7.7^{A}	13.5^{A}	12.9a^{A}	34^{A}	258^{A}	900^{A}	757a^{A}	124a^{B}	57^{A}
Pr > α	0.88	0.38	0.05	0.42	0.88	0.38	0.05	<0.038	0.52
					------Indiangrass------				
10	3.1a^{C}	4.5^{C}	3.1^{C}	11^{C}	104a^{C}	300^{C}	184^{C}	112^{AB}	41^{AB}
15	1.6b^{C}	3.9^{D}	3.5^{C}	9^{D}	54b^{C}	258^{D}	206^{C}	196^{AB}	64^{A}
20	1.9b^{C}	4.2^{C}	4.0^{C}	10^{D}	62b^{C}	283^{C}	235^{C}	173^{B}	56^{A}
Pr > α	0.04	0.69	0.54	0.53	0.04	0.69	0.54	0.15	0.48
					------Switchgrass------				
10	5.0^{BC}	9.5b^{AB}	6.2b^{B}	21b^{B}	167^{BC}	633c^{AB}	368b^{B}	137b^{A}	39^{B}
15	4.2^{B}	11.4b^{B}	9.6a^{A}	25ab^{B}	141^{B}	758b^{B}	566a^{A}	218ab^{A}	51^{A}
20	4.0^{B}	14.4a^{A}	11.0a^{A}	29a^{B}	133^{B}	958a^{A}	647a^{A}	302a^{A}	45^{A}
Pr > α	0.62	<0.001	<0.01	<0.01	0.62	<0.001	<0.01	0.02	0.23

[†]Average increase in forage dry matter (DM) day^{-1}. [‡]Increase in forage DM day^{-1} g^{-1}of the respective preceding harvest weight. [‡‡]Numbers 1-3 indicate order of three sequential harvests following a 30-, 15-, and 17-d long regrowth period, respectively. [§]Within a column, means of the same species followed by the same lowercase letter or the same cut height across species followed by the same uppercase letter are not significantly different at α = .05. [¶]The probability of difference between means of the same species within the column.

During the second year, all species had consistently greater forage yields ($P <.001$) for the 20-cm treatment than the other two (Table 4). Consistently also, forage biomass values were the least for the corresponding 10-cm treatment although not significantly different from the 15-cm ones for the second and third BB harvests or the second of GG ($P > .05$). For each species, however, cumulative forage biomass was significantly greater for the 20- and least for 10-cm treatment ($P < .001$). In fact, in all species, the 20-cm cutting-height produced >100% more forage DM than its 10-cm counterpart. The treatment differences in forage biomass were actually consistent with reported negative effects of severe defoliation on plant growth (Ferraro & Oesterheld, 2002). Usually, severely defoliated plants suffer irreversible tissue damages that eventually reflect in reduced subsequent yields. In fact, multiple defoliations have been found to reduce subsequent herbage biomass of warm-season grasses by over 60% (Mullahey, 1990; Forwood & Magai, 1992). This is so because proportions of photosynthetic tissue retained on defoliated plants usually influence how quickly they repair their damaged tissues (Oesterheld & McNaughton, 1991; Lee et al., 2000; Ferraro & Oesterheld, 2002). With severe defoliation, plants lack sufficient residual leaves to supply enough carbon for maintenance and regrowth. They remain in "negative carbon", consuming stored resources until their photosynthetic leaf areas are sufficiently restored (Richards, 1993).

The noted year differences in response to treatment clearly demonstrate the importance of sufficient recovery growth before plants experience subsequent defoliations. That allows the defoliated plants to restore their carbohydrate reserves, which usually influence stand persistence (Slepetys, 2008). During the first year of the current study, recovery growths towards the second and third harvests lasted only about two weeks, while, during the second year, the first and second regrowths took approximately three weeks and nearly five for the third. Based on the current results, three weeks recovery period seemed long enough for NWSG plants cut at 20 cm to effectively restore their photosynthetic capacities, provided soil moisture and nutrient supplies are not limiting.

Table 4. Effects of cutting-heights (cm) on mean forage productivity, belowground biomass, and root:total biomass (RTB) ratio of potted big bluestem, gamagrass, indiangrass, and switchgrass in an open-sided high tunnel from three consecutive harvests (Cut1-3) between mid-July and October

Height (cm)	Regrowth forage yield and growth rate									Belowground biomass		
	Forage yield				AGR[†]			RGR[‡]		Root & Crown		Ratio[§]
	Cut1[‡‡]	Cut2	Cut3	Total	Cut1	Cut2	Cut3	Cut2	Cut3	Crown	Root	RTB
	--------------- g DM pot^{-1} ---------------				------ mg d^{-1} pot^{-1} ------			--mg g^{-1}d^{-1}--		------ g pot^{-1}-----		
	--Big bluestem--											
10	4.5c^{BC¶}	2.6b^C	0.6b^B	7.71c^C	197c^{BC}	104b^C	14b^B	24^C	5^C	14.6c^B	9.9c^B	0.30^A
15	8.4b^B	4.2b^C	1.2b^B	13.7b^C	365b^B	167b^C	29b^B	21^C	7^C	24.5b^C	17.2b^B	0.31^A
20	14.5a^{B¶}	8.5a^C	2.7a^B	25.7a^C	629a^B	340a^C	68a^B	24^C	8^C	37.9a^C	28.4a^B	0.30^A
	---Gamagrass--											
10	5.6c^B	10.8b^B	8.6a^A	25.0b^B	242c^B	430b^B	216a^A	87^A	21^A	33.1b^B	11.5b^B	0.16^C
15	9.6b^B	14.6b^B	13.2b^A	37.5b^B	419b^B	585b^B	331b^A	64^A	24^A	41.5b^B	11.5b^C	0.13^D
20	15.6a^B	22.5a^B	18.7a^A	56.9a^B	680a^B	900a^B	469a^A	60^A	22^A	67.4a^B	17.7a^C	0.12^D
	--Indiangrass--											
10	2.6c^C	2.7c^C	1.2c^B	6.5c^C	114c^C	107c^C	30c^B	49^B	12^B	4.6c^C	2.6c^C	0.18^C
15	5.6b^C	5.4b^C	2.7b^B	13.7b^C	243b^C	217b^C	69b^B	40^B	13^B	9.75b^D	5.5b^D	0.18^C
20	8.6a^C	8.2a^C	4.1a^B	21.0a^C	376a^C	328a^C	102a^B	40^B	13^B	17.0a^D	9.4a^D	0.20^C
	--Swichgrass--											
10	11.1c^A	13.6c^A	7.5c^A	32.3c^A	483c^A	546c^A	187c^A	54a*^{B}	14^B	32.5c^A	19.5c^A	0.23^B
15	20.3b^A	20.7b^A	12.7b^A	53.8b^A	884b^A	829b^A	319b^A	41b^B	16^B	50.0b^A	30.2b^A	0.22^B
20	29.1a^A	28.7a^A	17.9a^A	75.7a^A	1267a^A	1150a^A	447a^A	41b^B	16^B	77.7a^A	49.9a^A	0.24^B
Pr > α#	<.001	<.001	<.001	<.001	<.001	<.001	<.001	>.1	>.1	<.001	<.001	>.1

[†]Average Increase in forage dry matter (DM) day^{-1}. [‡]Increase in forage DM day^{-1} g^{-1} of the respective preceding harvest weight. [‡‡]The number indicates the order in three sequential harvests following a 23-, 25-, and 40-d long regrowth period, respectively. [§]A ratio obtained by dividing the recovered root mass by the combined above- and belowground biomass (RTB). [¶]Within a column, means of the same species followed by the same lowercase letter or the same cut height across species followed by the same uppercase letter are not significantly different at α = .05. *For switchgrass only, treatment means differed significantly (P = .03). [#]The probability of difference between means of the same species within a column.

3.2.2 Daily Weight Gains and Relative Growth Rate

With respect to forage production, management decisions on appropriate harvest regimes are better based on the rate at which plants may recover from defoliation events. In the current study, the year1 AGRs, based on estimated daily weight gains (mg d^{-1}), showed that regrowth rates following the first (30-d) harvest were faster for the 10-cm than the 15- and 20-cm, for BB (204 mg d^{-1}) and IG (104 mg d^{-1}) (Table 3). The corresponding GG or SG values were not statistically different and averaged 265 and 147 mg d^{-1}, respectively. Towards the second harvest events, only SG exhibited treatment differences in AGRs, with mean daily gain for the 20-cm cutting-height (958 mg d^{-1}) being greater than for the 15- and 10-cm cutting-heights. Towards the third harvest, however, AGR for the 20-cm cut SG (662 mg d^{-1}) was 258 units faster than for the 10-cm, but statistically similar to the 15-cm one. Again, on the second year data, all species showed clear and consistent treatment differences in AGR (Table 4) with greater ($P < 0.001$) values for the 20-cm, than the 15-cm, and 10-cm cutting-heights, respectively.

Exceptions were BB towards the second and third harvests and GG towards the second harvest. Generally, AGRs for plants in the 20-cm were over 100% greater than those for 10-cm cutting-height. These results are consistent with the assertion that negative effects of severe defoliation on recovery growth are influenced by the proportions of their residual photosynthetic tissues (Crider, 1955; Ferraro & Oesterheld, 2002). Overall, plants tended to regrow faster towards the second harvest than their respective first and third harvests.

On hay fields, differences in pre-harvest energy reserves and stand vigor will influence the rates at which subsequent harvests could be realized. So, to appropriately assess response to defoliation, it is important that likely influence of initial plant size on recovery growth and/yield performance of defoliated grasses is also considered. In this section, therefore, possible effects of the pre-harvest plant sizes on regrowth yields calculated as daily forage biomass production per gram of the preceding harvest weights, as RGR estimates, are discussed.

The statistical analysis results on the respective RGR values for the second and third harvests are presented in Table 3. There was an increase in RGR for higher cutting-heights towards the second harvest, during year1. This increase in RGR was significant in all species except IG. However, towards the third harvest, RGR differences between cutting-heights, in year1, were only numerical. During the second year, RGR values towards the second harvest for BB, GG, and IG or the third for each species (Table 4) showed no treatment difference ($P > 0.1$). The fact that even for SG only one treatment differed from the rest ($P = 0.03$) and that this higher RGR value for the 10-cm was inconsistent with its forage biomass ranking makes it an isolated outlier. The observed declines in AGR and forage biomass with low cutting-heights clearly demonstrated the practical significance of appropriate harvest management in NWSG stands. Cutting NWSGs too low causes severe loss of growing points, thus leaving recovery more dependent on new sets of leaves (Briske 1986) and subsequently reduce forage biomass. At the same harvest frequency, plants cut too low take longer to re-establish sufficient photosynthetic leaf area and will, therefore, have relatively lower cumulative yields. So, the observed consistent treatment differences in derived yield responses, during the second year, are attributable to the longer recovery periods allowed.

3.2.3 Belowground Biomass

Cutting-height had significant effects ($P < 0.05$) on both root and crown weights (Table 4). In all species, the crown and root biomass produced for the 20-cm cutting-height were greater than for the other cutting-heights. In fact, it exceeded that for the 10-cm cutting-height by over 50%. It is only in GG that the differences between the 15-cm and 10-cm cutting-heights were not significant ($P > 0.05$). Over all, the magnitudes of the decrease in root and crown biomass were consistent with the severity of defoliation associated with the cutting-heights. The observed decline in belowground biomass weight agrees with reported negative impacts of multiple defoliations on root weight and their nonstructural carbohydrates content in grasses (Christiansen and Svejcar, 1987; Engel et al,. 1998). In fact, immediately following defoliation, grasses usually experience a stoppage in root growth that reflects the percentage of foliage removed and continues until recovery of the top growth is advanced (Crider, 1955). For example, in most C$_3$ and C$_4$ grasses, root growth ceases immediately following a $\geq 50\%$ leaf area removal (Richards, 1993; Turner et al., 1993). For shorter cutting-heights and/or harvest frequencies, therefore, reduction in respective root and crown biomass is expected, an observation also made in the current study. This is so because of preferential resource allocation to aboveground growth at the expense of roots, a scenario usually exhibited by plants recovering from defoliation (Richards, 1984; Turner et al., 1993; Turner et al., 2007).

The demonstrated decrease in crown and root biomass for the shorter cutting-heights has implications on how long a recovery growth should be allowed before plants can be considered ready for the next harvest. Subsequent defoliations that do not allow enough time for recovery may cause progressive decline in growth performance due to weakening of the plants, which may favor the growth of undesirables. Crown size is important as it is the origin of the adventitious roots, the main rooting system in warm-season grasses (Meltcalfe & Nelson, 1985). All species showed no effect of cutting-height in their derived RTB ratios, implying that defoliation affected the below and aboveground biomass production in similar proportions.

3.3 Species Response to Cutting-Heights

3.3.1 Species Forage Biomass

In the current study, the forage biomass means showed significant species differences (Table 3). During the first year, and for each cutting-height, IG had the least ($P < 0.001$) forage values at each harvest, as well as total biomass while GG generally had the greatest forage weight. For 15- and 20-cm cutting-heights, SG had third harvest biomass values similar to GG, while for the 10-cm cutting-height values were similar to those of BB (Table 3). At the second harvest, SG also had similar yield values to those of GG for the 20-cm cutting-height. During the second year, SG had greater ($P < 0.001$) per harvest and total forage weight values than any other species except GG at the third cut. Generally, GG appeared the second ranked species in forage biomass with both BB and IG showing more or less similar but the least values. The observed species biomass differences demonstrate variations in their abilities to recover from defoliation and the likelihood of altering the subsequent biomass proportions in mixed stands. Usually, plant species differ in their abilities to compensate for defoliation-imposed tissue damages (Dawson et al., 2000) as reflected in respective regrowth rates (van Staalduinen and Anten, 2005). In mixed stands, such differences may lead to subsequent changes in forage biomass (Temu et al., 2014), sward structure, and/species composition (Temu et al., 2015). Owing to differences, in species' ability to restore lost photosynthetic capacities, recovery growths often result in under/overcompensation of preceding tissue damages. Additionally, the notable similarities in BB and IG forage biomass response to intensities of defoliation suggest that harvesting may not drastically alter their proportional contributions to total biomass from shared mixed stands.

3.3.2 Species Average and Relative Growth Rates

Species differences were also observed in the rate at which the NWSGs produced the recorded forage biomass. During the first year, mean AGR values were greater for GG than for any other species except SG at the second and third harvests (Table 3). Towards each harvest, the rate of increase in forage biomass for IG was the least compared to any other species. For all harvests, the AGR values for GG and SG were over 100% greater than those of IG. The RGR for the second cut were greatest for SG, although not significantly different from that for the 10- or 15-cm cut IG. All species generally showed comparable RGR for the third harvest. The second ranked RGR value was for GG and BB that were significantly similar to each other. However, there was no species difference in RGR towards the third harvest. In the second year experiment, AGR values (Table 4) also showed notable differences in species rankings. For every cutting-height, SG had greater AGR values than any other species ($P < 0.001$) towards the first and second harvests, but similar to GG towards the third. There was also no significant AGR difference between BB and IG towards the second or third harvest. The second year species RGR values were greater for GG and the least for BB with no significant difference between IG and SG. These RGR values are comparable to previously reported research results of other NWSGs (Coyne and Bradford, 1995). Over all, SG and GG were the first and second most productive of the four NWSGs, respectively.

3.3.2 Species Belowground Biomass

On the second year data, crown weights were greater ($P < 0.001$) in SG than any other species except for the 10-cm cut GG (Table 4). For each cutting-height, SG also had greater root biomass (up to nearly 50 g) than any other species whose values ranged from as low as 2.6 g in IG to nearly 28.4 g in BB. Except for the 10-cm cut GG, the least crown and root weights were in IG followed by BB and GG, in ascending order. There were significant species differences in their RTB ratios with values in the order of BB > SG > IG > GG. Although comparable diversities in species response to defoliation attributable to differences in compensatory mechanisms have been reported (Meyer, 1998; Smith, 1998; Gutman et al., 2001), the magnitudes of the RTB ratio values, in the current study, suggest that root systems of the more severely impacted species had better efficiencies of obtaining soil-based resources for growth. Plants with better resource uptake and/utilization efficiencies are more likely to survive severe defoliation events, even when their forage productivity declines, which may later on be reversed by appropriate management.

4. Conclusions

There being no effect of root-trimming on all parameters determined, data indicate that, trimming outgrown and/entwined roots to facilitate seedling placement in mechanized planting of NWSGs may not negatively impact their growth or biomass production during establishment, under similar soil moisture and nutrient supply. That with comparable growing conditions, mowing transplanted NWSGs at 20-cm, during establishment, may not negatively impact their growth performance provided they are allowed a ≥ 3-week recovery period. Because the belowground and forage biomass response to treatments were in similar trends, data indicate that severe defoliation during establishment may impact stand persistence, negatively. The demonstrated greater species susceptibilities of BB and IG to tissue damages associated with intensive defoliation suggest that their proportions in frequently harvested mixed stands with GG and/or SG may be drastically reduced.

Acknowledgments

The authors are grateful to the USDA Evans Allen program for funding the study, the management of the

Agricultural Research Station in the College of Agriculture at Virginia State University for housing the project, as well as providing logistical and material support to the research team. The authors are also grateful to Kevin Kidd and Christos Galanopoulos for their help with trial management and data collection during the research. This article is a publication No. 330 of the Agricultural Research Station, Virginia State University.

References

Blake, A. K. (1935). Viability and germination of seeds and early life history of prairie plants. *Ecol. Monogr., 5*(4), 408-460. http://dx.doi.org/10.2307/1943035

Briske, D. D. (1986). Plant response to defoliation: morphological considerations and allocation priorities. p. 425-427. In P. J. Joss, P. W. Lynch & O. B. Williams (Eds.), *Rangelands: a resource under siege*. Proc. second Int. Rangeland Cong. Adelaide, Australia.

Christiansen, S., & Svejcar, T. (1987). Grazing effects on the total nonstructural carbohydrate pools in Caucasian Bluestem. *Agron. J., 79*, 761-764.

Coyne, P. I., & Bradford, J. A. (1985). Morphology and growth in seedlings of several C4, perennial grasses. *J. Range Manage., 38*, 504-512.

Crider, F. J. (1955). *Root-growth stoppage resulting from defoliation of grass* (No. 1102). US Department of Agriculture.

Dawson, L. A., Grayston, S. J., & Paterson, E. (2000). Effects of grazing on the roots and rhizosphere of grasses. In G. Lemaire, et al. (eds.), *Grassland ecophysiology and grazing ecology* (pp. 61-84). CAB Publishing.

Engel, R. K., Nichols, J. T., Dodd, J. L., & Brummer, J. E. (1998). Root and shoot responses of Sand Bluestem to defoliation. *J. Range Manage., 51*, 42-46. http://dx.doi.org/10.2307/4003562.

Ferraro, D. O., & Oesterheld, M. (2002). Effect of defoliation on grass growth. A quantitative review. *Oikos, 98*, 125-133. http://dx.doi.org/10.1034/j.1600-0706.2002.980113.x.

Forwood, J. R., & Magai, M. M. (1992). Clipping frequency and intensity effects on big bluestem yield, quality, and persistence. *J. Range Manage., 45*, 554-559.

Gutman, M., Noy-Meir I., Pluda D., Seligman N., Rothman S., & Sternberg, M. (2002). Biomass partitioning following defoliation of annual and perennial Mediterranean grasses. *Conserv Ecol., 5*, 1. Retrieved from http://www.consecol.org/vol5/iss2/art1

Lee, W. G., Fenner, M., Loughnan, A., & Lloyd, K. M. (2000). Long-term effects of defoliation: Incomplete recovery of a New Zealand Alpine Tussock Grass, *Chionochloa Pallens*, After 20 Years. *J. Appl. Ecol., 37*(2), 348-55. Retrieved from http://www.jstor.org/stable/2655915

Metcalfe D. S., & Nelson, C. J. (1985). The botany of grasses and legumes. In M. E. Heathet et al. (eds.), *Forages* (4th ed., pp. 52-63). Iowa State Univ. Press, Ames.

Meyer, G. A. (1998). Mechanisms promoting recovery from defoliation in goldenrod (Solidago altissima). *Can. J. Bot., 76*, 450-459.

Meyer, G. C., Melvin III, N. C., Turner, T. R., & Swartz H. J. (1999). Native warm-season grass establishment as affected by weed control in Maryland coastal plains (pp. 212-221). In *Proceedings of the 2nd Eastern Natives Symposium, Baltimore, MD*.

Miller, C. F., & Dickerson, J. A. (1999). The use of native warm-season grasses for critical area stabilization. In *Proceedings of the 2nd Eastern Native Grass Symposium, Baltimore, MD*. November 1999. Retrieved from https://efotg.sc.egov.usda.gov/references/public/VT/WSGguide.pdf

Mullahey, J. J., Waller, S. S., & Moser, L. E. (1990). Defoliation effects on production and morphological development of little bluestem. *J. Range Manage., 43*, 497-500.

Oesterheld, M. (1992). Effect of defoliation intensity on aboveground and belowground relative growth rates. *Oecol., 92*, 313-316.

Oesterheld, M., & McNaughton, S. J. (1991). Effect of stress and time for recovery on the amount of compensatory growth after grazing. *Oecol., 85*, 305-313.

Rhoden, E. G., Reeves III, J. B., Krizek, D. T., Ritchie, J. C., & Foy, C. D. (2000). Influence of root removal on shoot regrowth and forage quality of greenhouse-grown Eastern gamagrass. In *Proceedings of the Second Eastern Native Grass Symposium* (p. 276). DIANE Publishing.

Richards, J. H. (1984). Root growth response to defoliation in two Agropyron bunchgrasses: field observations with an improved root periscope. *Oecol., 64,* 21-25.

Richards, J. H. (1993). February. Physiology of plants recovering from defoliation. In *Proceedings of the XVII international grassland congress* (vol. 1993, pp. 85-94). Palmerston North, New Zealand: New Zealand Grassland Association.

SAS Institute Inc. (n.d.). Cary, NC, USA.

Satellite, N. O. A. A. (2013). Information Service. National Climatic Data Center. *US Dept of Commerce.*

Slepetys, J., & Šterne, D. (2008). The productivity and persistency of pure and mixed forage legume swards. *Latvian Journal of Agronomy, 11,* 276-281.

Smith, S. E. (1998). Variation in response to defoliation between populations of Bouteloua curtipendula var. caespitosa (Poaceae) with different livestock grazing histories. *American Journal of Botany, 85,* 1266.

Springer, T. L. (2005). Germination and early seedling growth of chaffy-seeded grasses at negative water potentials. *Crop Sci., 45,* 2075-2080.

Temu, V. W., Baldwin, B. S., Reddy, K. R., & Riffell, S. K. (2015). Harvesting Effects on Species Composition and Distribution of Cover Attributes in Mixed Native Warm-Season Grass Stands. *Environments, 2*(2), pp.167-185. http://dx.doi.org/10.3390/environments2020167

Temu, V. W., Kering, M. K., & Rutto, L. K. (2016). Effects of Planting Method on Enhanced Stand Establishment and Subsequent Performance of Forage Native Warm-Season Grasses. *Journal of Plant Studies, 5*(1), p38. http://dx.doi.org/10.5539/jps.v5n1p38

Temu, V. W., Rude, B. J., & Baldwin, B. S. (2014). Yield response of native warm-season forage grasses to harvest intervals and durations in mixed stands. *Agron., 4,* 90-107.

Turner, C. L., Seastedt, T. R., & Dyer, M. I. (1993). Maximization of Aboveground Grassland Production: The Role of defoliation frequency, intensity, and history. *Ecological Applications, 3,* 175-186.

Turner, L. R., Donaghy, D. J., Lane, P. A., & Rawnsley, R. P. (2007). Patterns of leaf and root regrowth, and allocation of water-soluble carbohydrate reserves following defoliation of plants of prairie grass (Bromus willdenowii Kunth.). *Grass Forage Sci., 62,* 497-506. http://dx.doi.org/10.1111/j.1365-2494.2007.00607.x

van Staalduinen, M., & Anten, N. (2005). Differences in the compensatory growth of two co-occurring grass species in relation to water availability. *Oecol., 146,* 190-199.

5

Belowground Influence of *Rhizobium* Inoculant and Water Hyacinth Composts on Yellow Bean Infested by *Aphis fabae* and *Colletotrichum lindemuthianum* under Field Conditions

Victoria Naluyange[1], Dennis M. W. Ochieno[1], Philip Wandahwa[2], Martins Odendo[3], John M. Maingi[4], Alice Amoding[5], Omwoyo Ombori[4], Dative Mukaminega[6] & John Muoma[1]

[1] Department of Biological Sciences, Masinde Muliro University of Science and Technology (MMUST), Kakamega, Kenya

[2] School of Agriculture, Veterinary Science and Technology, Masinde Muliro University of Science and Technology (MMUST), Kakamega, Kenya

[3] Socio-Economics and Statistics Division, Kenya Agricultural and Livestock Research Organization (KALRO), Kakamega, Kenya

[4] Department of Plant and Microbial Sciences, Kenyatta University, Nairobi, Kenya

[5] Department of Soil Science, Makerere University, P. O. Box 7062, Kampala, Uganda

[6] Faculty of Applied Sciences, Kigali Institute of Science and Technology (KIST), P.O. Box 3900, Kigali, Rwanda

Correspondence: Victoria Naluyange, Department of Biological Sciences, Masinde Muliro University of Science and Technology (MMUST), P.O. Box 190-50100, Kakamega, Kenya. E-mail: vicluy@gmail.com

Abstract

Rhizobium inoculant has been developed for bean production in Lake Victoria basin. Two types of compost have been developed, water hyacinth compost with cattle manure culture (H+CMC) or with effective microbes (H+EM). Influence of *Rhizobium* and composts on *Aphis fabae* and *Colletotrichum lindemuthianum* were investigated in the field. *Rhizobium* and hyacinth composts increased nodulation (×2 to 5); while *Aphis fabae* population increased (×2) on *Rhizobium*-inoculated plants with H+EM. Incidence of *C. lindemuthianum* was high in *Rhizobium*-inoculated plants. Plants that received diammonium phosphate (DAP) fertilizer had few nodules, reduced germination, slow growth and low yields. In conclusion, the water hyacinth composts contain beneficial microbes that promote root nodulation by *Rhizobium*, which is necessary for nitrogen fixation, while enhancing tolerance to aboveground infestations by *A. fabae* and *C. lindemuthianum*. We raise questions on our results to stimulate research, considering that bean breeding programs in Africa have mainly focused on microbial pathogens, and not insect pests.

Keywords: anthracnose, compost, manure, soil fertility, sustainability

1. Introduction

Common bean *Phaseolus vulgaris* is an important food security crop, and the major source of plant protein within the Lake Victoria basin in East Africa (David & Sperling, 1999). Beans complement the shortage of animal protein in East Africa, especially in the prevailing situation whereby fish production has been greatly impeded by the disastrous spread of water hyacinth in Lake Victoria (Ntiba et al., 2001; Hecky et al., 2010). The leguminous crop is also very important in agro-ecosystems, as it symbiotically fixes nitrogen through endophytic *Rhizobium* species (Bala et al., 2011; Devi et al., 2013). However, bean production has been declining to levels that are too low to meet the demand in East Africa (Mauyo et al., 2007). The main causes of declines in bean production are inferior germplasms, low soil fertility, pests and diseases (David & Sperling, 1999; Danielsen et al., 2013; Tittonell & Giller, 2013). Depletion of soil nutrients such as nitrogen and phosphorus has been a growing problem for bean production in East Africa (Kimani et al., 2007, 2008; Ayuke et al., 2011). Insect pests such as the black bean aphid *Aphis fabae* have been transmitting viral diseases (Beebe, 2012; Were et al., 2013), while fungal pathogens such as *Colletotrichum lindemuthianum* cause anthracnose disease of beans in East

Africa (Beebe, 2012; Kharinda, 2013). These factors have been complicated by the fact that local cultivars that are widely grown by smallholder farmers in East Africa have been succumbing to a complex of biotic and abiotic stresses (Otsyula et al., 2004; Ojiem et al., 2006; Kharinda, 2013).

There have been efforts to enhance sustainable crop production on farmlands in the Lake Victoria basin (Mireri et al., 2007; de Graaff et al., 2011), while conserving the lake for fish production (Lung'ayia et al., 2001; Nunan, 2013). Among the strategies, *Rhizobium* inoculants are being developed to enhance legume production by fixing nitrogen in the Lake Victoria basin (Bala et al., 2011). At the same time, nutrient-rich water hyacinth in the heavily eutrophied lake is being processed into compost, and transferred onto nutrient depleted farmlands for crop production (Naluyange et al., 2014). Removal of water hyacinth from Lake Victoria restores conditions that are favorable for fishing. Furthermore, water hyacinth compost contains phosphorus (Gunnarsson & Petersen, 2007; Naluyange et al., 2014), which is necessary for nodulation and nitrogen fixation in *Rhizobium*-inoculated bean seeds (Ssali & Keya, 1983), through processes such as enhancement of plant growth (Robson et al., 1981), improvement of shoot metabolism (Jakobsen, 1985) and specific roles in nodule initiation, growth and function (Israel, 1987). Such specific roles of phosphorus include ATP synthesis for nodule development and function (Ribet & Drevon, 1996), as well as for signal transduction and cell membrane biosynthesis (Graham & Vance, 2003). *Rhizobium*-inoculated beans are being promoted for improved yields and nitrogen fixation in the Lake Victoria region (Kihara et al., 2010; Thuita et al., 2012).

Rhizobium inoculants, when applied on legumes, have belowground effects such as enhanced root nodulation associated with better plant growth (Graham & Vance, 2000). Such beneficial processes of *Rhizobium* are affected by biotic factors like rhizosphere microbes (van Veen et al., 1997) as well as abiotic factors like soil fertility (Rotaru & Sinclair, 2009). Belowground colonization of roots by *Rhizobium* has been found to interact with aphids and other aboveground herbivores (Kempel et al., 2009; Katayama et al., 2010, 2011; Martinuz et al., 2012). For example, root colonization by *Rhizobium* has been found to promote plant resistance to insect pests (Thamer et al., 2011). However, *Rhizobium* colonization of roots has also been related to an increase in aphid and fungal incidences on leguminous shoots, which has been attributed to improved nutritive suitability of the host plant due to nitrogen fixation (Dean et al., 2009, 2014; Naluyange et al., 2014). Such effects of *Rhizobium* on legumes are modified by soil fertility amendments (El-Wakeil & El-Sebai, 2009; Dean et al., 2009). For instance, *Rhizobium* when applied using water hyacinth compost as an inoculant carrier improves the growth of faba bean (Mohamed & Abdel-Moniem, 2010).

In the Lake Victoria basin, combined application of water hyacinth compost and commercial *Rhizobium* inoculant had some positive effects on performance of the commercial Rosecoco bean cultivar, depending on the water hyacinth compost formulation (Naluyange et al., 2014). However, farmers in the Lake Victoria basin, especially in Western Kenya, mostly rely on local bean cultivars obtained from the market, including the yellow bean '*Mugasa*' (David & Sperling, 1999; Otsyula et al., 2004). Bean seeds from the local markets attain higher germination percentage than the certified commercial varieties (Otsyula et al., 2004). Furthermore, smallholder farmers in the Lake Victoria region rarely coat their seeds with fungicides that are always present in the commercial seeds. The absence of fungicides in the local bean seeds makes them ideal for *Rhizobium* inoculation, since such chemicals are potentially harmful to the inoculants (Graham & Vance, 2000; Stoddard et al., 2010). However, unlike commercial bean cultivars, the yellow bean is among local cultivars that are yet to be studied, especially in terms of *Rhizobium* nodulation and pest infestations under the influence of soil fertility amendments such as water hyacinth compost.

The objective of this study was to determine the influence of *Rhizobium* inoculant and water hyacinth composts on the performance of yellow bean in terms of growth and yields, and how these applications affect natural infestation of the plants by *A. fabae* and *C. lindemuthianum*. We hypothesize that water hyacinth composts and *Rhizobium* inoculant contain plant growth promoting microbes that improve belowground nutrient acquisition for yellow bean growth and yields, enabling the plants to tolerate aboveground infestations by *A. fabae* and *C. lindemuthianum*.

2. Materials and Methods

2.1 Experimental Design

The field experiment was conducted at the Masinde Muliro University of Science and Technology farm (N 00 17.104', E 034° 45.874'; altitude 1561m a.s.l.). Soils in this region have been classified as dystro-mollic Nitisols (FAO, 1974; Rota et al., 2006). Nutrient composition for the soil was; total phosphorus (18.9 ppm), total nitrogen (0.26 %), organic carbon (2.5 %), potassium (0.41 $cmol_c$ kg^{-1}), sodium (0.1 $cmol_c$ kg^{-1}), calcium (2.3 $cmol_c$ kg^{-1}), magnesium (0.8 $cmol_c$ kg^{-1}), zinc (1.9 ppm) and iron (0.37 ppm), with acidic pH of 4.2 (Naluyange et al., 2014).

The experiment was laid out in a randomized block design comprising 2×4 factorial treatments with *Rhizobium* inoculum factor having two levels (with or without inoculation) and fertility factor with four levels i.e. no fertilizer (Non), diammonium phosphate fertilizer-DAP (18-46-0), water hyacinth compost + cattle manure culture (H+CMC), and water hyacinth compost + effective microbes (H+EM). Each of the resulting 8 treatment combinations (plots) had 25 plants (n) in 3 blocks (i.e. N=600). Each plot was in form of a row containing the 25 plants spaced at 20 cm, with a distance of 40 cm between the plots, without border rows. The treatment rows were completely randomized to minimize non-experimental bias in sampling for natural infestations of aphids and anthracnose disease on bean plants. This experiment was conducted during the long rain season between 20th April to 30th July 2012, and then repeated between 30th May and 31st August 2012 (Figure 1).

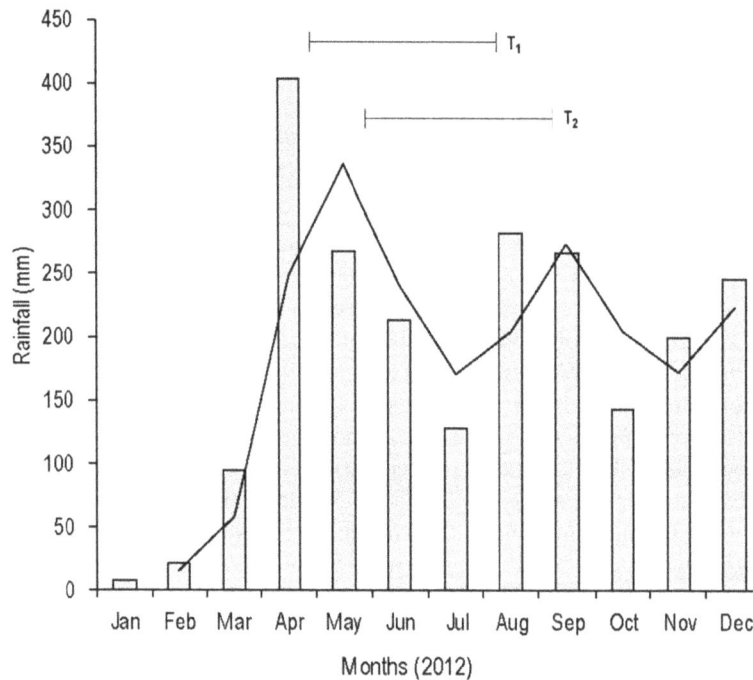

Figure 1. Rainfall in Kakamega county of Western Kenya in the year 2012 during the first field trial (T 1) between 20th April to 30th July 2012, and the second field trial (T 2) between 30th May and 31st August 2012. Total rainfall during the first trial was 1012 mm with a 3-month average of 253 mm. Total rainfall during the second trial was 622 mm with a 3-month average of 207 mm (Source: Naluyange 2013; Courtesy of the Kenya Agricultural and Livestock Research Organization (KALRO), Kakamega, Kenya)

2.2 Water Hyacinth Composts

Two formulations of compost made from water hyacinth + cattle manure culture (H+CMC) and water hyacinth + effective microbes (H+EM) were prepared using aboveground closed aerobic heap design (Naluyange et al., 2014). The H+CMC compost formulation was prepared by mixing chopped and dried water hyacinth material with a culture of decomposed cattle manure to supply saprophytic microbes. The H+CMC compost had a density of 58 g / 100 cm^3, with a nutrient concentration of total phosphorus (375 ppm), total nitrogen (1.1 %), organic carbon (13.4 %), potassium (21 cmol$_c$ kg^{-1}), sodium (1.9 cmol$_c$ kg^{-1}), calcium (22.3 cmol$_c$ kg^{-1}), magnesium (12 cmol$_c$ kg^{-1}), zinc (2 ppm) and iron (1.9 ppm), with alkaline pH of 8.1 (Naluyange et al., 2014).

The H+EM compost formulation was prepared by mixing dried and chopped water hyacinth material with Effective Microorganisms solution (EM™), containing photosynthetic bacteria (*Rhodopseudomonas palustris*), lactic acid bacteria (*Lactobacillus plantarum* and *L. casei*), yeast (*Saccharomyces cerevisae*), molasses and water (EM Technologies Ltd, Embu, Kenya). The H+EM compost had a density of 62 g / 100 cm^3, with the nutrient concentration of total phosphorus (270 ppm), total nitrogen (1 %), organic carbon (13.5 %), potassium (24.5 cmol$_c$ kg^{-1}), sodium (1.7 cmol$_c$ kg^{-1}), calcium (27.5 cmol$_c$ kg^{-1}), magnesium (15.3 cmol$_c$ kg^{-1}), zinc (4 ppm) and iron (1.7 ppm), with alkaline pH of 8.4 (Naluyange et al., 2014).

2.3 Seed Inoculation and Planting

Seeds of the local yellow bean cultivar '*Mugasa*' were purchased from the Kakamega town market in Western Kenya. These are among uncertified seeds that are widely grown by farmers (Otsyula et al., 2004). The bean seeds were inoculated with *Rhizobium* inoculant powder as per manufacturer's directions (BIOFIX®, MEA Ltd, Kenya). The seeds (250 g) were mixed in gum Arabic solution (0.5 gum Arabic/ 5 mL of sterile lukewarm water). The gum Arabic-coated seeds (250 g) were mixed with the *Rhizobium* inoculant powder (1 g). Controls were coated with gum Arabic solution only.

Planting holes of ~200 cm³ volume (i.e. ~5 cm diameter and ~10 cm deep) were dug using a shovel. The water hyacinth composts were applied using containers of 150 mL volumes per hole (i.e. ~90 g) as per the respective treatments and mixed with soil. Therefore, each planting hole received approximately 0.03 g phosphorus and 0.99 g nitrogen for the H+CMC compost; or 0.02 g phosphorus and 0.90 g nitrogen in case of the H+EM compost. For the DAP treatment, one leveled teaspoon (4.7 g) was mixed with soil in the planting hole (Naluyange et al. 2014). DAP fertilizer contains nitrogen (18 %) and phosphorus pentoxide P_2O_5 (46 %), with phosphorus (P) constituting 20 % of the total mass. Hence, every planting hole in the DAP treatment received 0.94 g phosphorus and 0.85 g nitrogen. One bean seed was sown in every planting hole at a depth of ~2 cm.

2.4 Data Collection

Data was recorded as described by Naluyange et al. (2014). The emergence date of every seedling was recorded independently, and used to determine the duration for germination. The number of seedlings that germinated out of the total number of seeds that were planted was used to calculate the germination percentage within 20 days from the planting date. When the first trifoliate leaves were fully formed in ~80% of the seedlings, plant height (stem base to petiole), length of the middle leaf (base to apex) and its width (widest part) were recorded. The date when the first flower of every plant appeared was recorded and used to calculate the duration for flowering. Ten days from the onset of flowering, 5 bean plants were randomly selected from each treatment per block for the estimation of number of root nodules associated with *Rhizobium* colonization. The bean plants were dug out with their root system still holding rhizosphere soil and transported to the laboratory in plastic bags. Rhizosphere soil from each bean plant was gently removed onto a white paper. Detached root nodules in the soil and those still attached to the roots were carefully quantified using a tally counter. This method maximized the collection and counting of root nodules. At harvest, the pods from every plant were packed in separate paper packets and sun dried for a period of five days; the weight of bean seeds per plant was recorded.

Aphid infestations on bean plants were recorded at the vegetative and flowering stage of bean growth. Three screw-capped vessels each containing 10 mL of 70% ethanol were placed on every treatment row of 25 plants. Aphids from every 8 plants per row were collected into each container using a camel hair brush from leaves and stems. The collected aphids were identified under a dissection microscope (Model Z45E, Leica Inc., USA) at × 10 magnifications using the features described by Martin (1983) and Holman (1998), and their absolute counts recorded using a tally counter. These insects have already been identified as the black bean aphid *A. fabae* (Naluyange et al., 2014). At the vegetative stage, the bean plants were also scored for incidence of anthracnose disease (*C. lindemuthianum*) i.e. the proportion of plants having anthracnose symptoms, characterized by dark brown to black lesions on leaves (Hagedorn & Inglis, 1986; Buruchara et al., 2010).

2.5 Statistical Analysis

Statistical analyses were conducted using SAS 9.1 software (SAS Institute Inc.) at $p < 0.05$ confidence level. Proc Means was used in the generation of descriptive statistics such as means and standard errors for nodule counts, aphid population, plant growth (duration and size) and yields (pod counts and seed weight). These data were graphically checked for normality using Proc Univariate; while Proc Transreg was used to find appropriate Box-Cox power transformations for normalization of data. Data for aphid population were log-transformed, while untransformed data were used for nodule counts. Frequencies for germination (%) and anthracnose disease incidence (%) were generated using Proc Freq. For plant growth and aphid population, Analyses of Variance (ANOVA) between treatment means for the fertility, *Rhizobium* and trial factors were done by Proc Mixed using the three blocks as random effects and the two trials (or plant growth stage for aphid data) as repeated measures. Means for plant growth and aphid population were separated using Ls-means when treatment effects were significant ($p < 0.05$). Anthracnose disease incidences and germination percentages were analyzed by Proc Genmod (χ^2 test; binomial) and percentages compared using Proc Multtest. Germination percentage data for the two seasons were combined because they were similar. Percentage increase and decrease in nodulation, aphid populations, plant growth and yields were calculated using the formula:-

$$\frac{Xi - Xc}{Xc} \times 100 \%$$

Whereby Xc = mean for untreated controls (Non); and Xi = mean for other treatments.

3. Results and Discussion

In this study, it was expected that plants treated with water hyacinth composts and *Rhizobium* inoculant would exhibit improved nodulation, growth and yields, while expressing tolerance to infestations by *A. fabae* and *C. lindemuthianum*. The average number of root nodules in the untreated plants was seven (mean = 7), but significantly increased in plants grown using the water hyacinth composts without *Rhizobium*, with H+CMC attaining an extra 200 % (i.e. +14 nodules) and H+EM scoring extra 278 % (i.e. +19 nodules), when compared to the untreated controls (df = 7, χ^2 = 1138, p < 0.0001) (Figure 2). Because these plants had not been inoculated,

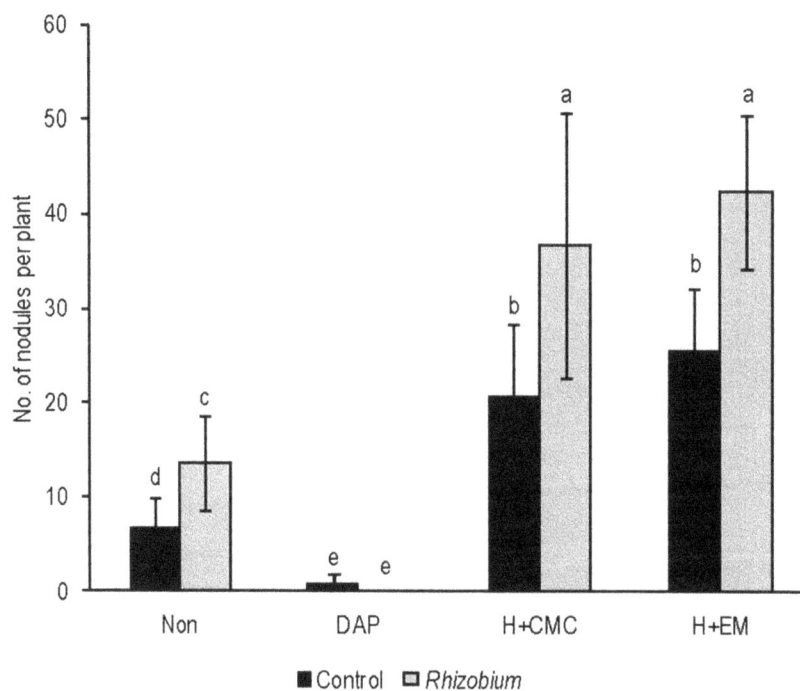

Figure 2. Number of root nodules in local yellow bean plants (var. *Mugasa*) as affected by commercial *Rhizobium* inoculant and soil fertility amendments; without fertilizer (Non), diammonium phosphate fertilizer (DAP), water hyacinth compost + cattle manure culture (H+CMC) and water hyacinth compost + effective microbes (H+EM). Bars with the same letter(s) are not significantly different (χ^2 test, p > 0.05)

the nodules are likely to have been formed by native strains of *Rhizobium* species (Dean et al., 2009; Naluyange et al., 2014). The *Rhizobium* inoculant, which significantly increased nodulation of plants grown without fertilizer by 100 % (i.e. +7 nodules), also performed better in combination with the water hyacinth composts, attaining high increases in nodulation at 440 % (i.e. +30 nodules) for H+CMC and 524 % (i.e. +36 nodules) in H+EM (Figure 2). These results on root nodulation are in line with the expectations that the water hyacinth composts and *Rhizobium* inoculant can promote root nodulation in the yellow bean, which indicates their compatibility (Naluyange et al., 2014). Microbes in the two water hyacinth composts may have enhanced root nodulation by *Rhizobium*, probably through phosphate solubilization (Argaw, 2012; Messele & Pant, 2012). Bean plants that received DAP had the lowest number of root nodules (Figure 2), despite the fertilizer being rich in phosphorus required for nodulation (Ssali & Keya, 1983; Graham & Vance, 2003). Reduction in nodule counts for DAP treatments ranged between -87 % to -100 % (i.e. zero nodules). The reduced nodulation has been linked to the chemical composition of inorganic fertilizers that limits the survival of Rhizobia (Peterson & Kremer, 1989). It is likely that the DAP inhibited *Rhizobium* nodulation, probably through acidification (Thawornchaisit & Polprasert, 2009). The soil was also already acidic with a pH of 4.2. Based on nutrient concentration, it is unlikely that nitrogen in DAP played a significant role in inhibiting root nodulation, because the levels of N supplied to every plant in the three fertility treatments were similar i.e. DAP (0.85 g), H+CMC compost (0.99 g) and H+CMC compost (0.90 g). However, the ammonium form of nitrogen in DAP could have inhibited root nodulation and diazotrophic activities of *Rhizobium* inoculant (Huss-Danell et al., 1982; Mendoza et al., 1995). The form of nitrogen and phosphorus in the two water hyacinth composts was not established (Naluyange et al., 2014).

Composts and other soil fertility amendments can influence seed germination through properties such as water holding capacity (Celik et al., 2004) and chemical activities including phytotoxicity (Hartz et al., 1996; Kabir et al., 2010). Influence of the two water hyacinth composts (H+CMC and H+EM) on seed germination percentage and duration were not evident, as there was no difference with those grown without fertilizer (p > 0.05) (Figure 3). This was also the case in the commercial Rosecoco bean variety (Naluyange et al., 2014). Overall germination percentage for bean seeds in the first trial (74.8%) was significantly lower than in the second trial (81.4%) (χ^2 = 6.93; p = 0.0085). However, there was no difference in germination percentage between the two trials when the DAP treatment was excluded in the statistical analysis (df = 1, χ^2 = 0.31, p = 0.58). Variations in germination percentage between the two trials can be explained by the large difference in germination percentages of seeds grown with DAP, which were -87 % in the first trial and -56 % in the second trial. It is likely that the differences in climatic factors such as rainfall (Figure 1), may have affected the efficacy of DAP through soil moisture related processes (Olson & Dreier, 1956; Hoeft et al., 1975; Hartz et al., 1996; Salvagiotti et al., 2013). Rhizobia are known to improve seed germination by secretion of Nod-factors and phytohormones (Prithiviraj et al., 2003; Cassan et al., 2009). In the current study, there was no indication that the *Rhizobium* inoculant stimulated seed germination, but the inoculated seeds even exhibited further germination suppression when grown with DAP (Figure 3). This is an issue that requires investigation. Plants grown with DAP took longer period to germinate (Table 1), an effect that has been reported in other studies (Naluyange et al., 2014).

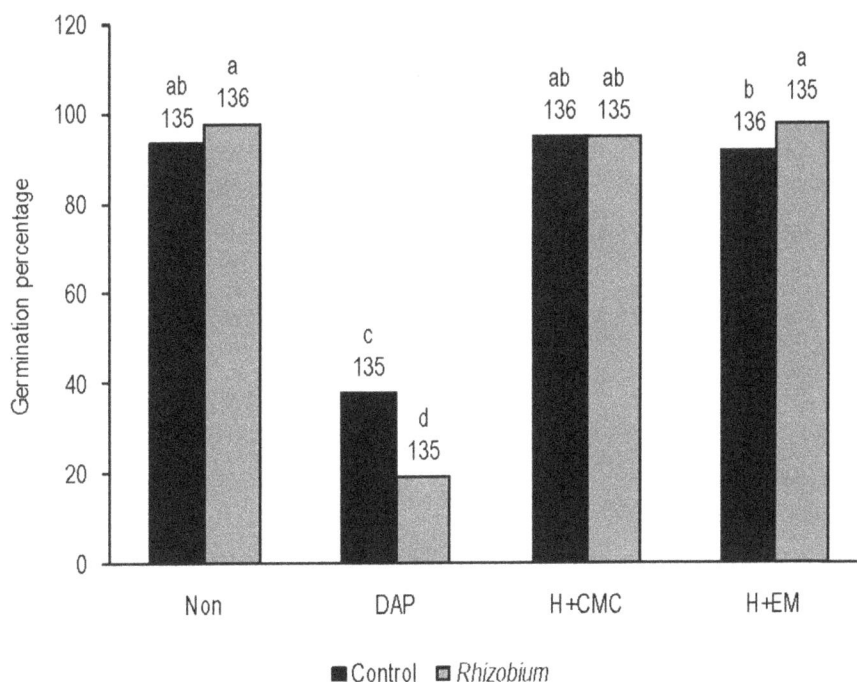

Figure 3. Germination percentage of local yellow bean seeds (var. *Mugasa*) as affected by commercial *Rhizobium* inoculant and soil fertility amendments; without fertilizer (Non), diammonium phosphate fertilizer (DAP), water hyacinth compost + cattle manure culture (H+CMC) and water hyacinth compost + effective microbes (H+EM). Numbers on top of bars represent sample sizes. Bars with the same letter(s) are not significantly different (χ^2 test, p > 0.05)

Beans grown with the two types of water hyacinth compost and those without fertilizer were not different in terms of plant size, days to flowering, as well as pod counts and seed weight per plot (p > 0.05) (Table 1). Furthermore, *Rhizobium*-related effects on the previously mentioned parameters were not detected (Table 1). In the Rosecoco bean, plants grown with the two water hyacinth composts were large in size, and exhibited improved growth and yields in the second trial (Naluyange et al., 2014). Differences between the yellow bean and the Rosecoco in terms of response towards the two types of compost may be cultivar-related (Naluyange, 2013); although methodological differences such as statistical analysis approach need to be considered in this judgment. Yellow bean plants treated with DAP took the longest time to emerge and flower, while exhibiting reduced yields per unit area in terms of pod count (-64 %) and seed weight (-67 %) (Table 1). The low yields are a result of the low germination percentage in the DAP treatment (Figure 3). Despite reduction in germination percentage, individual mature plants from DAP treatments become large in size indicating that inhibitive effects of DAP are temporary (Naluyange et al., 2014). Such negative effects of DAP may be due to edaphic and climatic conditions specific to study locations (Ghizaw et al., 1999), because DAP has been reported to enhance bean yields in many other regions around the world (Zhang et al., 2008; Zafar et al., 2013). Bean plants in the first trial took shorter time to flower, were taller and produced higher number of pods with greater seed weight per unit area compared to the second trial (Table 1). The primary reason for this observation is the relatively more rainfall during the first trial (1012 mm) than in the second trial (622 mm) in which the relatively young bean plants were exposed to rain shortages in the month of July 2012 (Figure 1).

Table 1. Plant growth and yields of the local yellow bean '*Mugasa*' as influenced by *Rhizobium* inoculant, water hyacinth compost containing cattle manure culture (H+CMC) or effective microbes (H+EM) and diammonium phosphate (DAP) fertilizer under field conditions in Western Kenya

		Means of means						
		Emergence days	Flowering days	Leaf length (cm)	Leaf width (cm)	Plant height (cm)	Pods (counts)	Yield (g)
Overall mean		6.90±0.24	40.32±0.46	8.37±0.32	5.44±0.18	5.16±0.12	125.29±12.58	125.2±13.06
First trial		6.87±0.14a	37.89±0.28 b	8.49±0.23a	5.76±0.12a	5.72±0.16 a	214.98±16.4 a	236.66±13.5 a
Second trial		6.81±0.10a	42.19±0.65 a	8.28±0.76a	5.22±0.47a	4.72±0.05 b	46.29±6.0 b	25.7±1.34 b
Fertilizer	Non	6.55±0.07 b	39.86±0.35a	8.0±0.27a	5.17±0.16a	4.95±0.18 b	132±12.12 a	136.41±14.75 a
	DAP	8.66±0.37 a	43.36±0.99 a	7.39±1.11a	4.89±0.59a	4.72±0.28 b	47.17±26.12 b	45.63±28.68b
	H+CMC	6.26±0.17 b	38.88±0.35 b	9.28±0.21a	6.04±0.13a	5.72±0.17 a	175.17±14.56 a	167.94±15.82 a
	H+EM	6.15±0.13 b	39.18±0.25 b	8.8±0.25a	5.68±0.19a	5.24±0.06 ab	146.83±7.18 a	150.85±7.83a
Inoculum	Control	6.86±0.37a	39.98±0.49a	8.17±0.41a	5.34±0.24a	5.23±0.18a	113.75±15.19a	112.13±16.12a
	Rhizobium	6.95±0.31a	40.66±0.78a	8.56±0.49a	5.55±0.27a	5.09±0.16a	136.83±20.18a	138.28±20.56a
Source of variation	df	F values						
Trial	1,4	1.78	41.82**	0.07	1.25	30.03**	93.16***	836***
Fertilizer	3,16	28.58***	19.21***	1.52	2.06	3.93*	10.56***	33.57***
Inoculum	1,16	0.82	1.85	0.39	0.33	0.29	1.86	1.27
Fertilizer × Inoculum	3,16	0.82	1.68	0.17	0.04	0.12	0.39	0.27

Without fertilizer (Non), diammonium phosphate fertilizer (DAP), water hyacinth compost + cattle manure culture (H+CMC) and water hyacinth compost + effective microbes (H+EM). Asterisk indicates the significant effect, ***$P \leq .001$, ** $P \leq 0.01$, * $P \leq 0.05$. Means with the same letter(s) are not significantly different; those with more than one letter are intermediate.

Populations of the black bean aphid *A. fabae* were generally low i.e. fewer than 1-2 aphids per plant in most treatments. *Aphis fabae* population was significantly higher in *Rhizobium* inoculated plants grown with hyacinth + effective microbes (H+EM) than the other treatments ($F_{7, 37}$; p = 0.0036) (Figure 4). This increase in *A. fabae* population by ~230 % may be linked activities of the *Rhizobium* inoculant, although microbes contained in the compost (H+EM) may have also contributed, as this effect was not evident in the other compost (H+CMC). It also implies that host-mediated effects of *Rhizobium* on aphids depend on the composition of microbial communities the legume symbiont interacts with in the rhizosphere (Whipps, 2001; Bais et al., 2006; Raaijmakers et al., 2009); as microbial communities that exist in cattle manure culture (Chachkhiani et al., 2004; Maeda et al., 2010), are unlikely to be similar to those in Effective Microbes® (Naluyange et al., 2014). Increase in aphid populations due to *Rhizobium* inoculation has also been reported in soybean and in the commercial Rosecoco bean (Dean et al., 2009; Naluyange et al., 2014). Since *Rhizobium* fixes nitrogen in legumes, then increased *A. fabae* density may be linked to high organic nitrogen content that determines nutritive suitability of host plants (Mattson, 1980; Dean et al., 2009; Ballhorn et al., 2013). It is also possible that *Rhizobium* inoculated plants emit more attractive volatiles to *A. fabae*, a suggestion that requires investigation. However, in the Lima bean *Phaseolus lunatus*, the application of *Rhizobium* minimized olfactory attraction of the Mexican bean beetle (*Epilachna varivestis*) towards plants induced to produce attractive volatiles using jasmonic acid (Ballhorn et al., 2013). The *A. fabae* population did not vary between the vegetative and flowering stages of bean growth ($F_{1,37}$; p = 0.33).

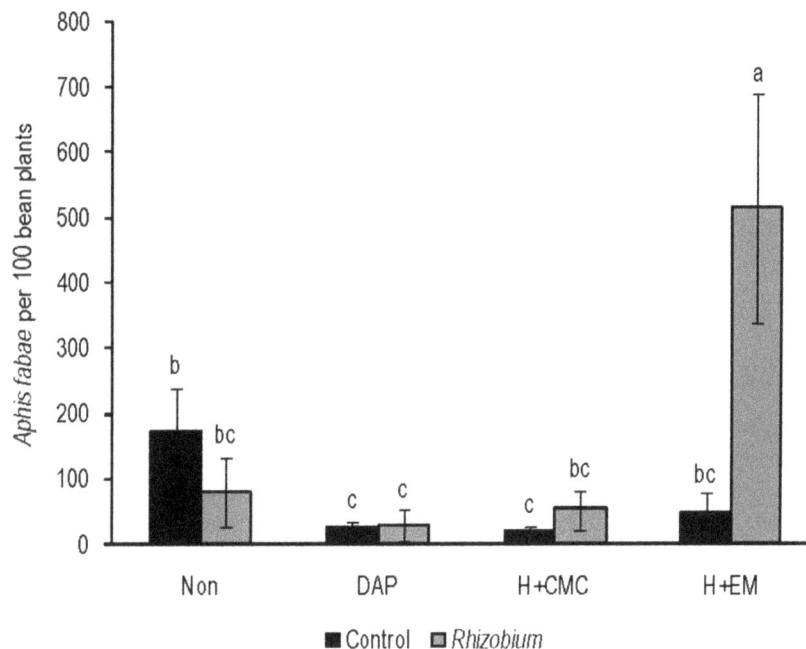

Figure 4. Population of *Aphis fabae* on local yellow bean plants (var. *Mugasa*) as affected by commercial *Rhizobium* inoculant and soil fertility amendments; without fertilizer (Non), diammonium phosphate fertilizer (DAP), water hyacinth compost + cattle manure culture (H+CMC) and water hyacinth compost + effective microbes (H+EM). Bars with the same letter(s) are not significantly different (F test, p > 0.05)

Incidence of anthracnose disease caused by *C. lindemuthianum* was significantly higher in *Rhizobium*-inoculated plants than in the non-inoculated plants, particularly in the first trial (Figure 5). It is likely that *C. lindemuthianum* infestations are dependent on *Rhizobium* activities, which can be supported by the similarity of trends in root nodule counts (Figure 2) and anthracnose incidence across fertility treatments (Figure 5). Just like *A. fabae*, this effect can also be linked to enhanced nitrogen content due to N_2 fixation. This is because *Colletotrichum* species that cause anthracnose disease and many other phytopathogens have high affinity for nitrogen in host plants (Nam et al., 2006; Tavernier et al., 2007; Lobato et al., 2009; Ochieno, 2010). On average, anthracnose incidence in the first trial (~62%) was higher than in the second trial (~7%). This is because the first trial had more rainfall (Figure 1), which favours *C. lindemuthianum* infestations on beans that are more prevalent under moist conditions (Kumar et al., 1999).

Apart from symbiotic nitrogen fixation, there could be other tripartite host-mediated interactions between *Rhizobium* species, the black bean aphid *A. fabae* and the anthracnose pathogen *C. lindemuthianum* that need to be established (Stout et al., 2006). First, feeding by aphids induces *Rhizobium* nodulation in leguminous roots (Heath & Lau, 2011). This is because the number of root nodules was higher in plants that had high *A. fabae* population, particularly in the H+EM compost; which is reverse of the assumption that *Rhizobium* is the one that causes increase in aphid infestation (Dean et al., 2009; Naluyange et al., 2014). This also contradicts scenarios in which *Rhizobium* reduces aphid population on crops through induced plant resistance (El-Wakeil & El-Sebai, 2009; Martinuz et al., 2012). *Rhizobium* may actually suppress plant immunity instead of boosting induced resistance (Mithofer, 2002; Luo & Lu, 2014). Secondly, feeding wounds inflicted by *A. fabae* stylets may have facilitated the entry of *Colletotrichum* hyphae into plant tissues. Since *Rhizobium* inoculated plants had more *A. fabae* and hence higher number of stylet wounds, then more *Colletotrichum* hyphae may have penetrated

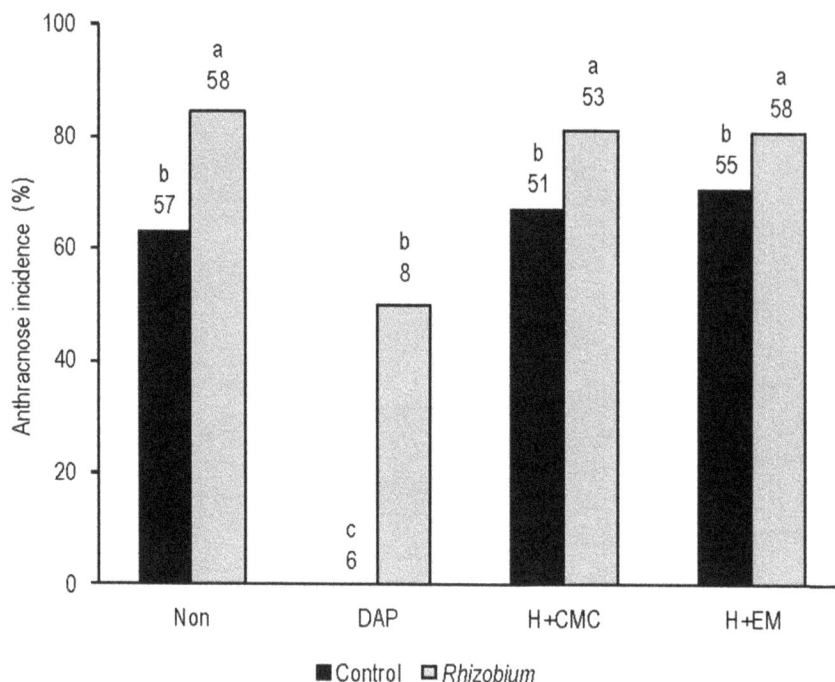

Figure 5. Anthracnose incidences in local yellow bean plants (var. *Mugasa*) as affected by commercial *Rhizobium* inoculant and soil fertility amendments; without fertilizer (Non), diammonium phosphate fertilizer (DAP), water hyacinth compost + cattle manure culture (H+CMC) and water hyacinth compost + effective microbes (H+EM). Numbers on top of bars represent sample sizes. Bars with the same letter(s) are not significantly different (χ^2 test, p > 0.05)

resulting in high anthracnose incidence. However, aphid stylet wounds are too small to permit the entry of fungal pathogens into plant tissues (Mitchell, 2004; Will & van Bell, 2006). Also, there are reports indicating that feeding by some aphid species inhibits plant tissue colonization by *Colletotrichum* species (Russo et al., 1997; Stout et al., 2006). Third, high number of aphids on *Rhizobium* inoculated plants may have secreted vast amounts of honeydew, which may have facilitated saprophytic colonization of bean surfaces by *Colletotrichum* hyphae prior to pathogenic penetration. However, this scenario is quite unlikely considering the low number of aphids (1-2) per plant. Fourth, both the aphid and the fungal pathogen may have compromised the immune system of the host plant to enhance their colonization (Stout et al., 2006). However, this interpretation may not fully stand, because plants have separate defense mechanisms against *Colletotrichum* and insect pests (Ajlan & Potter, 1991). Finally, plant infection by *Colletotrichum* may have influenced the production of aphid-attracting volatile organic compounds, because some endophytic fungi influence the production of insect attractive compounds (Cardoza et al., 2003; Jallow et al., 2008). However, this cannot be generalized, because *Colletotrichum* exhibits both biotrophic and necrotrophic infection phases (Bhadauria et al., 2011), which are associated with increase or decrease in *A. fabae* infestations on beans, respectively (Al-Naemi & Hatcher, 2013). Relationships between *Rhizobium*, *Colletotrichum* and aphids require further investigation.

Despite the high infestations by *A. fabae* and *C. lindemuthianum* in *Rhizobium*-inoculated plants grown with water hyacinth compost (H+EM), the growth and yields were not negatively affected, while the desirable root nodulation was enhanced. This indicates that the plants were tolerant to the insect pest *A. fabae* and the fungal pathogen *C. lindemuthianum* (Naluyange et al., 2014); although there are possibilities that plant growth promoting benefits of *Rhizobium* were cancelled by costs towards plant defensive mechanisms (Thaler et al., 1999; Heil et al., 2000).

In conclusion, the water hyacinth composts contain beneficial microbes that promote root nodulation by *Rhizobium*, which is necessary for nitrogen fixation, while enhancing tolerance to aboveground infestations by *A. fabae* and *C. lindemuthianum*. We raise questions on our results to stimulate research, considering that bean breeding programs in Africa have mainly focused on microbial pathogens, and not insect pests.

Acknowledgements

This work was conducted under the sponsorship of the Lake Victoria Research (VicRes) Initiative, a regional collaborative research programme of the Inter-University Council for East Africa (IUCEA). Research funds were provided by the Government of Sweden through the Swedish International Development Cooperation Agency (Sida), under the framework of the Lake Victoria Development Partnership (LVDP) Programme.

References

Ajlan, A. M., & Potter, D. A. (1991). Does immunization of cucumber against anthracnose by *Colletotrichum lagenarium* affect host suitability for arthropods? *Entomologia Experimentalis et Applicata, 58*, 83-91. http://dx.doi.org/10.1111/j.1570-7458.1991.tb01455.x

Al-Naemi, F., & Hatcher, P. E. (2013). Contrasting effects of necrotrophic and biotrophic plant pathogens on the aphid *Aphis fabae*. *Entomologia Experimentalis et Applicata, 148*, 234-245. http://dx.doi.org/10.1111/eea.12091

Argaw, A. (2011) Evaluation of co-inoculation of *Bradyrhizobium japonicum* and phosphate solubilizing *Pseudomonas* spp. effect on soybean (*Glycine max* L. Merr.) in Assossa Area. *Journal of Agricultural Science and Technology, 14*, 213-224.

Ayuke, F. O., Brussaard, L., Vanlauwe, B., Six, J., Lelei, D. K., Kibunja, C. N., & Pulleman, M. M. (2011). Soil fertility management: Impacts on soil macrofauna, soil aggregation and soil organic matter allocation. *Applied Soil Ecology, 48*, 53-62. http://dx.doi.org/10.1016/j.apsoil.2011.02.001

Bais, H. P., Weir, T. L., Perry, L. G., Gilroy, S., & Vivanco, J. M. (2006). The role of root exudates in rhizosphere interactions with plants and other organisms. *Annual Review of Plant Biology, 57*, 233-266. http://dx.doi.org/10.1146/annurev.arplant.57.032905.105159

Bala, A., Karanja, N., Murwira, M., Lwimbi, L., Abaidoo, R., & Giller, K. (2011). *Production and use of Rhizobial inoculants in Africa*. Retrieved from www.N2Africa.org

Ballhorn, D. J., Kautz, S., & Schadler, M. (2013). Induced plant defense via volatile production is dependent on rhizobial symbiosis. *Oecologia, 172*, 833-846. http://dx.doi.org/10.1007/s00442-012-2539-x

Beebe, S. (2012). Common bean breeding in the Tropics. In J. Janick (Ed.), *Plant Breeding* Reviews (Vol. *36*, pp 357-426). John Wiley & Sons, Inc., Hoboken, NJ, USA. http://dx.doi.org/10.1002/9781118358566.ch5

Bhadauria, V., Banniza, S., Vandenberg, A., Selvaraj, G., & Wei, Y. (2011). EST mining identifies proteins putatively secreted by the anthracnose pathogen *Colletotrichum truncatum*. *BMC Genomics, 12*, 327. http://dx.doi.org/10.1186/1471-2164-12-327

Buruchara, R., Mukankusi, C., & Ampofo, K. (2010). *Bean Disease and Pest Identification and Management: Handbooks for small-scale seed producers* no.04. International Center for Tropical Agriculture (CIAT); Pan-Africa Bean Research Alliance (PABRA).

Cardoza, Y. J., Teal, P. E. A., & Tumlinson, J. H. (2003). Effect of peanut plant fungal infection on oviposition preference by *Spodoptera exigua* and on host-searching behavior by *Cotesia marginiventris*. *Environmental Entomology, 32*, 970-976. http://dx.doi.org/10.1603/0046-225X-32.5.970

Cassan, F., Perrig, D., Sgroya, V., Masciarellia, O., Penna, C., & Luna, V. (2009). *Azospirillum brasilense* Az39 and *Bradyrhizobium japonicum* E109, inoculated singly or in combination, promote seed germination and early seedling growth in corn (*Zea mays* L.) and soybean (*Glycine max* L.). *European Journal of Soil Biology, 45*, 28-35. http://dx.doi.org/10.1016/j.ejsobi.2008.08.005

Celik, I., Ortas, I., & Kilic, S. (2004). Effects of compost, mycorrhiza, manure and fertilizer on some physical properties of a Chromoxerert soil. *Soil and Tillage Research, 78*, 59-67. http://dx.doi.org/10.1016/j.still.2004.02.012

Chachkhiani, M., Dabert, P., Abzianidze, T., Partskhaladze, G., Tsiklauri, L., Dudauri, T., & Godon, J. J. (2004). 16S rDNA characterisation of bacterial and archaeal communities during start-up of anaerobic thermophilic digestion of cattle manure. *Bioresource Technology, 93*, 227-232. http://dx.doi.org/10.1016/j.biortech.2003.11.005

Danielsen, S., Boa, E., Mafabi, M., Mutebi, E., Reeder, R., Kabeere, F., & Karyeija, R. (2013). Using plant clinic registers to assess the quality of diagnoses and advice given to farmers: a case study from Uganda. *The Journal of Agricultural Education and Extension, 19*, 183-201. http://dx.doi.org/10.1080/1389224X.2012.741528

David, S., & Sperling, L. (1999). Improving technology delivery mechanisms: lessons from bean seed systems in eastern and central Africa. *Agriculture and Human Values, 16*, 381-388.

http://dx.doi.org/10.1023/A:1007603902380

de Graaff, J., Kessler, A., & Nibbering, J. W. (2011). Agriculture and food security in selected countries in Sub-Saharan Africa: diversity in trends and opportunities. *Food Security, 3,* 195-213. http://dx.doi.org/10.1007/s12571-011-0125-4

Dean, J. M., Mescher, M. C., & de Moraes, C. M. (2009). Plant-rhizobia mutualism influences aphid abundance on soybean. *Plant and Soil, 323,* 187-196. http://dx.doi.org/10.1007/s11104-009-9924-1

Dean, J. M., Mescher, M. C., & de Moraes, C. M. (2014). Plant dependence on Rhizobia for nitrogen influences induced plant defenses and herbivore performance. *International Journal of Molecular Sciences, 15,* 1466-1480. http://dx.doi.org/10.3390/ijms15011466

Devi, M. J., Sinclair, T. R., Beebe, S. E., & Rao, I. M. (2013). Comparison of common bean (*Phaseolus vulgaris* L.) genotypes for nitrogen fixation tolerance to soil drying. *Plant and Soil, 364,* 29-37. http://dx.doi.org/10.1007/s11104-012-1330-4

El-Wakeil, N. E., & El-Sebai, T. N. (2009). Role of biofertilizer on faba bean growth, yield, and its effect on bean aphid and the associated predators. *Archives of Phytopathology and Plant Protection, 42,* 1144-1153. http://dx.doi.org/10.1080/03235400701650882

FAO. (1947). *Soil Map of the World 1: 5,000, 000* (vol. 1, p. 72). Legend, FAO/UNESCO.

Ghizaw, A., Mamo, T., Yilma, Z., Molla, A., & Ashagre, Y. (1999). Nitrogen and phosphorus effects on faba bean yield and some yield components. *Journal of Agronomy and Crop Science, 182,* 167-174. http://dx.doi.org/10.1046/j.1439-037x.1999.00272.x

Graham, P. H., & Vance, C. P. (2003). Legumes: importance and constraints to greater use. *Plant Physiology, 131,* 872-877. http://dx.doi.org/10.1104/pp.017004

Graham, P. H., & Vance, C. P. (2000). Nitrogen fixation in perspective: an overview of research and extension needs. *Field Crops Research, 65,* 93-106. http://dx.doi.org/10.1016/S0378-4290(99)00080-5

Gunnarsson, C. C., & Petersen, C. M. (2007). Water hyacinths as a resource in agriculture and energy production: A literature review. *Waste Management, 27,* 117-129. http://dx.doi.org/10.1016/j.wasman.2005.12.011

Hagedorn, D. J., & Inglis, D. A. (1986). *Handbook of Bean Diseases* (p. 24). University Wisconsin-Madison Extension Bulletin A3374.

Hartz, T. K., Costa, F. J., & Schrader, W. L. (1996). Suitability of composted green waste for horticultural uses. *HortScience, 31,* 961-964.

Heath, K. D., & Lau, J. A. (2011). Herbivores alter the fitness benefits of a plant-*Rhizobium* mutualism. *Acta Oecologica, 37,* 87-92. http://dx.doi.org/10.1016/j.actao.2010.12.002

Hecky, R. E., Mugidde, R., Ramlal, P. S., Talbot, M. R., & Kling, G. W. (2010). Multiple stressors cause rapid ecosystem change in Lake Victoria. *Freshwater Biology, 55,* 19-42. http://dx.doi.org/10.1111/j.1365-2427.2009.02374.x

Heil, M., Hilpert, A., Kaiser, W., & Linsenmair, K. E. (2000). Reduced growth and seed set following chemical induction of pathogen defence: does systemic acquired resistance (SAR) incur allocation costs? *Journal of Chemical Ecology, 88,* 645-654. http://dx.doi.org/10.1046/j.1365-2745.2000.00479.x

Hoeft, R. G., Walsh, L. M., & Liegel, E. A. (1975). Effect of seed-placed fertilizer on the emergence (germination) of soybeans (*Glycine max* L.) and snapbeans (*Phaseolus vulgaris* L.). *Communications in Soil Science and Plant Analysis, 6,* 655-664. http://dx.doi.org/10.1080/00103627509366601

Holman, J. (1998). Species of the genus *Aphis* (Sternorrhyncha: Aphidoidea) living on Hieracium (Asteraceae: Cichorieae). *European Journal of Entomology, 95,* 383-394.

Huss-Danell, K., Sellstedt, A., Flower-Ellis, A., & Sjöström, M. (1982). Ammonium effects on function and structure of nitrogen-fixing root nodules of *Alnus incana* (L.) Moench. *Planta, 156,* 332-340. http://dx.doi.org/10.1007/BF00397471

Israel, D. W. (1987). Investigation of the role of phosphorus in symbiotic dinitrogen fixation. *Plant Physiology, 84,* 835-840. http://dx.doi.org/10.1104/pp.84.3.835

Jakobsen, I. (1985). The role of phosphorus in nitrogen fixation by young pea plants (*Pisum sativum*). *Physiologia Plantarum, 64,* 190-196. http://dx.doi.org/10.1111/j.1399-3054.1985.tb02334.x

Jallow, M. F. A., Dugassa-Gobena, D., & Vidal, S. (2008). Influence of an endophytic fungus on host plant selection by a polyphagous moth via volatile spectrum changes. *Arthropod-Plant Interactions, 2*, 53-62. http://dx.doi.org/10.1007/s11829-008-9033-8

Kabir, E., Bell, R. W., & Johansen, C. (2010). Triple superphosphate placement affects early growth of chickpea. In *19th World Congress of Soil Science, Soil Solutions for a Changing World, Brisbane, Australia* (pp. 276-279).

Katayama, N., Nishida, T., Zhang, Z. Q., & Ohgushi, T. (2010). Belowground microbial symbiont enhances plant susceptibility to a spider mite through change in soybean leaf quality. *Population Ecology, 52*, 499-506. http://dx.doi.org/10.1007/s10144-010-0207-8

Katayama, N., Zhang, Z. Q., & Ohgushi, T. (2011). Community-wide effects of below-ground rhizobia on above-ground arthropods. *Ecological Entomology, 36*, 43-51. http://dx.doi.org/10.1111/j.1365-2311.2010.01242.x

Kempel, A., Brandl, R., & Schadler, M. (2009). Symbiotic soil microorganisms as players in aboveground plant-herbivore interactions-the role of Rhizobia. *Oikos, 118*, 634-640. http://dx.doi.org/10.1111/j.1600-0706.2009.17418.x

Kharinda, B. T. M. (2013). *Phenotypic characterization of common bean genotypes for resistance to the pathogen Colletotrichum lindemuthianum constraining bean production in western Kenya.* MSc. Thesis, Masinde Muliro University of Science and Technology, Kakamega, Kenya.

Kihara, J., Vanlauwe, B., Waswa, B., Kimetu, J. M., Chianu, J., & Bationo, A. (2010). Strategic phosphorus application in legume-cereal rotations increases land productivity and profitability in Western Kenya. *Experimental Agriculture, 46*, 35-52. http://dx.doi.org/10.1017/S0014479709990810

Kimani, J. M., Kimani, P. M., Githiri, S. M., & Kimenju, J. W. (2007). Mode of inheritance of common bean (*Phaseolus vulgaris* L.) traits for tolerance to low soil phosphorus (P). *Euphytica, 155*, 225-234. http://dx.doi.org/10.1007/s10681-006-9324-z

Kimani, J. M., & Tongoona, P. (2008). The mechanism of genetic control for low soil nitrogen (N) tolerance in common beans (*Phaseolus vulgaris* L.). *Euphytica, 162*, 193-203. http://dx.doi.org/10.1007/s10681-007-9499-y

Kumar, A., Sharma, P. N., Sharma, O. P., & Tyagi, P. D. (1999). Epidemiology of bean anthracnose *Colletotrichum lindemuthianum* under sub-humid mid-hills zone of Himachal Pradesh. *Indian Phytopathology, 52*, 393-397.

Lobato, A. K. S., Gonçalves-Vidigal, M. C., Vidigal-Filho, P. S., Costa, R. C. L, Lopes, M. J. S., Cruz, A. S., ... Gonçalves, A. M. O. (2009). Nitrogen compounds responses in two cultivars of common bean inoculated with *Colletotrichum lindemuthianum*. *Research Journal of Biological Sciences, 4*, 293-297.

Lung'ayia, H., Sitoki, L., & Kenyanya, M. (2001). The nutrient enrichment of Lake Victoria (Kenyan waters). *Hydrobiologia, 458*, 75-82. http://dx.doi.org/10.1023/A:1013128027773

Maeda, K., Hanajima, D., Morioka, R., & Osada, T. (2010). Characterization and spatial distribution of bacterial communities within passively aerated cattle manure composting piles. *Bioresource Technology, 101*, 9631-9637. http://dx.doi.org/10.1016/j.biortech.2010.07.057

Martin, J. H. (1983). The identification of common aphid pests of tropical agriculture. *Tropical Pest Management, 29*, 395-411. http://dx.doi.org/10.1080/09670878309370834

Martinuz, A., Schouten, A., Menjivar, R. D., & Sikora, R. A. (2012). Effectiveness of systemic resistance toward *Aphis gossypii* (Hom., Aphididae) as induced by combined applications of the endophytes *Fusarium oxysporum* Fo162 and *Rhizobium etli* G12. *Biological Control, 62*, 206-212. http://dx.doi.org/10.1016/j.biocontrol.2012.05.006

Mattson, W. J. (1980). Herbivory in relation to plant nitrogen content. *Annual Review of Ecology and Systematics, 11*, 119-161. http://dx.doi.org/10.1146/annurev.es.11.110180.001003

Mauyo, L. W., Okalebo, J. R., Kirkby, R. A., Buruchara, R., Ugen, M., & Maritim, H. K. (2007). Spatial pricing efficiency and regional market integration of cross-border bean (*Phaseolus Vulgaris* L.) marketing in East Africa: The case of Western Kenya and Eastern Uganda. In A. Bationo, B. Waswa, J. Kihara & J. Kimetu (Eds.), *Advances in Integrated Soil Fertility Management in sub-Saharan Africa: Challenges and Opportunities* (pp. 1027-1034). Springer, Netherlands. http://dx.doi.org/10.1007/978-1-4020-5760-1_100

Mendoza, A., Leija, A., Martínez-Romero, E., Hernández, G., & Mora, J. (1995). The enhancement of ammonium assimilation in *Rhizobium elti* prevents nodulation of *Phaseolus vulgaris. Molecular Plant-Microbe Interactions, 8*, 584-592. http://dx.doi.org/10.1094/MPMI-8-0584

Messele, B., & Pant, L.M. (2012). Effects of inoculation of *Sinorhizobium ciceri* and phosphate solubilizing bacteria on nodulation, yield and nitrogen and phosphorus uptake of Chickpea (*Cicer arietinum* L.) in Shoa Robit Area. *Journal of Biofertilizers & Biopesticides, 3*, 129. http://dx.doi.org/10.4172/2155-6202.10000129

Mireri, C., Atekyereza, P., Kyessi, A., & Mushi, N. (2007). Environmental risks of urban agriculture in the Lake Victoria drainage basin: A case of Kisumu municipality, Kenya. *Habitat International, 31*, 375-386. http://dx.doi.org/10.1016/j.habitatint.2007.06.006

Mitchell, P. L. (2004). Heteroptera as vectors of plant pathogens. *Neotropical Entomology, 33*, 519-545. http://dx.doi.org/10.1590/S1519-566X2004000500001

Mohamed, H. M., & Abdel-Moniem, A. (2010). Evaluation of water hyacinth and sugarcane bagasse composts as a carrier for *Rhizobium* inoculants and their effects on faba bean. *Research Journal of Agriculture and Biological Sciences, 6*, 1022-1028.

Naluyange, V. (2013). *Compatibility of Rhizobium inoculant with water hyacinth manure formulations in common bean and consequences on Aphis fabae and Colletotrichum lindemuthianum infestations*. MSc. Thesis, Masinde Muliro University of Science and Technology, Kakamega, Kenya.

Naluyange, V., Ochieno, D. M. W., Maingi, J. M., Ombori, O., Mukaminega, D., Amoding, A., ... Muoma, J. V. O. (2014). Compatibility of *Rhizobium* inoculant and water hyacinth compost formulations in Rosecoco bean and consequences on *Aphis fabae* and *Colletotrichum lindemuthianum* infestations. *Applied Soil Ecology, 76*, 68-77. http://dx.doi.org/10.1016/j.apsoil.2013.12.011

Nam, M. H., Jeong, S. K., Lee, Y. S., Choi, J. M., & Kim, H. G. (2006). Effects of nitrogen, phosphorus, potassium and calcium nutrition on strawberry anthracnose. *Plant Pathology, 55*, 246-249. http://dx.doi.org/10.1111/j.1365-3059.2006.01322.x

Ntiba, M. J., Kudoja, W. M., & Mukasa, C. T. (2001). Management issues in the Lake Victoria watershed. *Lakes and Reservoirs Research and Management, 6*, 211-216. http://dx.doi.org/10.1046/j.1440-1770.2001.00149.x

Nunan, F. (2013). Wealth and welfare? Can fisheries management succeed in achieving multiple objectives? A case study of Lake Victoria, East Africa. *Fish and Fisheries.* http://dx.doi.org/10.1111/faf.12012

Ochieno, D. M. W. (2010). *Endophytic control of Cosmopolites sordidus and Radopholus similis using Fusarium oxysporum V5w2 in tissue culture banana*. Doctoral thesis and Propositions, Wageningen University and Research Centre, The Netherlands.

Ojiem, J. O., de Ridder, N., Vanlauwe, B., & Giller, K. E. (2006). Socio-ecological niche: a conceptual framework for integration of legumes in smallholder farming systems. *International Journal of Agricultural Sustainability, 4*, 79-93.

Olson, R. A., & Dreier, A. F. (1956). Fertilizer placement for small grains in relation to crop stand and nutrient efficiency in Nebraska. *Soil Science Society of America Journal, 20*, 19-24. http://dx.doi.org/10.2136/sssaj1956.03615995002000010005x

Otsyula, R., Rachier, G., Ambitsi, N., Juma, R., Ndiya, C., Buruchara, R., & Sperling, L. (2004). *The use of informal seed producer groups for diffusing root rot resistant varieties during period of acute stress. Addressing Seed Security in Disaster Response: Linking relief with development* (pp. 69-89). Cali, Colombia: International Center for Tropical Agriculture (CIAT).

Peterson, H. L., & Kremer, R. J. (1989). Compatibility of Rhizobia and fertilizers combined as legume inocula. *World Journal of Microbiology and Biotechnology, 5*, 95-104. http://dx.doi.org/10.1007/BF01724964

Prithiviraj, B., Zhou, X., Souleimanov, A., Kahn, W. M., & Smith, D. L. (2003). A host-specific bacteria-to-plant signal molecule (Nod factor) enhances germination and early growth of diverse crop plants. *Planta, 216*, 437-445.

Raaijmakers, J. M., Paulitz, T. C., Steinberg, C., Alabouvette, C., & Moënne-Loccoz, Y. (2009). The rhizosphere: a playground and battlefield for soilborne pathogens and beneficial microorganisms. *Plant and Soil, 321*, 341-361. http://dx.doi.org/10.1007/s11104-008-9568-6

Ribet, J., & Drevon, J. J. (1996). The phosphorus requirement of N2 fixing and urea-fed *Acacia mangium*. *New Phytologist, 132*, 383-390. http://dx.doi.org/10.1111/j.1469-8137.1996.tb01858.x

Robson, A. D., O'hara, G. W., & Abbott, L. K. (1981). Involvement of phosphorus in nitrogen fixation by subterranean clover (*Trifolium subterraneum* L.). *Australian Journal of Plant Physiology, 8*, 427-436. http://dx.doi.org/10.1071/PP9810427

Rota, J. A., Wandahwa, P., & Sigunga, D. O. (2006). Land evaluation for soybean (*Glycine max* L. Merrill) production based on kriging soil and climate parameters for the Kakamega district, Kenya. *Journal of Agronomy, 5*, 142-150. http://dx.doi.org/10.3923/ja.2006.142.150

Rotaru, V., & Sinclair, T. R. (2009). Interactive influence of phosphorus and iron on nitrogen fixation by soybean. *Environmental and Experimental Botany, 66*, 94-99. http://dx.doi.org/10.1016/j.envexpbot.2008.12.001

Russo, V. M., Russo, B. M., Peters, M., Perkins-Veazie, P., & Cartwright, B. (1997). Interaction of *Colletotrichum orbiculare* with thrips and aphid feeding on watermelon seedlings. *Crop Protection, 16*, 581-584. http://dx.doi.org/10.1016/S0261-2194(97)00024-0

Salvagiotti, F., Barraco, M., Dignani, D., Sanchez, H., Bono, A., Vallone, P., ... Gudelj, V. J. (2013). Plant stand, nodulation and seed yield in soybean as affected by phosphate fertilizer placement, source and application method. *European Journal of Agronomy, 51*, 25-33. http://dx.doi.org/10.1016/j.eja.2013.06.006

Ssali, H., & Keya, S. O. (1983). The effect of phosphorus on nodulation, growth and dinitrogen fixation by beans. *Biological Agriculture and Horticulture, 1*, 135-144. http://dx.doi.org/10.1080/01448765.1983.9754387

Stoddard, F. L., Nicholas, A. H., Rubiales, D., Thomas, J., & Villegas-Fernandez, A. M. (2010). Integrated pest management in faba bean. *Field Crops Research, 115*, 308-318. http://dx.doi.org/10.1016/j.fcr.2009.07.002

Stout, M. J., Thaler, J. S., & Thomma, B. P. H. J. (2006). Plant-mediated interactions between pathogenic microorganisms and herbivorous arthropods. *Annual Review of Entomology, 51*, 663-689. http://dx.doi.org/10.1146/annurev.ento.51.110104.151117

Tavernier, V., Cadiou, S., Pageau, K., Laugé, R., Reisdorf-Cren, M., Langin, T., & Masclaux-Daubresse, C. (2007). The plant nitrogen mobilization promoted by *Colletotrichum lindemuthianum* in *Phaseolus* leaves depends on fungus pathogenicity. *Journal of Experimental Botany, 58*, 3351-3360. http://dx.doi.org/10.1093/jxb/erm182

Thaler, J. S., Fidantsef, A. L., Duffey, S. S., & Bostock, R. M. (1999). Trade-offs in plant defense against pathogens and herbivores: a field demonstration of chemical elicitors of induced resistance. *Journal of Chemical Ecology, 25*, 1597-1609. http://dx.doi.org/10.1023/A:1020840900595

Thamer, S., Schadler, M., Bonte, D., & Ballhorn, D. J. (2011). Dual benefit from a belowground symbiosis: nitrogen fixing Rhizobia promote growth and defense against a specialist herbivore in a cyanogenic plant. *Plant and Soil, 341*, 209-219. http://dx.doi.org/10.1007/s11104-010-0635-4

Thawornchaisit, U., & Polprasert, C. (2009). Evaluation of phosphate fertilizers for the stabilization of cadmium in highly contaminated soils. *Journal of Hazardous Materials, 165*, 1109-1113. http://dx.doi.org/10.1016/j.jhazmat.2008.10.103

Thuita, M., Pypers, P., Herrmann, L., Okalebo, R.J., Othieno, C., Muema, E., & Lesueur, D. (2012). Commercial rhizobial inoculants significantly enhance growth and nitrogen fixation of a promiscuous soybean variety in Kenyan soils. *Biology and Fertility of Soils, 48*, 87-96. http://dx.doi.org/10.1007/s00374-011-0611-z

Tittonell, P., & Giller, K. E. (2013). When yield gaps are poverty traps: The paradigm of ecological intensification in African smallholder agriculture. *Field Crops Research, 143*, 76-90. http://dx.doi.org/10.1016/j.fcr.2012.10.007

van Veen, J. A., van Overbeek, L. S., & van Elsas, J. D. (1997). Fate and activity of microorganisms introduced into soil. *Microbiology and Molecular Biology Reviews, 61*, 121-135.

Were, H. K., Kabira, J. N., Kinyua, Z. M., Olubayo, F. M., Karinga, J. K., Aura, J., ... Torrance, L. (2013). Occurrence and distribution of potato pests and diseases in Kenya. *Potato Research, 56*, 325-342. http://dx.doi.org/10.1007/s11540-013-9246-9

Whipps, J. M. (2001). Microbial interactions and biocontrol in the rhizosphere. *Journal of Experimental Botany, 52*, 487-511. http://dx.doi.org/10.1093/jexbot/52.suppl_1.487

Will, T., & van Bel, A. J. E. (2006). Physical and chemical interactions between aphids and plants. *Journal of*

Experimental Botany, 57, 729-737. http://dx.doi.org/10.1093/jxb/erj089

Zafar, M., Abbasi, M. K., Rahim, N., Khaliq, A., Shaheen, A., Jamil, M., & Shahid, M. (2013). Influence of integrated phosphorus supply and plant growth promoting rhizobacteria on growth, nodulation, yield and nutrient uptake in *Phaseolus vulgaris. African Journal of Biotechnology, 10*, 16781-16792.

Zhang, W., Ma, W., Ji, Y., Fan, M., Oenema, O., & Zhang, F. (2008). Efficiency, economics, and environmental implications of phosphorus resource use and the fertilizer industry in China. *Nutrient Cycling in Agroecosystems, 80*, 131-144. http://dx.doi.org/10.1007/s10705-007-9126-2

6

Chlorococcum humicola (Nageli) Rabenhorst as a Renewable Source of Bioproducts and Biofuel

Santhoshkumar K.[1], Prasanthkumar S.[1] & J. G. Ray[1]

[1] Laboratory of Ecology & Ecotechnology, School of Biosciences, Mahatma Gandhi University, Kottayam, Kerala, India

Correspondence: J. G. Ray, Laboratory of Ecology & Ecotechnology, School of Biosciences, Mahatma Gandhi University, Kottayam, Kerala, India. E-mail: jgray@mgu.ac.in

Abstract

Among the diverse new generation biomass yielding species, green algae are the most promising organisms. Compared to biomass production of other organisms, production of algae is less laborious, quite fast, and more economical. Moreover, eutrophicated waters get naturally purified in the cultivation process of algae. Algal biomass from monoculture of specific species, which are rich in carbohydrates, proteins and lipids, is considered a good source of diverse bio-products and feed-stock for food, feeds and bio-fuels. Quantity and quality of algal biomass for specific products depend on the species and strains as well as environmental conditions of cultivation. In this connection, biomass productivity and oil-yield of a local strain of *Chlorococcum humicola* (Nageli) Rabenhorst was assessed in Bold's Basal Medium. Long-term storage capacity of the alga was tried by entrapping the algal cells in sodium alginate beads, which showed viability up to 14 months. Estimation of total carbohydrate, protein, lipid and chemical characterization of oil as well as the feasibility of its conversion to biodiesel revealed the industrial potential of this local strain as a source of food and biofuel. Fatty acid profiling of the extracted oil showed that 70% are mono-saturated and 12.2 % are nutritionally important polyunsaturated fatty acids. The oil could be effectively trans-esterified to methyl esters and the conversion was confirmed by FTIR spectroscopy. Further standardization of the mass production of the alga in natural environmental conditions for biomass and oil is progressing to optimize its value as globally competent food, nutraceutical and biofuel resource.

Keywords: *Chlorococcum humicola*, algal biomass, algal biodiesel, algal bioproducts, synthetic seed

1. Introduction

Green algae are significant new biomass resource for the production of natural bioactive compounds and renewable energy. They have unique nutritional quality (Becker, 2007) that can add value to conventional food preparations for humans and animals. Because of the high protein content, biomass from many species of microalgae is generally considered as the potential source of proteins (Spolaore, Joannis-Cassan, Duran, & Isambert, 2006) for the future. In addition to proteins, green algae are good source of carbohydrates and lipids for food and fuels. Among the known commercially and industrially amenable green resources, algae can provide the highest and cheapest biomass per unit light and area. Simultaneously they can remediate nutrient load or degrade other toxic pollutants in water (Chiu et al., 2015) and reducing atmospheric CO_2 level through carbon fixation process (Morais & Costa, 2007). Moreover, several pharmaceutical products are derived from algal biomass (Yamaguchi, 1997), especially from that of Chlorophytes.

Biopharmaceutical industries are in search of low cost biomaterials for production of therapeutics in a sustainable manner (Johnson, 2008). *Chlorococcum humicola* (Nageli) Rabenhorst (Figure 1) is a freshwater unicellular green alga coming under the class Chlorophyta. Even though *C. humicola* is proved to be a rich source of structurally novel and biologically active metabolites (Bhagavathy, Sumathi & Jancy Sherene Bell , 2011), biomass-productivity of its specific strains are not well known.

Long-term storage of algal stock in sodium alginate beads in pure culture form is useful for stock culture management (Gaudin, Lebeau, & Robert , 2006) as well as long term storage of the algal seeds (Faafeng, Donk, & Källqvist, 1994) for biomass production and production of secondary metabolites (Moreno-Garrido, 2008). Immobilized algae can also be used in wastewater treatment (Travieso et al., 1996) and removal of heavy metals from waste water (Becker, 1994; Murugesan, Maheswari, & Bagirath, 2008). However, specific methods to develop a seed material of this alga for convenient mass-cultivation remain quite unexplored.

Protein content of green algae varies depending upon the species and strains (Fleurence, 1999; Gatenby et al., 2003). In addition to proteins, carbohydrates present in algae are also variable that have important value as food and fuels. Algal carbohydrates are easily digestible compounds that have high demand in the preparation of conventional foods, pharmaceutical and nutraceutical compounds (Becker, 2007). Moreover, the residual biomass rich in carbohydrates after the extraction of lipids or proteins is used for the production of ethanol (Gao, Shimamura, Ishida, & Takahashi, 2012). Some of the algal lipids such as omega-3 fatty acid and DHA are nutritionally valuable (Spolaore, Joannis-Cassan, Duran, & Isambert, 2006). Polyunsaturated fatty acids, especially omega-3 and omega-6 in algal oils are used as medicines for health of heart and brain (Ignarro, Balestrieri, & Napoli, 2007). Essential fatty acids present in algae are used as dietary supplements in many pharmaceutical products (Benatti, Peluso, Nicolai, & Calvani, 2004). Algal biomass rich in oils is also used as a natural source of bio-fuels, which is highly cost effective, environmental-friendly and renewable source of liquid fuel (Scott & Bryner, 2006). However, magnitude of oil and biomass production of an alga depends on the cultural conditions and biomass productivity (Olofsson et al., 2012). Naturally, chemical characterization of the bio-oil as well as the general biomass of hitherto unexplored species and local strains of green algae becomes quite meaningful. Such kinds of data are universally significant to assess the industrial potential of new species or strains of algae.

Assessment of biomass productivity, total proteins, carbohydrates and lipid content as well as lipid characterization of a local strain of C. humicola was the major objective of the current investigation. Since trials of long-term storage are essential to ensure continuous industrial production of algae, synthetic seed preparation and its viability in the 'seed form' became another objective. Overall, the present investigation point to the significance of a local strain of C. humicola as a potential feed-stock for food and fuel. Synthetic seeds of this alga could be preserved for more than a year.

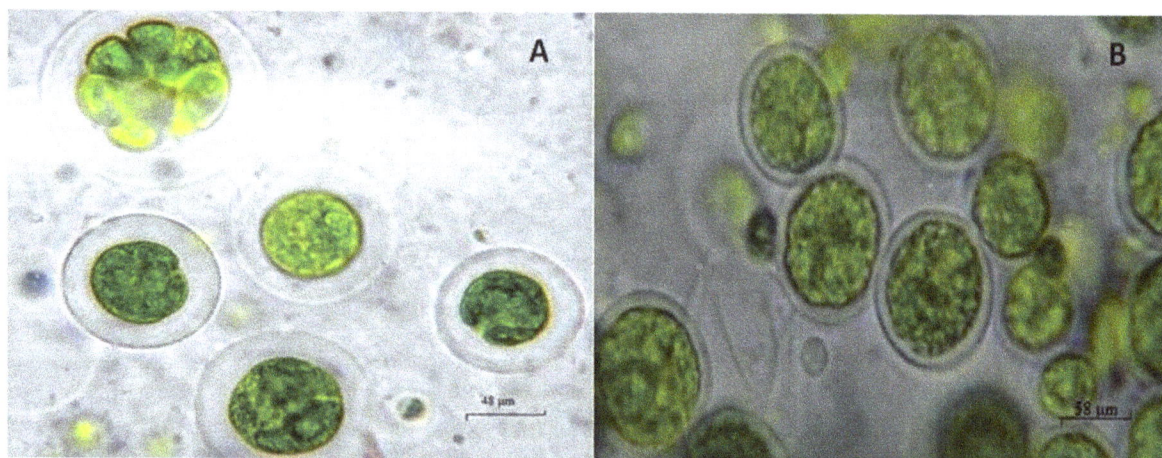

Figure 1. *Chlorococcum humicola* showing solitary cells (A) and colonial cells (B)

2. Materials and methods

2.1 In vitro Culture of Algae

C. humicola is unicellular, non-motile, spherical cells having smooth cell walls. Cells are seen in colonies or solitary and are varied in sizes (48 μm- 58 μm) having a 'hollow sphere like chloroplast', completely filling each cell, with a lateral notch and a single pyrenoid (Phillipose, 1967). In the current investigation, viable cells of a local strain of C. humicola were isolated from a fresh water temple pond (9°45'02.9"N 76°23'45.6"E) of Kottayam District of Kerala, India. Pure culture of the strain in BBM is maintained in the algal culture facility centre in the Ecotechnology Laboratory, School of Biosciences, Mahatma Gandhi University.

Temperature and pH of the collected water samples were measured. 100 mg of fresh biomass after centrifugation of the pure culture was inoculated and cultured in one litre flasks using Bold's Basal Medium (BBM) (Andersen, 2005) in triplicate (Figure 2). All the culture vessels were incubated under controlled conditions of light (8000 Lux), temperature (24 ± 2^0C) and pH (7.30). Productivity was measured on completion of 30 days of growth. On completion of the incubation period, biomass was collected by centrifugation and the solid biomass was further air dried. Percentage increase of biomass per day per litre was calculated as productivity of algae as per the formula:

$$\% \text{ of Biomass productivity of algae } /L/day = \frac{\text{Final dry weight}}{\text{Initial dry weight } \times \text{ total no. of culturing days x Vol.}} \times 100\%$$

Figure 2. Culture of *C. humicola* for biomass production of alga in BBM – intensity of colour reveals growth intensity

2.2 Synthetic Seed Preparation and Test for Long Term Storage Viability

Synthetic seeds of algae were prepared by using sodium alginate. Sodium alginate (4%) were prepared in sterilized BBM media and stirred continuously at 60^0C in a water bath for one hour. One gram of uniculture of algae sample was washed out with sterile water and mixed with prepared sodium alginate slurry in the ratio of 1 volume of cell inoculums: 2 volume of sodium alginate. It was then dropped into 0.2 M calcium chloride solution using a pipette to form calcium alginate beads with algal cells entrapped. The beads were kept for 30 minutes to solidify. They were then washed 3 to 4 times with distilled water. Algal seeds were stored under dark condition in a refrigerator at 4 degree Celsius. At every three month interval, viability of the beads was tested by culture (5 beads for 100 mL culture medium) in BBM to assess their viability.

2.3 Quantitative Estimation of Carbohydrates

Carbohydrate content was determined as per the method of Dubois et al. (1956), using glucose as standard; 100 mg of the lyophilized algal sample was hydrolyzed by keeping it in boiling water bath for three hours with 5 mL of 2.5 N HCl and cooled to room temperature. The total volume was made up to 100 mL by using double distilled water and centrifuged at 2723g for 5 minutes at 4^0C. The supernatant was used for carbohydrate estimation.

2.4 Quantitative Estimation of Total Proteins

Total protein content was determined as per the method of Lowry (Lowry, Rosebrough, Lewis, & Randall, 1951), using bovine serum albumin as standard; 5 mg of freeze dried algal sample was mixed with 5 mL of 80% acetone (Rotek vortex mixture: 1331) for 1 minute and centrifuged at 5000 rpm for 5 minutes at 4^0C and the pellet was homogenized with 0.2 mL of 24% TCA (w/v) and centrifuged at 7000 rpm for 5 minutes at 4^0C. The homogenate was incubated at 95 ^0C for 15 minute in a water bath and cooled to room temperature. 600µL of ultra pure water was added and centrifuged at 15000 rpm at 4^0C for 2 minutes. The pellet was collected and re-suspended in 0.5 mL Lowry D reagent and it was then incubated at 55^0C for 3 hours. The sample was cooled to room temperature ($27^0C – 30^0C$) and centrifuged at 15000 rpm for 20 minutes at 4^0C. The supernatant was collected and used for estimation of total proteins.

2.5 Extraction of Algal Oil

Total lipids were extracted as per the method of Bligh and Dyer (1959) using Soxhlet; 10 gm of freeze dried biomass was taken into a round bottom flask and added 100 mL of chloroform: methanol (2:1 v/v) mixture into the biomass. The biomass was then kept soaked in the organic solvents for 4 hrs under continuous shaking in a rotary shaker at 750 rpm; afterwards the mixture was centrifuged at 6000 rpm for 5 minutes at room temperature (27^0C - 30^0C). Residual biomass was separated from the extract and then the oil along with the solvent was transferred in to a separating funnel. About 40 mL of distilled water was added to this mixture to separate the oil from the solvent. The oil got separated as an organic phase in bottom layers; this was then collected into a bottle. The separated biomass and the oil were made free of the solvent by using rotary evaporator.

The air dried residual biomass free of the solvent was further subjected to hot method of extraction for collection of the remaining neutral lipids. The biomass was taken in to Soxhlet extractor with 75 mL of hexane, refluxed under 70^0 C for 2 hours. The extracted oil components were collected and the oil was made-free of the solvent by using rotary evaporator. Finally, the two extracted oil samples were mixed together to get the total oil.

2.6 Chemical Characterization of Algal Oil

Chemical characterization of the oil was carried out using the advanced Government of India analytical facility at CARE Kerala, Chalakkudy. Exactly 50 mg of algal oil was saponified with 1 mL of saturated KOH-CH_3OH solution at 50^0 C for 10 minutes and then followed by methanolysis with 5% HCl in methanol at 60^0C for another 10 minutes in screw capped test tubes. The methyl fatty acids were separated by adding 2 mL of water into it and fatty acid phase was recovered. GC-MS (Agilent make 7890A- 5975C) instrument was used for the fatty acid profiling. 1 mL of methyl fatty acid sample was injected to the GC column. Helium was used as carrier gas at flow rate of 54 mL/min. Chromatographic data was recorded and compared using Agilent data analysis software.

2.7 Transesterification of Algal Oil and the Production of Biodiesel

400 mg of algal oil extracted were taken into a round bottom flask and mixed with 15 mL of methanolic sulphuric acid containing 2% sulphuric acid in methanol (v/v) and refluxed at 60^0 C for 4 hours with continuous shaking. The reaction was monitored by thin layer chromatography (TLC) with the solvent system, Hexane: Ethyl acetate/ hexane: Toluene at the ratio of 9:1. The reaction was continued till the oil spot was disappeared on TLC plate. After the completion of reaction (2-4 hr), the contents were transferred to separating funnel and 25 mL water was added to it. The aqueous layer was extracted twice with ethyl acetate (25 mL each) and pooled the ethyl acetate layer. The extract was dried over anhydrous Na_2SO_4 and concentrated under vacuum.

2.8 FTIR Analysis of Algal biodiesel

FTIR characterization for 'biodiesel' samples produced was carried out (IS10 FTIR, Thermo Scientific) in transmission mode in 400-4000 cm^{-1} wave number range.

Fatty acid composition of algal oil

Percentage of oil in algae (%) was calculated using the formula (Abubakar, Mutie, & Muhoho, 2012)

$$= \frac{\text{Weight of oil (g)}}{\text{Dry weight of sample (g)}} \times 100$$

3. Results and Discussion

3.1 Biomass Productivity of Alga in Vitro Culture Media

Since BBM is known to enhance maximum production of protein and chlorophyll in green algae (Sankar & Ramasubramanian, 2012), the same medium was used for the assessment of biomass productivity of *C. humicola* (Table 1) in the current experimentation. Productivity of 73.8% mg/ L/ day obtained suggests the alga to be a suitable candidate for high yield of biomass and other derivatives, easily amenable to industrial trials.

Table 1. Biomass productivity, carbohydrates, proteins and lipids contents in the biomass of *C. humicola*

Cultural conditions	Medium (BBM)
Quantity of the medium	1 L
pH	7.30
Temp in ^{0}C	24±2
Light intensity (Lux)	8000
Duration of days	30
Fresh Weight of inoculums (mg)	100
Dry weight of inoculums (mg)	19.3
Dry weight of biomass after 30 days (mg)	427.57 ± 4
Growth of alga mg/L/day	73.8%
Carbohydrates mg/gm of biomass	22.4%
Proteins in mg/gm of biomass	25.5%
Lipids in mg/gm of biomass	13%

3.2 Experimentation on Long Term Storage and Viability

In general, encapsulation of micro algae in alginate as 'synthetic-seed-material' (SSM) can be considered a profitable method, to reduce the cost of long-term storage of pure culture or stock maintenance. Synthetic seed material of algae has several other industrial applications such as phycoremediation (Rai & Mallick, 1992) hydrogen production (Das, 2001) and maintenance of aseptic specimens of algae during culture transportation. In the present experimentation, successful preparation of SSM of *C. humicola* and its long-term maintenance is achieved (Figure 3). The SSM of *C. humicola* was stored at 4^0C in a usual laboratory refrigerator for about 14 months, and the same was successfully cultured in BBM at every three-month intervals with quite same viability till the 14th month (Figure 4; Table 2). This fact is evidential to viability of the SSM for further duration. Even though, it is well known that alginate encapsulation method of algae maintains ultra structural integrity and normal physiological activities (Corrêa et al., 2009; Dainty, Goulding, Robinson, Simpkins, & Trevan, 1986) during sufficiently long period of time, this is the first demonstration of retention of green-algae in alginate beads over a year with quite good viability.

Figure 3. Synthetic seed materials (SSM) of Algal cells after preparation

Figure 4. Viability testing by culturing of *C. humicola* in Bold's basal medium: A. after 3 months; B. after 6 months; C. after 9 months; D. after 12 months; E. after 14 months – intensity of the colour reveals equal viability

Table 2. Growth rate of *C. humicola* in terms of number of cells/100 mL after culture of SSM in BBM at different intervals

No. of cells in one seeds/ mL of inoculum	After 3 month cells / mL	After 6 month cells / mL	After 9 month cells / mL	After 12 month cells / mL	After 14 month cells / mL
1.84±0.039 x10^6	13.3±0.07x10^7	13.3±0.04x10^7	13.4±0.02x10^7	13.4±0.02x10^7	13.3±0.02x10^7

3.3 Total Proteins, Carbohydrates and Lipids

Chemical characterization of the biomass of the strain showed 22.4 % carbohydrates, 25.5 % proteins and 13 % of lipids (Table 1). Uma et al. (2015) reported more or less similar protein content for *C. humicola* cultured in outdoor environments in CFTRI medium enriched with NPK fertilizer; even higher protein content is known for *Chlorella sp.* cultured in BBM (Sankar & Ramasubramanian, 2012). But the lipid content observed in the present experimentation using the local strain of *C. humicola* is found to be higher than that of the previous reports. Since protein and lipid content of algae depend not only on the species, but also on diverse environmental conditions (Morris , Smith & Glover,1981), further standardization for optimum yield of proteins and lipids is essential in the assessment of *C. humicola* as a protein or lipid-rich algal resource.

3.4 Transesterification of Oil and FTIR Confirmation of Biodiesel

Chemical profile of the oil from this alga (Table 3) has shown 70 % monosaturated fatty acids, 17.4 % monounsaturated fatty acids and 12.2 % polyunsaturated fatty acids. Commonly used lipids for biodiesel productions have C16:0 and C18:1fatty acids (Knothe, 2005). Usually algal oil with saturated and poly unsaturated fatty acids containing 14-18 carbon molecules such as C14:0, C16:0, C16:1, C18:1, C18:2, C18:3 are used as the feed-stock for biodiesel productions (Duong , Li , Nowak & Schenk, 2012; Stansell, Gray & Sym, 2012). Chemical characterization of the oil extracted from *C. humicola* obtained in the current investigation revealed that 95.4% of it is C14-18 fatty acids.

According to the American Society for Testing and Materials (ASTM) D6751 and European EN 14214 standards, monosaturated fatty acids are given preferences for the production of good quality biodiesels (Knothe, 2005). Moreover, algal oil containing fatty acids such as palmitic, stearic, oleic and Linoleic acids are considered good for biodiesel (Knothe, 2008). Therefore, the oil extracted from *C. humicola* was subjected to transesterification trials. The IR spectra (Figure 5) peak 1741.72cm^{-1} of transesterified algal oil confirmed the formation of biodiesel. Since the quality of the oil remains the same irrespective of the medium used (Mahmah, Chetehouna, & Mignolet, (2011), this alga may be considered as a good oil resource for biodiesel production; however, further standardization of environmental conditions and media is required to assure the optimum oil yield.

Observation of 12.2 % of the total fatty acids to be of the two important essential fatty acids in this local strain, it may be considered a nutraceutically valuable alga. Only a very low concentrations (1.41- 4.04 %) of Linoleic acid and Linolenic acid (0.19- 0.67%) is known for Chlorella sp (Hempel, Petrick, & Behrendt, 2012). Since, *Chlorella* species are the major group of micro algae used for industrial production of essential fatty acids (Pulz & Gross 2004), this local strain of *C. humicola* with significant amount of essential fatty acids may be considered one of the best green algal resources for essential fatty acids.

Table 3. GCMS fatty acid profile of *Chlorococcum humicola*: *monosaturated fatty acids (70%), ** monounsaturated fatty acids (17.4%), *** polyunsaturated fatty acids (12.2%)

Fatty acid profile	Result in 1 ml (ppm)	% of fatty acids
Caproicacid (C6:0) *	66.04	11.2
Undecanoicacid (C11:0)*	36.51	6.2
Lauricacid (C12:0)*	68.42	11.6
Myristicacid (C14:0)*	66.59	11.3
Pentadecanoicacid (C15:0)*	33.17	5.6
Palmiticacid (C16:0)*	80.29	13.6
Palmitoleic acid (C16:1 cis)**	34.46	5.8
Stearicacid (C18:0)*	64.86	10.9
Oleic acid (C18:1 cis)**	68.37	11.6
Linoleic acid (C18:2 cis)***	34.47	5.8
Linolenic acid (C18:3 cis)***	37.79	6.4

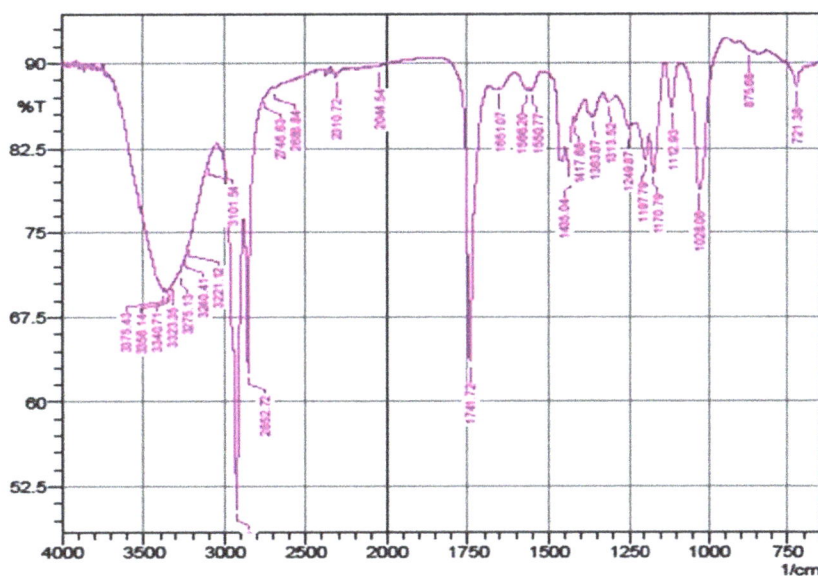

Figure 5. FTIR spectrum of algal biodiesel – peak at 1741.72cm^{-1} confirm the formation of biodiesel

4. Conclusion

Chemical analysis of the biomass of *C. humicola* has shown that it is rich in carbohydrate, protein and lipids, especially of nutraceutically significant compounds in its oil, which indicate industrial value of this alga. High percentage of monosaturated fatty acids in its oil indicates this local strain of *C. humicola* as a good candidate for biofuel feedstock. Sodium alginate encapsulation method is found quite feasible for maintaining a stock culture of this alga for sufficiently long period of 14 months. Further explorations on lipid production potential of this alga in different media under varied levels of diverse nutrients and heterotrophic conditions are essential for ensuring its industrial applications.

Acknowledgment

Authors wish to acknowledge the support received for UV Spectrum analytical facilities at institute for intensive research in basic sciences (IIRBS) and FTIR facility of Department of Biotechnology, Government of India at School of Biosciences, Mahatma Gandhi University.

References

Abubakar, L. U., Mutie, A. M., & Muhoho, A. (2012). Characterization of Algae Oil (Oilgae) and its Potential As Biofuel in Kenya. *Journal of Applied Phytotechnology in Environmental Sanitation, 1*(4), 147-153.

Andersen, R. A. (2005). *Algal Culturing Techniques* (p. 578). UK: Elsevier Academic Press.

Becker, E. W. (1994). *Microalgae: Biotechnology and Microbiology* (p. 262). Cambridge, NY: Cambridge Univ. Press.

Becker, E. W. (2007). Micro-algae as a source of protein. *Biotechnology Advances, 25*, 207-210. http://dx.doi.org/10.1016/j.biotechadv.2006.11.002

Benatti, P., Peluso, G., Nicolai, R., & Calvani, M. (2004). Polyunsaturated fatty acids: biochemical, nutritional and epigenetic properties. *Journal of the American College of Nutrition, 23*, 281-302. http://dx.doi.org/10.1080/07315724.2004.10719371

Bhagavathy, S., Sumathi, P., & Jancy Sherene Bell, I. (2011). Green algae Chlorococcum humicola- a new source of bioactive compounds with antimicrobial activity. *Asian Pacific Journal of Tropical Biomedicine, 1*, S1–S7. http://dx.doi.org/10.1016/S2221-1691(11)60111-1

Bligh, E., & Dyer, W. (1959). A Rapid Method of Total Lipid Extraction and Purification. *Canadian Journal of Biochemistry and Physiology, 37*, 911-917. http://dx.doi.org/10.1139/o59-099

Chiu, S., Kao, C., Chen, T., Chang, Y., Kuo, C., & Lin, C. (2015). Cultivation of microalgal Chlorella for biomass and lipid production using wastewater as nutrient resource. *Bioresource Technology, 184*, 179-189. http://dx.doi.org/10.1016/j.biortech.2014.11.080

Corrêa, X. R., Tamanaha, M. S., Horita, C. O., Radetski, M. R., Corrêa, R., & Radetski, C. M. (2009). Natural impacted freshwaters: In situ use of alginate immobilized algae to the assessment of algal response. *Ecotoxicology, 18*, 464-469. http://dx.doi.org/10.1007/s10646-009-0301-x

Dainty, A. L., Goulding, K. H., Robinson, P. K., Simpkins, I., & Trevan, M. D. (1986). Stability of alginate-immobilized algal cells. *Biotechnology and Bioengineering, 28*, 210-216. http://dx.doi.org/10.1002/bit.260280210

Das, D. (2001). Hydrogen production by biological processes: a survey of literature. *International Journal of Hydrogen Energy, 26*, 13-28. http://dx.doi.org/10.1016/S0360-3199(00)00058-6

Dubois, M., Gilles, K. A., Ton, J. K. H., Rebers, P. A., & Smith, F. (1956). Colorimetric Method for Determination of Sugars and Related Substances. *Analytical Chemistry, 28*, 350-356. http://dx.doi.org/10.1021/ac60111a017

Duong, V. T., Li, Y., Nowak, E., & Schenk, P. M. (2012). Microalgae isolation and selection for prospective biodiesel production. *Energies, 5*, 1835-1849. http://dx.doi.org/10.3390/en5061835

Faafeng, B. A., Donk, E., & Källqvist, S. T. (1994). In situ measurement of algal growth potential in aquatic ecosystems by immobilized algae. *Journal of Applied Phycology, 6*, 301-308. http://dx.doi.org/10.1007/BF02181943

Fleurence, J. (1999). Seaweed proteins.*Trends in Food Science & Technology, 10*, 25-28. http://dx.doi.org/10.1016/S0924-2244(99)00015-1

Gao, M. T., Shimamura, T., Ishida, N., & Takahashi, H. (2012). Investigation of utilization of the algal biomass residue after oil extraction to lower the total production cost of biodiesel. *Journal of Bioscience and Bioengineering, 114*, 330-333. http://dx.doi.org/10.1016/j.jbiosc.2012.04.002

Gatenby, C. M., Orcutt, D. M., Kreeger, D., Parker, B. C., Jones, V. A., & Neves, R. J. (2003). Biochemical composition of three algal species proposed as food for captive freshwater mussels. *Journal of Applied Phycology, 15*, 1-11. http://dx.doi.org/10.1023/A:1022929423011

Gaudin, P., Lebeau, T., & Robert, J. M. (2006). Microalgal cell immobilization for the long-term storage of the marine diatom Haslea ostrearia. *Journal of Applied Phycology, 18*, 175-184. http://dx.doi.org/10.1007/s10811-006-9092-0

Hempel, N., Petrick, I., & Behrendt, F. (2012). Biomass productivity and productivity of fatty acids and amino acids of microalgae strains as key characteristics of suitability for biodiesel production. *Journal of Applied Phycology, 24*, 1407-1418. http://dx.doi.org/10.1007/s10811-012-9795-3

Ignarro, L. J., Balestrieri, M. L., & Napoli, C. (2007). Nutrition, physical activity and cardiovascular disease: an update. *Cardiovascular Research, 73*, 326–340. http://dx.doi.org/10.1016/j.cardiores.2006.06.030

Johnson, F. X. (2008). Industrial Biotechnology and Biomass Utilisation. *Futur Prospect for Industrial Biotechnology*, 196.

Knothe, G. (2005). Dependence of biodiesel fuel properties on the structure of fatty acid alkyl esters. *Fuel Processing Technology, 86*, 1059-1070. http://dx.doi.org/10.1016/j.fuproc.2004.11.002

Knothe, G. (2008). "Designer" biodiesel: Optimizing fatty ester composition to improve fuel properties. *Energy and Fuels, 22*, 1358-1364. http://dx.doi.org/10.1021/ef700639e

Lowry, O. H., Rosebrough, N. J., Lewis, A. F., & Randall, R. J. (1951). The folin by oliver. *Readings, 193*, 265-275. http://dx.doi.org/10.1016/0304-3894(92)87011-4

Mahmah, S. C. B., Chetehouna, K., & Mignolet, E. (2011). Biodiesel production using Chlorella sorokiniana a green microalga. *Revue des Energies Renouvelables, 14*, 21–26.

Morais, M. G., & Costa, J. A. V, (2007). Biofixation of carbon dioxide by Spirulina sp. and Scenedesmus obliquus cultivated in a three-stage serial tubular photobioreactor. *Journal of Biotechnology, 129*, 439-445. http://dx.doi.org/10.1016/j.jbiotec.2007.01.009

Moreno-Garrido, I. (2008). Microalgae immobilization: Current techniques and uses. *Bioresource Technology, 99*, 3949-3964. http://dx.doi.org/10.1016/j.biortech.2007.05.040

Morris, I., Smith, A. E., & Glover, H. E. (1981). Products of photosynthesis in phytoplankton off the Orinoco River and in the Caribbean Sea. *Limnology and Oceanography, 26*, 1034-1044. http://dx.doi.org/10.4319/lo.1981.26.6.1034

Murugesan, A. G., Maheswari, S., & Bagirath, G. (2008). Biosorption of cadmium by live and immobilized cells of Spirulina platensis. *International Journal of Environmental Research, 2*, 307-312.

Olofsson, M., Lamela, T., Nilsson, E., Bergé, J. P., Del Pino, V., Uronen, P., & Legrand, C. (2012). Seasonal variation of lipids and fatty acids of the microalgae Nannochloropsis oculata grown in outdoor large-scale photobioreactors. *Energies, 5*, 1577-1592. http://dx.doi.org/10.3390/en5051577

Phillipose, M. T. (1967). *Chlorococcales, ICAR, Monograph on Algae* (pp. 73-74). New Delhi,

Pulz, O., & Gross, W. (2004). Valuable products from biotechnology of microalgae. *Applied Microbiology and Biotechnology, 65*, 635-648. http://dx.doi.org/10.1007/s00253-004-1647-x

Rai, L. C., & Mallick, N. (1992). Removal and assessment of toxicity of Cu and Fe to Anabaena doliolum and Chlorella vulgaris using free and immobilized cells. *World Journal of Microbioly & Biotechnology, 8*, 110-4. http://dx.doi.org/10.1007/BF01195827

Sankar, M., & Ramasubramanian, V. (2012). Biomass production of commercial algae Chlorella vulgaris on different culture media. *E-Journal of Life Sciences, 1*, 56-60.

Scott, A., & Bryner, M. (2006, December). Alternative fuels: Rolling out next generation technologies. Chemical week (pp. 20-27).

Spolaore, P., Joannis-Cassan, C., Duran, E., & Isambert, A. (2006). Commercial applications of microalgae. *Journal of Bioscience and Bioengineering, 101*, 87-96. http://dx.doi.org/10.1263/jbb.101.87

Stansell, G. R., Gray, V. M., & Sym, S. D. (2012). Microalgal fatty acid composition: Implications for biodiesel quality. *Journal of Applied Phycology, 24*, 791-801. http://dx.doi.org/10.1007/s10811-011-9696-x

Travieso, L., Benitez, F., Weiland, P., Sánchez, E., Dupeyrón, R., & Dominguez, A. R. (1996). Experiments on immobilization of microalgae for nutrient removal in wastewater treatments. *Bioresource Technology, 55*, 181-186. http://dx.doi.org/10.1016/0960-8524(95)00196-4

Uma, R., Sivasubramanian, V., & S. N. D (2015). C. humicola in fertilizer based outdoor cultivation. *Indian Journal of Pharmaceutical Science & Research, 5*, 19-22.

Yamaguchi, K. (1997). Recent advances in microalgal bioscience in Japan, with special reference to utilization of biomass and metabolites: a review. *Journal of Applied Phycology, 8*, 487-502.

7

Inhibitory Activity of *Citrus Madurensis* Ripe Fruits Extract on Antigen-induced Degranulation in RBL-2H3 Cells

Kimihisa Itoh[1], Kazuya Murata[2], Megumi Futamura-Masuda[2], Takahiro Deguchi[2], Yuko Ono[2], Marin Eshita[2], Masahiko Fumuro[1], Morio Iijima[1, 3] & Hideaki Matsuda[1, 2]

[1]The Experimental Farm, Kindai University, Wakayama, Japan

[2]Faculty of Pharmacy, Kindai University, Osaka, Japan

[3]Faculty of Agriculture, Kindai University, Nara, Japan

Correspondence: Hideaki Matsuda, Faculty of Pharmacy, Kindai University, 3-4-1 Kowakae, Higashiosaka, Osaka 577-8502, Japan. E-mail: matsuda@phar.kindai.ac.jp

Abstract

The purpose of this study was to search edible ripe *Citrus* fruits which are applicable for functional food materials as juice, tea and/or jam with sweet taste and rich aroma. A fifty percent ethanolic extract (CMR-ext) obtained from the edible ripe fruit of *Citrus madurensis* exhibited an inhibitory activity of antigen-induced degranulation in anti-dinitrophenyl (DNP) IgE antibody sensitized rat basophilic leukemia (RBL) -2H3 cells. The inhibitory effect of the CMR-ext on degranulation in RBL-2H3 cells was attributable to 3',5'-di-C-β-glucopyranosylphloretin (**1**) which is a constituent of *C. madurensis*. The effect of **1** on Akt and mitogen-activated protein kinases (MAPK) phosphorylation was examined in RBL-2H3 cells. Western blot analysis revealed that **1** (50 μM) inhibited the degranulation by suppression of Akt and p38 phosphorylation.

Keywords: *Citrus madurensis*, 3',5'-di-C-β-glucopyranosylphloretin, degranulation inhibition

1. Introduction

Our previous studies (Kubo et al., 1989, Matsuda et al., 1991, Itoh et al., 2009) on *Citrus* fruit have indicated that the extracts of unripe fruits of *C. unshiu* MARKOVICH and *C. hassaku* HORT ex. T. TANAKA showed potent anti-allergic and melanogenesis inhibitory activities, whereas ripe *Citrus* fruit extracts of them had poor activities. It was found that the active constituents of these unripe *Citrus* fruit extracts were several flavonoids, such as hesperidin and narirutin from *C. unshiu* fruit and naringin and neohesperidin from *C. hassaku* fruit. These findings suggested that these unripe *Citrus* fruit may be useful ingredients for anti-allergic agents and/or skin-whitening cosmetics. Recently, on the basis of our investigations, (Fujita et al., 2008, Itoh et al., 2009, Murata et al., 2013, Futamura et al., 2016) several products originated from some unripe *Citrus* fruits are launched in the functional food market with expectation for anti-allergic and tyrosinase inhibitory activities. However, unripe *Citrus* fruits have bitter taste and aren't edible for functional food with sweet taste. On the other hand, there is another market for edible functional food as juice, tea and/or jam with sweet taste and/or rich aroma. Therefore, the purpose of this study was to search edible ripe *Citrus* fruit, which are applicable for functional food materials with sweet taste. Our previous chemotaxonomic report on several *Citrus* fruits indicated that fruit of *C. madurensis* LOUREIRO scarcely contain above mentioned flavanone glycosides (Kubo et al., 2004). Ogawa et al. (2001) reported peels, juice sacs and leaves of *C. madurensis* contain 3',5'-di-C-β-glucopyranosylphloretin (**1**). Since *C. madurensis* is a perpetual *Citrus* breed which gives ripe and unripe fruit together, the plant has the advantage of being collectable ripe fruit throughout the year. Thus, we focused on ripe *C. madurensis* fruits.

RBL-2H3 cells originated from rat basophilic leukemia (RBL) have been frequently used to evaluate inhibitory activity of different compounds on type I allergic reaction, and to study IgE-Fcε receptor interactions in relation to intracellular signaling pathways in the process of degranulation (Ortega et al., 1988, Ikawati et al., 2001, Funaba et al., 2003, Choi et al., 2012, Murata et al., 2013). When granules in mast cells or basophils degranulate, an enzyme, β-hexosaminidase, is released along with histamine. Therefore, the enzyme is commonly used as the marker of mast cell degranulation or histamine release (Cheong et al., 1998). In fact, several constituents with type I allergic inhibitory activity have been isolated from several natural resources, such as *Mentha × piperita* Linne var. *citrata* Briq leaves (Sato & Tamura, 2015), *Coix lachryma-jobi* Linne var. *ma-yuen* Stapf bran (Chen et al., 2012), *Arachis hypogaea* Linne skins (Tomochika et al., 2011) and *Caesalpinia sappan* Linne root and heartwood (Yodsaoue et al., 2009) by using degranulation inhibitory activity assay in RBL-2H3 cells. In the present work, we evaluate the degranulation inhibitory activity of 50% ethanolic extract of edible ripe fruit of *C. madurensis* (CMR-ext) and **1** isolated from CMR-ext by use of RBL-2H3 cells. In several reports (Mastuda et al., 2002, Murata et al., 2013) concerning with anti-allergic constituent of plant resources, it was noticed that a flavonoid, baicalein, exhibited anti-degranulation activity. Thus we used baicalein as a positive control agent.

2. Materials and Methods

2.1 Reagents

Monoclonal mouse IgE anti-dinitrophenyl (anti-DNP IgE) was purchased from Yamasa Corporation (Tokyo, Japan). DNP-labeled human serum albumin (DNP-HSA) was purchased from Sigma-Aldrich (St. Louis, MO, USA). Fetal bovine serum (FBS) was purchased from Nichirei Bioscience (Tokyo, Japan). All primary antibodies were obtained from Cell Signaling Technology (Danvers, MA, USA), including Extracellular signal Regulated Kinase (ERK) (#4695), P-ERK (#4376), p38 (#9212), P-p38 (#9215), Akt (#4691), P-Akt (#2965) and β-actin (#4967). A horseradish peroxidase- labeled secondary antibody (anti-rabbit IgG HRP linked whole antibody, NA934-1ML) and chemiluminescent (ECL) kit were obtained from GE Healthcare (Tokyo, Japan). Baicalein was obtained from Wako Pure Chemical Industries, Ltd. (Osaka, Japan). Other chemical and biochemical reagents were of reagent grade and were purchased from Wako Pure Chemical Industries, Ltd. (Osaka, Japan) and/or Nacalai Tesque, Inc. (Kyoto, Japan) unless otherwise noted.

2.2 Plant Materials and Extraction

Fruits of *C. madurensis* (cv. Shikikitsu in Japanese) were collected in the Experimental Farm, Kindai University (34° 2′ N, 135° 11′ E, 17 m ASL), located in Wakayama Prefecture, Japan in April, 2012. The *C. madurensis* trees are grown in the ground for the purpose of the genetic resources preservation. Ripe and unripe fruits were collected from two trees which were propagated by grafting (the height of trees, 2.5 m; canopy width, 3.6 m; the age of trees, 40 years old; the life span of trees, 50-70 years). The data of cultivation environment are as follows: annual mean temperature, 17.6°C; maximum temperature, 34.5°C and 28.8°C (soil); minimum temperature, -0.1°C and 8.3°C (soil); annual rainfall, 2,283 mm/year. The collected fruits were visually classified by the color of fruit; i.e., fruits with whole yellow and whole green appearance were defined as ripe and unripe fruits, respectively (Figure 1). Physical data of ripe and unripe fruits (n = 20) was as follows: diameters of fruits; 34.6 ± 4.3 mm (ripe fruits), 24.6 ± 3.6 mm (unripe fruits), fresh weight of fruits; 14.9 ± 5.5 g (ripe fruits), 7.2 ± 2.4 g (unripe fruits). The samples were identified by the Experimental Farm, Kindai University, air-dried at 50°C for 72 h in an automatic air-drying apparatus (Vianove Inc., Tokyo, Japan), and powdered. Voucher specimens of ripe and unripe fruits (Shikikitsu Ripe Fruits: CMR201204 and Shikikitsu Unripe Fruits: CMU201204) are deposited in the Experimental Farm, Kindai University. The each fruits powder (10 g) was extracted with 50% ethanol (EtOH) (100 ml) for 2 h under reflux. The extract was evaporated under reduced pressure and then lyophilized to give the 50% EtOH extract of ripe fruits (CMR-ext) in 46.7% yield. The yield of 50% EtOH extract of unripe fruits was 30.8%.

2.3 Anti-degranulation Activity

Anti-degranulation activity was examined according to the method of Murata et al. (2013). RBL-2H3 cells (Japan Health Sciences Foundation, Osaka, Japan) were cultured in Enhanced Eagle's Minimum Essential Medium (EMEM) supplemented with 10% FBS and antibiotics (100 U/ml penicillin and 100 μg/ml streptomycin) in a CO_2 incubator (37°C, 5% CO_2). Cells were inoculated in a 24-well plate at 2×10^5 cells/ well with 400 μl of culture medium. After 24 h incubation in a CO_2 incubator, 100 μl of anti-DNP IgE dissolved with EMEM (final concentration 0.45 μg/ml) was added to each well to start cell sensitization, followed by incubation in a CO_2 incubator for 24 h. Cells were washed twice with 500 μl of siraganian buffer {NaCl 119 mM, KCl 5 mM, $MgCl_2$ 0.4 mM, piperazine-1,4-bis(2-ethanesulfonic acid) (PIPES) 25 mM, NaOH 40 mM; pH 7.2} and 180 μl of siraganian (+) buffer (siraganian buffer supplemented with 5.6 mM glucose, 0.1% BSA and 1 mM calcium chloride) was added.

The test sample was dissolved with dimethyl sulfoxide (DMSO) and diluted with siraganian (+) buffer to a final DMSO concentration of 0.1% v/v. In the control group, DMSO solution diluted with siraganian (+) buffer to 0.1% of final concentration was used instead of the sample solution. An aliquot of 20 μl of test solution were added to each well, and then incubated for 0.5 h in a CO_2 incubator. Degranulation was induced by adding 50 μl of DNP-HSA dissolved with siraganian (+) buffer (final concentration 0.01 μg/ml), followed by incubation in a

CO_2 incubator for 0.5 h. A portion of the supernatant (50 µl) was transferred to a 96-well plate and 50 µl of β-hexosaminidase substrate, p-nitrophenyl-N-acetyl-β-D-glucosamide dissolved with 0.1 M citric acid aqueous solution (final concentration 0.5 mM), was added. After incubation in a CO_2 incubator for 1 h, 100 µl of alkaline buffer (0.05 M $NaHCO_3$ and 0.05 M Na_2CO_3, pH 10) was added to terminate the reaction, and absorbance at 405 nm was measured using a microplate reader (Sunrise Rainbow Thermo, Tecan Japan Co., Ltd., Kanagawa, Japan). Baicalein was used as a positive control agent.

To evaluate the cytotoxic effect of test sample against RBL-2H3 cells, the cell viability was determined by a 2-(2-methoxy-4-nitrophenyl)-3-(4-nitrophenyl)-5-(2,4-disulfophenyl)-2H-tetrazolium (WST-8) assay using a commercial kit (Cell count reagent SF). RBL-2H3 cells (2.2×10^3 cells/well) were exposed to control or the sample solutions in 96-well plates for 48 h. DMSO solution diluted with siraganian (+) buffer (0.1%, v/v) served as the solvent control. After control or the sample solutions were exposed, 10 µl of Tetra Color ONE solution was added to each well, and the 96-well plate was continuously incubated at 37 °C for 4 h, then the OD values for each well were measured at wavelength 450 nm using a microplate reader.

The inhibition (%) of the release of β-hexosaminidase by the test samples was calculated by the following equation,

$$Inhibition\ (\%) = [1-(T-B-N)/(C-N)] \times 100$$

Control (C): DNP-BSA (+), test sample (-); Test (T): DNP-BSA (+), test sample (+); Blank (B): DNP-BSA (-), test sample (+); Normal (N): DNP-BSA (-), test sample (-).

2.4 Isolation and Identification Procedure for 1

The following isolation procedure for **1** was carried out according to the methods of Ogawa et al. and Sato et al. (2001, 2006). A suspension of CMR-ext (100 g) in water (500 ml) was extracted successively with hexane (500 ml × 2), ethyl acetate (AcOEt, 500 ml × 5), and butanol (BuOH, 500 ml × 5). Evaporation of the solvent gave a hexane-soluble fraction (0.07 g; yield from CMR-ext: 0.07%), an AcOEt-soluble fraction (2.6 g; 2.6%), and a BuOH-soluble fraction (26.4 g; 26.4%). The aqueous layer was evaporated under reduced pressure and then lyophilized to give a water-soluble fraction (70.8 g; 70.8%). A part of the BuOH-soluble fraction (25.3 g) was loaded on a Diaion HP-20 (6.5 × 27 cm, Mitsubishi Chemical) column. Elution with MeOH and water in increasing proportions monitored with TLC [Merck No. 1.15685 silica gel 60 RP-18 F_{254}S, water: MeOH (1:1, v/v), detection; UV and 10% H_2SO_4 followed with heating] gave 22 chromatographic fractions of 500 ml each. TLC analysis of the collected fractions allowed us to assemble them into three fractions (Fr. 1 to 3). Fr. 1 [fr.1 to 6, elution solvent; water / MeOH (1:1), yield; 18.0 g], Fr. 2 [fr. 7 to 17, water to water / MeOH (1:1), 4.6 g], and Fr. 3 [fr. 18 to 22, water/MeOH (1:1) to MeOH, 1.1 g]. Fr. 2 (4.5 g) was submitted to column chromatography over 225 g of silica gel (Merck No. 1.09385 silica gel 60, 4.6 × 27 cm). Elution with chloroform ($CHCl_3$) and MeOH in increasing proportions monitored with TLC [Merck No. 1.05715 silica gel 60 F_{254}, $CHCl_3$/MeOH/H_2O (6:4:1, v/v), detection; UV and 10% H_2SO_4 followed with heating] gave 21 chromatographic fractions of 450 ml of each. TLC analysis of the collected fractions allowed us to assemble them into five fractions (Fr. 2-1 to 2-5). Fr. 2-1 [fr. 1-8, elution solvent; $CHCl_3$/MeOH (10:1) to (4:1), 0.8 g], Fr. 2-2 [fr. 9, $CHCl_3$/MeOH (7:3), 0.28 g], Fr. 2-3 [fr. 10-12, $CHCl_3$/MeOH (7:3), 1.14 g], Fr. 2-4 [fr. 13-16, $CHCl_3$/MeOH (7:3), 0.63 g], and Fr. 2-5 [fr. 17-21, $CHCl_3$/MeOH (7:3) to MeOH, 1.29 g]. Fr. 2-4 showed a single spot (*Rf* -value: 0.30) on TLC [$CHCl_3$/MeOH/H_2O (6:4:1, v/v). Purification of Fr. 2-4 was carried out by preparative HPLC [SunFire Prep C18 OBD, 19 i.d. × 250 mm, mobile phase: 18% acetonitrile (CN_3CN) containing 0.1% trifluoroacetic acid, 15 ml/min, detection UV 280 nm]. Identification of the compound was carried out by the comparison of the physicochemical data of the purified compound with those of the reported data (Ogawa et al., 2001, Sato et al., 2006) for the known **1** (Figure 2).

2.5 HPLC Determination of 1 in CMR-ext

Content of **1** in CMR-ext was determined by HPLC analysis described in previous report (Itoh et al., 2009) with minor modification. An accurately weighed CMR-ext (50 mg) was added in a volumetric 100 ml flask. After addition of MeOH up to 100 ml, the sample of CMR-ext was extracted by ultrasonic radiation for 30 min at room temperature. After filtration with a membrane filter (0.45 µm, GL Sciences Inc. Tokyo, Japan), an aliquot of 10 µl of the sample solution was injected into the HPLC system. The HPLC system consisted of a Shimadzu SCL-10Avp (Shimadzu, Kyoto) with a Shimadzu pump unit LC-20AT, Shimadzu UV-Vis detector SPD-10Avp and Chromato-PRO (Run Time Corporation, Kanagawa, Japan). The TSK gel ODS-120T (4 µm, 250 × 4.6 mm i.d.) column (Tosoh Co., Tokyo) was used at 37ºC. The mobile phase was a gradient system of a solution A [0.1% H_3PO_4 in distilled water: CH_3CN (9:1 v/v)] and solution B [0.1% H_3PO_4 in distilled water: CH_3CN (2:8 v/v)] in the following ratio 0 min, solution A: solution B 9:1; for 30 min, 3:7 v/v. The flow rate was 0.8 ml/min; detection was at UV 280 nm; and the t_R for **1** was 15.3 min. The peak area ratios *versus* concentrations of **1** ($r = 0.9999$) yielded straight-line relationships in the range of 0.625-40 µg/ml with the above correlation coefficients. Under this condition, narirutin, naringin, hesperidin and neohesperidin were eluted at the t_R of 16.8, 17.4, 17.8, and 18.4 min, respectively.

2.6 Western Blot Analysis

RBL-2H3 cells were treated using the same method as described above. The harvested cells were lysed with a lysis buffer (Cell Signaling Technology) and centrifuged at $9,200 \times g$ for 10 min. The protein concentration of the supernatant was determined with a Protein Assay (Bio-Rad, Hercules, CA, USA). A solution of the same protein concentration was prepared and subjected to electrophoresis followed by transfer of protein to PVDF membranes at 60 V for 4 h. The resulting membranes were blocked with Blocking One-P solution for 20 min at room temperature for phosphorylated protein and 5% skimmed milk solution (TBST: in mM; NaCl 137, KCl 2.7, Tris 25 and 0.05% Tween, pH 7.4) for 1 h at room temperature for non-phosphorylated proteins (β-actin). Antibodies were diluted with blocking buffers in the ratio antibody: blocking buffer (1:1,000). The membrane was treated with the antibody solution and the membrane was washed with TBST. Anti-rabbit IgG HRP linked whole antibody were diluted in the same manner as the primary antibody and the solution was used to immerse each membrane. After washing the membrane with TBST, the proteins on membranes were visualized using ECL detection system.

2.7 Statistical Analysis

The experimental data were evaluated for statistical significance using Bonferroni/Dunn's multiple-range test with GraphPad Prism for Windows, Ver. 5 (GraphPad Software Inc., 2007).

3. Results and Discussion

Inhibitory activity of CMR-ext on antigen-induced degranulation was determined by measuring inhibitory activity of β-hexosaminidase release in RBL-2H3 cells according to the method of Murata et al. (2013). Dinitrophenyl-labeled human serum albumin (DNP-HSA) was used as an antigen. As shown in Table 1, treatment of CMR-ext (3.1 to 200 µg/ml) significantly inhibited degranulation induced by DNP-HSA from RBL-2H3 cells sensitized with anti-DNP IgE. Cytotoxicity of test samples against RBL-2H3 cells were evaluated by measuring cell proliferation using a commercial kit. CMR-ext didn't show any significant effects on cell proliferation at the concentration of 3.1 to 200 µg/ml. Baicalein inhibited degranulation without any significant effects on cell proliferation at 50 µM. Since *C. madurensis* is a perpetual *Citrus* breed which gives ripe and unripe fruit together, both fruit were collectable at the same time. Therefore, we compared the degranulation inhibitory activity of the ripe fruit extract (CMR-ext) with that of the unripe fruit extract. Unripe fruit extract significantly inhibited degranulation without any significant effects on cell proliferation at 12.5, 50 and 200 µg/ml, however the unripe extract showed less activity than CMR-ext (Table 1). These results suggested that ripe *C. madurensis* fruit may be applicable for functional food materials as juice, tea and/or jam with sweet taste and rich aroma. Thus, we focused on ripe *C. madurensis* fruits.

1-a; Ripe fruits   1-b; Unripe fruits

Figure 1. Photographs of ripe and unripe fruits of *C. madurensis*

Table 1. Inhibitory activities of CMR-ext and unripe *C. madurensis* fruit extract on DNP-HSA induced degranulation in RBL-2H3 cells

Samples	Concentration (μg/ml or μM)	Inhibition (%)	Cell proliferation (%)
Run 1			
Control			100.0±0.1
Ripe *C. madurensis*	3.1 μg/ml	30±5	101.9±2.8
fruit extract (= CMR-ext)	12.5	51±3	100.9±0.9
	50	53±3	103.5±0.7
	200	59±3	99.2±2.7
Baicalein	50 μM	76±1	107.6±2.2
Run 2			
Control			100.0±3.3
Unripe *C. madurensis*	3.1 μg/ml	4±5	96.8±1.8
fruit extract	12.5	24±3	97.9±0.7
	50	28±4	99.3±2.6
	200	34±4	107.5±3.0
Baicalein	50 μM	70±3	103.3±3.0

Each value in inhibition represents the mean ± S.D. of 3 experiments. Baicalein was used as a positive control agent. Each value in cell proliferation represents the mean ± S.D. of 3 experiments.

According to the report by Ogawa et al. (2001), *C. madurensis* are thought to originate from natural hybrids between the genera *Citrus* and *Fortunella*, and contain **1** in their peels, juice sacs and leaves (Figure 2) (Ogawa et al., 2001). Recently, **1**, a component of *C. madurensis* peels, was reported to have tyrosinase inhibitory activity (Lou et al., 2012) and antioxidant activity (Yu et al., 2013). To identify active constituents of CMR-ext, at first, we isolated **1** (pale yellow powder, isolation yield, 0.32% from CMR-ext) according to the method of Ogawa et al. and Sato et al. (2001, 2006), and identified its chemical structure on the basis of several NMR spectral data (^{1}H-NMR, ^{13}C-NMR, and DEPT) analysis, thereafter the degranulation inhibitory activity of **1** was examined. As shown in Table 2, **1** inhibited degranulation at the concentration of 50, 100 and 200 μM without any significant effects on cell proliferation. The content (mg/g of extract) of **1** in CMR-ext was determined by HPLC analysis. As a result, CMR-ext contained 11.0 mg/g of **1**. Thus, a part of the degranulation inhibitory activity of CMR-ext is attributable to **1**. Comparison of the inhibitory activity of CMR-ext with that of **1** gave a hypothesis that other constituents of CMR-ext may also contribute to the activity. The HPLC analysis revealed that the contents of several flavonoids in CMR-ext were as follows: hesperidin, 0.9 mg/g, neohesperidin, 0.9 mg/g, while narirutin and naringin were not detected. These data were in accordance with those of reports of Kawaii et al. (1999). Our previous paper (Murata et al., 2013) reported that hesperidin and neohesperidin showed a weak degranulation inhibitory activity in RBL-2H3 cells. To identify other active ingredients, further studies are required, and now undergoing.

Table 2. Inhibitory activity of 3',5'-di-C-β-glucopyranosylphloretin (**1**) on DNP-HSA induced degranulation in RBL-2H3 cells

Samples	Concentration (μM)	Inhibition (%)	Cell proliferation (%)
Control			100.0±0.1
1	50	10±2	98.7±1.2
	100	32±2	103.9±4.8
	200	46±2	106.7±1.9
Baicalein	50	83±0	107.6±2.2

Each value in inhibition represents the mean ± S.D. of 3 experiments. Baicalein was used as a positive control agent. Each value in cell proliferation represents the mean ± S.D. of 3 experiments.

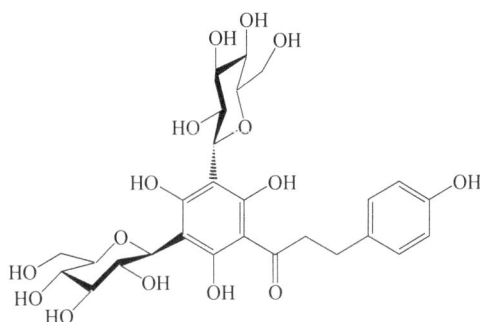

Figure 2. The chemical structure of 3',5'-di-C-β-glucopyranosylphloretin (**1**)

Type I allergy is defined as hypersensitive reaction. As illustrated in Figure 3, this type of allergy is known to be evoked by antigen-induced activation of Fcε receptors expressed on the surface of mast cells and basophils (Tomochika et al., 2011). Crosslinking of IgEs is essential, and is the first step triggering the signaling cascades such as Akt and MAPK that lead to degranulation of chemical mediators such as histamine, arachidonic acid metabolites and neutral proteases (Tomochika et al., 2011).

Therefore, to investigate the inhibition mechanism of degranulation by **1**, phosphorylation of Akt and relevant MAPKs (p38 and ERK) was examined by Western blot analysis. DNP-HSA induction at 0.01 µg/ml in RBL-2H3 cells led to phosphorylation of Akt, p38 and ERK (Figure 4). In our preliminary examination of DNP-HSA induction (0.01 µg/ml) on the phosphorylation of Akt, p38 and ERK, these were phosphorylated 15 min after induction, and phosphorylation peaked at 1 h (data not shown). In RBL-2H3 cells pretreated with **1** at 50 µM, phosphorylation of Akt and p38 was suppressed at 1 h, whereas phosphorylation of ERK was not suppressed (Figure 4). Significant effect was not observed in the expression level of Akt, p38, ERK and β-actin with or without DNP-HSA and/or **1**. These results suggest that **1** inhibits DNP-HSA-induced degranulation in RBL-2H3 cells by suppression of Akt and p38 phosphorylation as one of the degranulation inhibition mechanisms.

In conclusion, CMR-ext significantly inhibited DNP-HSA-induced degranulation in anti- DNP IgE antibody sensitized RBL-2H3 cells, without any effects on cell proliferation. It was revealed that a part of the degranulation inhibitory activity of the CMR-ext was attributable to **1**. To the best of our knowledge, this is the first report on degranulation inhibitory activity of **1**. Western blot analysis suggested that **1** inhibited degranulation by suppression of Akt and p38 phosphorylation.

Thus, ripe *C. madurensis* fruit may be applicable for functional food materials as juice, tea and/or jam with sweet taste and rich aroma in expectation of anti-type I allergic effect. There is an advantage that ripe fruits of this plant can be collected throughout the year due to a perpetual breed. However, further investigations are required to examine *in vivo* effects in animals and to reveal other active constituents.

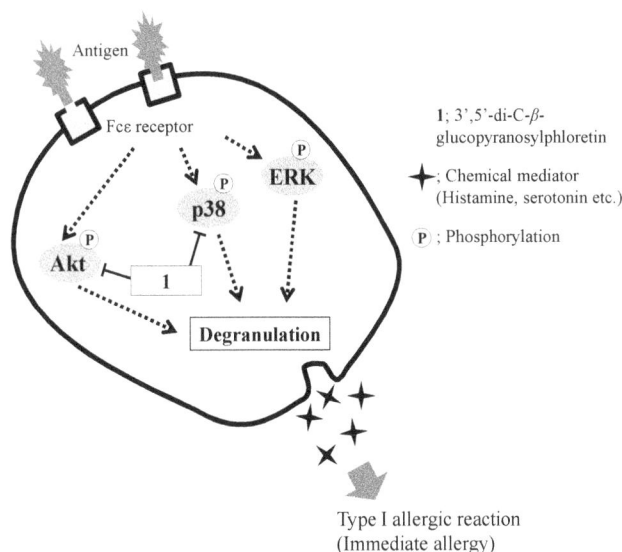

Figure 3. Schematic representation showing the inhibition of **1** on degranulation in RBL-2H3 Cells

P-Akt

Akt

P-p38

p38

P-ERK

ERK

β-Actin

| DNP-HSA (0.01 µg/ml) | - | + | + |
| 1 (50 µM) | - | - | + |

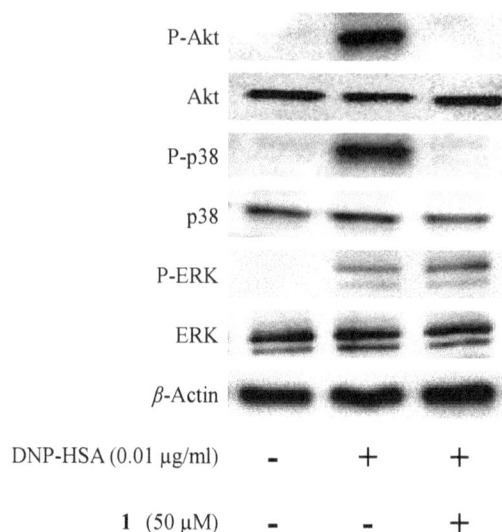

Figure 4. Effect of 1 on Akt, p38 and ERK phosphorylation in RBL-2H3 cells

RBL-2H3 cells were treated with 1 (50 µM) for 1 h, and degranulation was induced with DNP-HSA (0.01 µg/ml) for 0.5 h. Phosphorylation of Akt, p38 and ERK was determined with Western blot analysis.

Acknowledgment

This work was financially supported by the grant-in-aid for encouragement of research of Kindai University (project number; SR16), 2014-2015. We are grateful to all technical staffs of Yuasa Experimental Farm, Kindai University for the collection of C. madurensis fruit. I am deeply grateful to Dr. Shunsuke Naruto for his invaluable guidance and advice.

References

Chen, H. J., Lo, Y. C., & Chiang, W. (2012). Inhibitory effects of adlay bran (Coix lachryma-jobi L. var. ma-yuen Stapf) on chemical mediator release and cytokine production in rat basophilic leukemia cells. Journal of Ethnopharmacology, 141, 119-127. http://dx.doi.org/10.1016/j.jep.2012.02.009

Cheong, H., Choi, E. J., Yoo, G. S., Kim, K., M., & Ryu, S. Y. (1998). Desacetylmatricarin, an anti-allergic component from Taraxacum platycarpum. Planta Medica, 64(6), 577-578.

Choi, Y., Kim, M. S., & Hwang, J. K. (2012). Inhibitory effects of panduratin A on allergy-related mediator production in rat basophilic leukemia mast cells. Inflammation, 35(6), 1904-1915. http://dx.doi.org/10.1007/s10753-012-9513-y

Fujita, T., Shiura, T., Masuda, M., Tokunaga, M., Kawase, A., Iwaki, M., Gato, T., Fumuro, M., Sasaki, K., Utsunomiya, N., & Matsuda, H. (2008). Anti-allergic effect of a combination of Citrus unshiu unripe fruits extract and prednisolone on picryl chloride-induced contact dermatitis in mice. Journal of Natural Medicines, 62(2), 202-206. http://dx.doi.org/10.1007/s11418-007-0208-x

Funaba, M., Ikeda, T., & Abe, M. (2003). Degranulation in RBL-2H3 cells: regulation by calmodulin pathway. Cell Biology International, 27(10), 879-885. http://dx.doi.org/10.1016/S1065-6995(03)00177-X

Futamura-Masuda, M., Tokunaga, Y., Deguchi, T., Zentani, T., Enomoto, T., Murata, K., & Matsuda, H. (2016). Improving effects of the food containing dried powder of unripe Citrus unshiu fruits on nasal symptoms in the subjects with perennial nasal hypersensitivity. Allergology & Immunology, 23(7), 106-116.

Ikawati, Z., Wahyuono, S., & Maeyama, K. (2001). Screening of several Indonesian medicinal plants for their inhibitory effect on histamine release from RBL-2H3 cells. Journal of Ethnopharmacology, 75(2-3), 249-256.

Itoh, K., Hirata, N., Masuda, M., Naruto, S., Murata, K., Wakabayashi, K., & Matsuda, H. (2009). Inhibitory effects of Citrus hassaku extract and its flavanone glycosides on melanogenesis. Biological & Pharmaceutical Bulletin, 32(3), 410-415. http://dx.doi.org/10.1248/bpb.32.410

Itoh, K., Masuda, M., Naruto, S., Murata, K., & Matsuda, H. (2009). Antiallergic activity of unripe Citrus hassaku fruits extract and its flavanone glycosides on chemical substance-induced dermatitis in mice.

Journal of Natural Medicines, 63, 443-450. http://dx.doi.org/10.1007/s11418-009-0349-1

Kawaii, S., Tomono, Y., Katase, E., Ogawa, K., & Yano, M. (1999). Quantitation of flavonoid constituents in *Citrus* fruits. *Journal of Agricultural and Food Chemistry, 47*(9), 3565-3571. http://dx.doi.org/10.1021/jf990153+

Kubo, M., Fujita, T., Nishimura, S., Tokunaga, M., Matsuda, H., Gato, T., Tomohiro, N., Sasaki, K., & Utsunomiya, N. (2004). Seasonal variation in anti-allergic activity of *Citrus* fruits and flavanone glycoside content. *Natural Medicines, 58*(6), 284-294.

Kubo, M., Yano, M., & Matsuda, H. (1989). Pharmacological study on *Citrus* fruits. I. Anti-allergic effect of fruit of *Citrus unshiu* Markovich (1). *Yakugaku Zasshi, 109*(11), 835-842.

Lou, S. N., Yu, M. W., & Ho, C. T. (2012). Tyrosinase inhibitory components of immature calamondin peel. *Food Chemistry, 135*(3), 1091-1096. http://dx.doi.org/10.1016/j.foodchem.2012.05.062

Matsuda, H., Yano, M., Kubo, M., Iinuma, M., Oyama, M., & Mizuno, M. (1991). Pharmacological study on *Citrus* fruits. II. Anti-allergic effect of fruit of *Citrus unshiu* Markovich (2). On flavonoid components. *Yakugaku Zasshi, 111*(3), 193-198.

Mastuda, H., Morikawa, T., Ueda, K., Managi, H., & Yoshikawa, M. (2002). Structural requirements of flavonoids for inhibition of antigen-induced degranulation, TNF-α and IL-4 production from RBL-2H3 cells. *Bioorganic & Medicinal Chemistry, 10*, 3123-3128.

Murata, K., Takano, S., Masuda, M., Iinuma, M., & Matsuda, H. (2013). Anti-degranulating activity in rat basophil leukemia RBL-2H3 cells of flavanone glycosides and their aglycones in citrus fruits. *Journal of Natural Medicines, 67*, 643-646. http://dx.doi.org/10.1007/s11418-012-0699-y

Ogawa, K., Kawasaki, A., Omura, M., Yoshida, T., Ikoma, Y., & Yano, M. (2001). 3',5'-Di-*C*-*β*-glucopyranosylphloretin, a flavonoid characteristic of the genus *Fortunella*. *Phytochemistry, 57*, 737-742.

Ortega, E., Schweitzer-Stenner, R., & Pecht, I. (1988). Possible orientational constraints determine secretory signals induced by aggregation of IgE receptors on mast cells. *The EMBO Journal, 7*(13), 4101-4109.

Sato, S., Akiya, T., Nishizawa, H., & Suzuki, T. (2006). Total synthesis of three naturally occurring 6,8-di-*C*-glycosylflavonoids: phloretin, naringenin, and apigenin bis-*C*-*β*-D-glucosides. *Carbohydrate Research, 341*(8), 964-970. http://dx.doi.org/10.1016/j.carres.2006.02.019

Sato, A., & Tamura, H. (2015). High antiallergic activity of 5,6,4'-trihydroxy-7,8,3'-trimethoxyflavone and 5,6-dihydroxy-7,8,3',4'-tetramethoxyflavone from eau de cologne mint (*Mentha × piperita citrata*). *Fitoterapia, 102*, 74-83. http://dx.doi.org/10.1016/j.fitote.2015.02.003

Tomochika, K., Shimizu-Ibuka, A., Tamura, T., Mura, K., Abe, N., Onose, J., & Arai, S. (2011). Effects of peanut-skin procyanidin A1 on degranulation of RBL-2H3 cells. *Bioscience, Biotechnology, and Biochemistry, 75*(9), 1644-1648. http://dx.doi.org/10.1271/bbb.110085

Yodsaoue, O., Cheenpracha, S., Karalai, C., Ponglimanont, C., & Tewtrakul, S. (2009). Anti-allergic activity of principles from the roots and heartwood of *Caesalpinia sappan* on antigen-induced *β*-hexosaminidase release. *Phytotherapy Research, 23*, 1028-1031. http://dx.doi.org/10.1002/ptr.2670

Yu, M. W., Lou, S. N., Chiu, E. M., & Ho, C. T. (2013). Antioxidant activity and effective compounds of immature calamondin peel. *Food Chemistry, 136*(3-4), 1130-1135. http://dx.doi.org/10.1016/j.foodchem.2012.09.088

8

Morphological Diversity of Farmers' and Improved Potato (*Solanum tuberosum*) Cultivars Growing in Eritrea

Biniam M. Ghebreslassie[1,2], S. M. Githiri[2], Tadesse M.[1] & Remmy W. Kasili[3]

[1] Department of Horticulture, Hamelmalo Agricultural College, Eritrea

[2] Department of Horticulture, Jomo Kenyatta University of Agriculture and Technology, Nairobi, Kenya

[3] Institute of Biotechnology Research, Jomo Kenyatta University of Agriculture and Technology, Nairobi, Kenya

Correspondence: Biniam M. Ghebreslassie, Department of Horticulture, Hamelmalo Agricultural College, Eritrea; Department of Horticulture, Jomo Kenyatta University of Agriculture and Technology, Nairobi, Kenya. E-mail: bm95913@yahoo.com

Abstract

Farmers' and improved potato (*Solanum tuberosum* L.) cultivars growing in Eritrea are main sources of food and income to many growers. The current study was proposed to characterize 17 farmers' and 4 imported cultivars of potato using 33 morphological descriptors. Planting was done in two geographically distinct locations, HAC and Asmara, Eritrea. The experiment was laid out in a randomized complete block design with three replications having 18 plants per plot. Plants grown at HAC emerged early (24.52 days) and reach maturity (94.84 days) while at Asmara it took 43.77 and 123.59 days, respectively. However, yield was higher in Asmara (0.49 kg/plant) compared to HAC (0.37 kg/plant). An accession having many and longer stems was associated with more tuber production, but inversely related to yield. Similarly, accessions with higher stem thickness and tuber size were associated with high yields. The PCA analysis indicated that the first four components explained about 85% of the total variability among the studied materials. The PCA clustered the materials in to four main groups (GI, GII, GIII, GIV) mainly explained by flowering patterns and yield related descriptors. The work has provided useful information on morphological characteristics of the farmer's potato to avoid duplication of resources and identify promising materials for future breeding program.

Keywords: diversity, Eritrea, morphological characterization, potato, principal component

1. Introduction

Potato cultivation has long history in Eritrea, however, little is known about their diversity. Understanding the genetic diversity of a species can assist in the analysis of the taxonomic structure of potential breeding populations for crop improvement (Afuape, Okocha, & Njoku, 2011). The correct identification of a cultivar can be used to certify its pureness as an accession (Rosa, de Campos, de Sousa, Sforça, Torres, & de Souza, 2010); although many cultivars share the same traits.

A number of methods are currently available for characterization including pedigree, morphology, agronomic performance, biochemical, and molecular (DNA-based) data (Mohammadi & Prasanna, 2003). The typical approach to variety identification involves the observation and recording of morphological characters or descriptors (Nováková, Šimáčková, Bárta, & Čurn, 2010). Morphological assessment is based on phenotypic characteristics of a plant species that determine diversity and similarity between and within populations. These descriptors are traits such as leaf type, tuber shape and flower colour which can be analyzed during different developmental stages of the crop. According to Fongod, Mih, & Nkwatoh (2012) morphological descriptors could be suitable for use in distinguishing accessions. Several studies using morphological descriptors have been conducted on potato (Ahmadizadeh & Felenji, 2011; Arslanoglu, Aytac, & Oner, 2011; Felenji, Aharizad, Afsharmanesh, & Ahmadizadeh, 2011) and other crops e.g. sweet potato (Afuape et al., 2011; Fongod, Mih & Nkwatoh, 2012; Tairo, Mneney, & Kullaya, 2008), chilli (Del, Perez, Alvarez, Arrazate, Damian, & Lomeli, 2007), peanut (Upadhyaya, Ortiz, Bramel, & Singh, 2003), or rice (Gana, Shaba, & Tsado, 2013; Sinha & Mishra, 2013).

Few researches have been conducted to study the genetic diversity of potato materials growing in Eritrea. Landrace potato varieties are widely known by names based on their flower colour or place of origin/cultivation such as *Carneshim, Israel, Shashemanie, Asela, America* etc. Similar naming and duplication problem was reported by Tairo et al. (2008) in sweet potatoes growing in Tanzania. This makes it difficult and complicated to collect, identify and classify cultivars while avoiding duplication as many local names are conferred to the same cultivars and *vice versa*. The authors added that, varied cultivar naming system limits the proper identity of the cultivar and follow up in the field. Moreover, lack of accurate varietal identification limits conducting and/or evaluating varietal performance to select and improve suitable farmer-preferred varieties. For that reason, large study was carried out to assess diversity among potato accessions growing in Eritrea using morphological traits; this paper reports on the growth performance, correlation matrix between the variables, principal component analysis (PCA) and bi-plot while in another paper cluster analysis employing singe linkage and mean deviation of each group from the average was reported. The finding provided information on the basic characteristics of potato landraces and varieties growing in Eritrea to identify and avoid duplication of resources. The finding will also be resource information for breeders for future crop improvement strategy to increase productivity of potato in Eritrea.

2. Materials and Methods

2.1 Plant Materials

A total of 21 potato accessions collected from farmers and National Agricultural Research Institute (NARI) were used (Table A-1). The tubers from framers were freshly harvested while from NARI were stored for some time but not sprouted. The tuber seeds were allowed to sprout and later the same number and type of plant materials were planted in two locations under farmers' conditions to determine morphological similarity and diversity among themselves. The first site was located at the experimental field of Hamelmalo Agricultural College (HAC) with an altitude of about 1330 m above sea level with 15'53 N and 38'26 E coordinates. Since the crop is adapted to cool seasons, planting was done towards the end of winter when the temperature is relatively low. The second site was located in the high land (Asmara) with 2363 m above sea level altitude and 15'20N and 38'56E coordinates. The chemical property of the soil in both sites was analyzed in the soil laboratory of the National Agricultural Research Institute (NARI) Halhale (Biniam, Githiri, Tadesse, & Kasili, 2015).

The experiment was laid out in a randomized complete block design (RCBD) with three replications. Planting was made in furrows at inter- and intra-row spacings of 0.70 and 0.40 m, respectively. Three rows and six tubers per row (18 plants per plot) and 63 plots per site was used. At planting time, Di-ammonium Phosphate (DAP) fertilizer was applied at the rate of 200kg/ha. One month after emergence, Urea fertilizer was applied at the rate of 150 kg/ha. A total of 33 descriptors (14 vegetative; 6 flowering and 13 reproductive) were used (Table 1).

Table 1. The vegetative, flowering and reproductive descriptors used in this study

Vegetative	Flowering	Reproductive
Days to emergence (Emer_DAP)	Days to flowering (Flow_DAP)	Tuber diameter/size (T_size)
Emergence % (Emer_%)	No of flowers per plant (No_flow)	Primary tuber Skin color (PTSC)
No of primary stems (No PS)	Primary Flower Color (PFC)	Intensity of primary skin color (IPSC)
Stem color (SC)	Secondary Flower Color (SFC)	Secondary tuber skin color (STSC)
No of interjected leaf lets (No inter leaf)	Distribution of SFC (Diss. SFC)	Distribution of STSC (Diss_STSC)
No of lateral leaf lets (No later leaf)	Flower degree (Flow_dg)	Primary Tuber Flesh Color (PTFC)
Plant Height (PH)		Secondary Tuber Flesh Color (STFC)
Stem Thickness (ST)		Tuber shape (T_shape)
Stem Wing (SW)		Tuber Eye Depth (T_ED)
Growth Habit (GH)		Skin texture (S_texture)
Branching Habit (BH)		Tuber set (T_set)
Primary Sprout Colour (PSC)		Tuber weight (T_wt)
Secondary Sprout Colour (SSC)		Maturity time (Maturity)
Distribution of SSC (Diss. SSC)		

Data for morphological traits were recorded as per the descriptors reported by Huaman, Williams, Salhuana and Vincent (1977). The qualitative morphological characters were performed by numerical coding of each character, whereas for the quantitative traits, data collection was performed using measurement. In case of quantitative data, analysis of ANOVA was carried out to determine any significant variation among the accessions. Standardization of the data by means of the coefficient matrix function was made and then subjected to multivariate analysis of correlation coefficient, PCA and bi-plot analysis using SPSS (ver. 20) and GenStat (12th ed). The phenotypic correlation coefficient (r) values were calculated to measure the relationship between two sets of variables and correlation matrix was generated. Eigen values, percent variance, variance and cumulative percentage of each of the extracted factors were calculated and PCA analysis done on the basis of major factors.

3. Results

3.1 Plant Growth Characteristics

The data collected from the field in both sites indicated that there was a significant effect of location (environment) on the growth and performance of the plants. Data collected on the emergence date, plant branching habit, days to flowering, No of flowers, No of primary stems, stem thickness, growth habit, leaf interject, number of lateral leaves, maturity time and over all yield quality and quantity showed significant difference between the two regions (Table 2). The table showed that materials grown in HAC took fewer days to emerge (24.52) and reach maturity (94.84) as compared to plants grown in Asmara (43.77) and (123.59) days, respectively. Despite that, yield weight was higher at Asmara as compared to HAC (Table 2). Mean average of all the parameters in the two sites was used to calculate the PCA, coefficient matrix and bi-plot to minimize environmental effect. It was noted that plants grown in the high altitude (Asmara) failed and/or were late to produce flowers, the effect of photoperiod.

The separate cluster analysis of the landraces and varieties grown at Asmara and Hemelmalo generated each three groups at 84% similarity level. There was no sharp relationship in grouping of the materials in the two locations (Table 3).

Table 2. Statistical summary of the variables showed significant difference between the sites

Variables	Site		Statistical summary		
	HAC	Asmara	Mean	LSD	St Dv.
Emergence Date	24.52	43.77	33.66	0.96**	5.13
Branching Habit	1.07	1.40	1.23	0.09**	0.50
Growth Habit	1.57	1.84	1.75	0.11**	0.35
Stem Thickness	8.22	8.68	8.40	0.34*	1.24
No. of Primary Stems	3.23	2.30	2.78	0.27**	1.06
Flowering Date	41.07	73.74	57.01	1.95**	29.40
No. of Flowers	2.30	1.86	2.03	0.12**	1.29
Leaf interject	1.68	2.00	1.84	0.04**	0.39
No. of lateral eaves	4.03	3.94	3.99	0.08*	0.52
Maturity time	94.84	123.59	108.30	1.84**	9.49
Tuber set	8.35	9.85	8.99	0.83**	4.89
Tuber Size	38.00	40.20	39.42	1.21**	6.64
Tuber weight	0.37	0.49	0.43	0.04**	0.17

LSD = Least Significant Difference; StDv. = standard deviation; ** highly significant at < 0.01; * significant at <0.05

Table 3. Growth performance characteristic of the studied potato landraces and improved varieties in two geographically different locations (ASM and HAC)

Plant Material	Asmara								HAC							
	PH	ST	No PS	T_wt (Kg/plt)	Tuber set/plt	T_Size (cm)	Maturity (Days)	Cluster Grouping	PH	ST	No PS	T_wt (Kg/plt)	Tuber set/plt	T_Size	Maturity (Days)	Cluster Grouping
Yeha	22.1	7.5	3.5	0.311	14.08	31.0	132.0	GI	28.1	7.8	5.3	0.386	15.17	35.1	92.3	GI
Tsaeda_Embaba_I	23.1	7.6	3.1	0.328	19.42	27.3	131.0	GI	28.7	8.0	5.3	0.304	14.25	32.5	94.3	GI
Keyh_Embaba_I	16.4	6.9	1.7	0.163	6.00	34.3	116.3	GI	16.4	6.1	2.2	0.105	5.50	32.2	72.0	GIII
Tsaeda_Embab_II	24.8	8.2	4.3	0.472	18.67	30.3	132.0	GI	27.6	7.4	5.6	0.384	13.33	34.5	92.7	GI
Cameshim	22.2	10.8	2.0	1.050	6.92	56.8	126.7	GI	25.8	10.6	3.5	0.671	6.42	52.6	92.7	GII
Shashemanie_I	22.2	7.0	3.3	0.506	18.58	32.7	133.3	GI	33.0	7.7	5.3	0.406	13.08	40.0	93.0	GI
Zafira_I	20.6	7.4	1.8	0.364	6.67	39.4	122.3	GIII	20.9	8.0	3.1	0.408	6.67	43.3	90.0	GII
Round_Sudan	21.5	9.5	1.3	0.363	6.50	40.8	108.7	GI	12.8	9.2	1.2	0.363	5.33	41.8	95.0	GII
Oval_Sudan	22.5	11.0	2.1	0.413	6.83	38.4	128.3	GIII	22.1	9.7	2.7	0.254	4.75	32.7	84.7	GII
Keyh_Embaba_II	33.1	8.2	3.0	0.642	17.00	36.2	106.7	GII	35.9	7.6	5.5	0.560	16.00	34.1	89.0	GI
Tsaeda_Embaba_III	23.3	11.4	1.6	0.816	7.33	52.3	124.7	GI	22.5	9.8	2.0	0.633	7.50	41.4	95.67	GII
Banba	19.5	8.1	2.5	0.788	7.08	50.6	126.0	GI	19.8	8.2	2.2	0.292	4.50	41.4	111.0	GII
Baren	17.6	8.8	2.6	0.386	7.00	42.5	126.0	GI	16.3	9.3	2.4	0.393	4.83	40.4	117.7	GII
Orla	19.3	7.5	2.6	0.236	5.50	38.7	113.7	GIII	18.2	8.9	1.8	0.205	4.75	36.5	92.3	GII
Slaney	18.8	8.9	2.2	0.538	6.17	46.9	121.3	GI	16.3	9.2	3.8	0.351	5.00	38.8	114.7	GII
Shashemanie_II	21.3	10.2	2.0	0.613	18.33	35.4	128.7	GI	36.2	7.9	4.3	0.560	18.33	38.1	93.0	GI
Keyh Embaba_III	18.8	8.0	1.5	0.272	10.67	35.0	123.3	GI	11.1	4.6	2.2	0.148	7.17	20.0	62.3	GII
Ajeba	18.3	7.0	2.7	0.357	7.23	38.7	99.3	GII	18.9	6.7	3.1	0.348	5.33	41.9	87.0	GII
Zafira_II	22.3	10.1	1.4	0.622	4.25	53.6	128.0	GIII	18.2	9.9	1.8	0.362	5.67	39.6	115.0	GII
Safira	20.9	7.4	2.3	0.286	4.75	42.0	117.3	GI	16.8	6.7	2.7	0.333	5.17	41.4	87.0	GII
Grandnain	25.1	9.8	1.5	0.662	6.70	50.8	129.3	GI	15.6	8.5	1.8	0.341	5.00	41.0	116.3	GII
Grand mean	21.6	8.6	2.3	0.492	9.79	40.7	122.6		21.9	8.2	3.2	0.372	8.27	38.1	94.7	
LSD (5%)	5.1	1.5	0.8	0.210	4.21	5.9	6.20		4.5	1.5	1.5	0.077	1.24	4.6	8.1	

3.2 Correlation Coefficient Matrix

Pearson's correlation is a measure of strength of linear relationship between two variables (Sinha & Mishra, 2013). Correlation coefficient analysis of the studied traits helped to decide how many traits to select and remove the ineffective ones. Out of the total 33 descriptors used, 16 exhibited significant contribution to explain variation among the accession and were used for further classification analysis. It was reported by Felenji et al. (2011) that the factors which justify more percentage of variations are importance for further study. The pair wise correlation among the identified 16 traits was thus generated (Table 4). The table shows positive and negative correlation between the variables along with their magnitude. Generally, all variables associated with flowering exhibited a strong relationship to each other.

On the other hand, Emergence day showed strong negative relation with growth habit and No. of primary stems where as strong positive with the stem thickness. Similarly, No of primary stem showed strong positive relation with plant height and tuber set. In other word, tuber set has strong positive correlation with No. of primary stem, plant height, flower degree but negative correlation with tuber weight and size. On the other hand, tuber weight was strongly correlated with the tuber size and stem thickness, while negatively correlated with tuber set (numbers).

Table 4. Correlation coefficient matrix among the 16 descriptors of 21 accessions studied

Variables	Emer_DAP	SW	GH	No_PS	PH	ST	Flow_DAP	SFC	Dis_SFC	Flow_dg	No_ flow	Maturity	T_ shape	T_size	T_set
SW	-0.22	1.00													
GH	**-0.65**	0.02	1.00												
No_PS	**-0.72**	0.26	0.38	1.00											
PH	-0.38	0.06	-0.05	**0.74**	1.00										
ST	**0.62**	0.20	-0.54	-0.29	0.09	1.00									
Flow_DAP	0.28	-0.01	-0.15	0.12	0.12	0.04	1.00								
SFC	0.00	-0.36	-0.09	0.20	0.37	-0.09	**0.70**	1.00							
Dis_SFC	0.11	-0.15	-0.07	0.26	0.31	-0.05	**0.91**	**0.89**	1.00						
Flow_dg	-0.02	-0.06	0.03	0.47	0.39	-0.10	**0.90**	**0.75**	**0.92**	1.00					
No_flow	0.16	-0.06	-0.04	0.35	0.29	-0.02	**0.90**	**0.64**	**0.88**	**0.95**	1.00				
Maturity	0.29	0.27	-0.09	0.10	0.08	0.50	0.31	-0.01	0.18	0.22	0.25	1.00			
T_shape	0.53	0.13	-0.42	-0.58	-0.29	**0.62**	-0.33	-0.45	-0.45	-0.54	-0.40	0.11	1.00		
T_size	0.41	0.12	-0.39	-0.43	-0.17	**0.63**	-0.15	-0.17	-0.26	-0.37	-0.33	0.34	**0.79**	1.00	
T_set	-0.49	0.02	0.19	**0.82**	**0.82**	-0.23	0.39	0.49	0.53	**0.67**	0.56	0.04	**-0.69**	-0.56	1.00
T_wt	0.09	0.13	-0.37	0.12	0.50	**0.63**	0.14	0.24	0.15	0.13	0.05	0.33	0.37	**0.67**	0.16

Where: Emer_DAP= emergence date after planting; SW=stem wing; GH=growth habit; No PS= number primary stem; PH=plant height; ST= stem thickness; Flow_DAP= flowering day after planting; SFC= secondary flower colour; Dis_SFC= distribution of SFC; Flow_dg= flowering degree; No_flow=number flowers per plant; T_ =tuber.

3.3 Principal Component Analysis (PCA)

Based on the initial Eigen value ≥ 1 scored in the current study, four components were selected. The PCA gave comparable results in the two localities, however, the yield related parameters and PH were explained by PCA1

Table 5. PCA of the 16 morphological traits for 21 potato materials evaluated

Variables	Principal Components			
	1	**2**	**3**	**4**
Emergence date after planting (Emer DAP)	0.264	0.583	-0.684	-0.055
Stem Wing (SW)	-0.156	0.011	0.147	0.814
Growth Habit (GH)	-0.134	-0.706	0.183	0.187
No Primary Stems (No PS)	0.183	-0.337	0.843	0.257
Plant Height (PH)	0.205	0.177	0.914	-0.051
Stem thickness (ST)	0.012	0.853	-0.083	0.289
Flowering days after planting (Flow DAP)	0.963	0.050	-0.060	0.114
Secondary Flower Color (SFC)	0.785	0.071	0.257	-0.392
Distribution of _SFC (Diss. SFC)	0.958	-0.005	0.144	-0.085
Flower degree (Flow_dg)	0.936	-0.115	0.269	0.065
No of flowers per plant (No_flow)	0.938	-0.075	0.103	0.106
Maturity	0.287	0.348	-0.030	0.692
Tuber shape (T_shape)	-0.427	0.701	-0.373	0.111
Tuber size (T_size)	-0.270	0.813	-0.180	0.161
Tuber set (T_set)	0.477	-0.246	0.794	-0.008
Tuber weight/yield (T_wt)	0.075	0.813	0.462	0.090
Eigen-value/latent roots for each PC	6.0	3.85	2.37	1.32
Variation in Percentage (%) for each PC	37.50	24.08	14.08	8.25

in HAC while by PCA2 in Asmara. Moreover, high percentage of variability was explained by the first four components in Asmara (84.08%) as compared to HAC (80.58%) when data were analyzed separately. When data were pooled off together, the first four components explain about 85% of the total variability. The identified traits contained in these components exhibited significantly on the morphological expression of the accessions. PCA1 with eigenvalue of 6.0 contributed 37.50% of variation among the observed 21 potato accessions. The traits accounted to this component are Flowering day, secondary flower colour (SFC), Distribution of SFC, Flower degree and No of flowers per plant (Table 5); PC2 with 3.85 eigenvalue contributed 24.08% of the total variation relating to tuber shape, tuber size, tuber weight, growth habit and stem thickness; PC3 was corresponding to Emergence day, plant height, No. of primary stem and tuber set (Table 5).

3.4 Morphological Diversity

A principal component bi-plot of the 21 accessions and the 16 traits were carried out to show the underlying relationships among the accessions, accessions with traits and between the traits. The 21 accessions on axes representing the two rotated components PCA1 and PCA2 indicated that the plants are distributed in the first three quadrants, while the traits are scattered in I, II and IV quadrants widely (Figure 1).The scatter plot matrix score clustered the materials into four main groups (GI, GII, GIII and GIV) (Figure 1). Group I consisting of eight farmers' materials predominantly with flowering (white and/or red) coloured. They are characterized by intermediate emergence and intermediate maturing with a more than two and long primary stems (Biniam et al., 2015). Whereas GII consists of nine materials mainly made up of varieties of late emerging and late maturing with few and short primary stems, but relatively better yield. This group is positively associated with T_shape, T_size, T_wt, Maturity and Emer_DAP, while negatively related to GH, No_PS and T_set. GIII consists of three accessions (Orla, Zafira_1 and Oval_Sudan) characterize by early emerging and early maturing, semi rosette growth habit and intermediate yield quantity. Most of the materials in GII and GIII are characterized by non-flowering pattern. On the other hand, GIV is a single material (Ajiba) characterized by very early maturity and lower yield quantity. This classification is not in exact, but in accordance with the cluster analysis generated employing single linkage using the same data (Biniam et al., 2015).

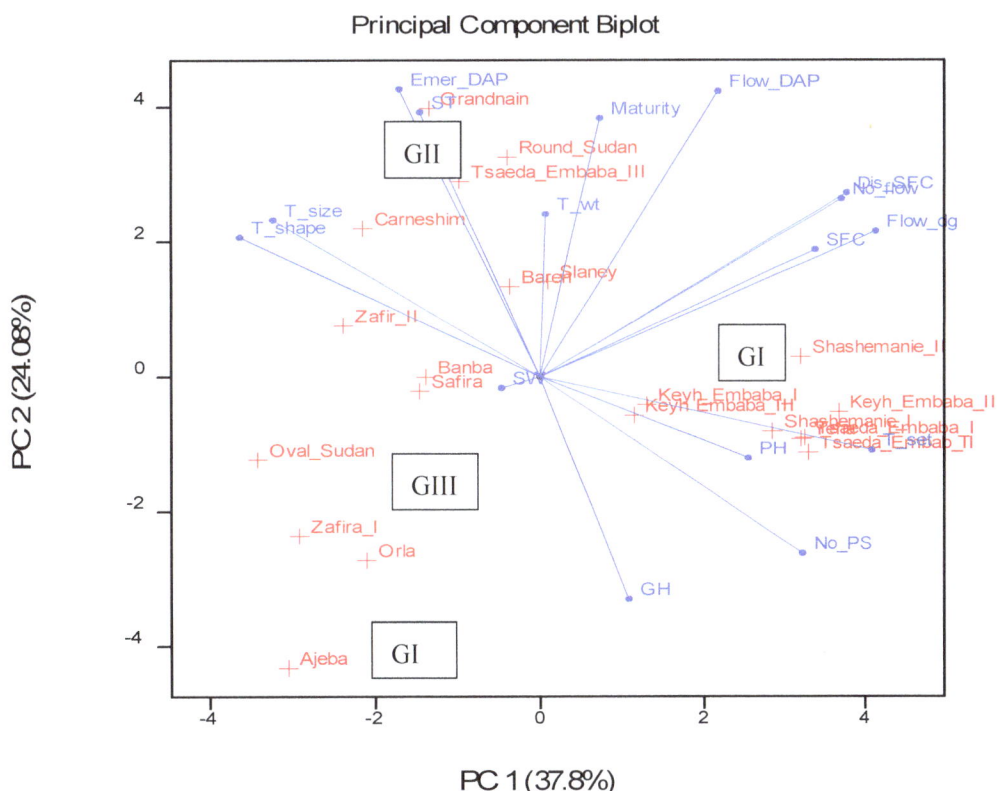

Figure 1. Scatter plot matrix of PC score of accessions and variables using of PCA1 and PCA2

4. Discussion

The growth characteristics study revealed that growth performance of plants is strongly affected by the environment and soil type. According to Crawford (2000) morphological descriptors unlike the molecular markers are not free from environmental and pleotropic effects. This is attributed to the climatic condition of the two sites, where HAC is subtropical with higher mean temperature (24°C) than Asmara (16°C) that accelerates the growth rate at the cost of yield quality and quantity. It was reported by Waniale, Wanyera, and Talwana (2014) that there were significant differences in all quantitative traits measured among the Mungbean varieties planted in two different locations in Uganda for characterization.

The preliminary PCA of the studied traits helped to decide how many traits to select and remove the ineffective ones. It was reported by Felenji et al. (2011) that the factors which justify more percentage of variations are importance for further study. The identified traits showed positive and negative correlation to each other with few remaining independent. It can then be inferred from the result that an accession having more and longer stems is likely to produce more tubers, but usually inversely relate to yield or tuber weight. This agrees with previous finding by Felenji et al. (2011) and Ahmadizadeh & Felenji (2011) where they observed positive correlation between number of stems, plant height and tuber numbers. Similarly, Rashidi (2009) reported that number of tubers has a high correlation with number of stems produced. Moreover, it was noted from the current study that accessions with higher stem thickness and tuber size are associated with high yields. This is in accordance with the report made by (Felenji et al., 2011). The negative correlation between tuber weight and tuber number is ascribed to the fact that the more number of the tubers per plant produced, the smaller the size of each, which can be related to the strong competition for resources. Previously, Felenji et al. (2011) drew the same conclusion out of their findings. According to Okii, Tukamuhabwa, Odong, Namayanja, Mukabaranga, Paparu, and Gepts (2014), it was inferred that the strong correlations between two traits allow to make simultaneous selections and use of the related traits interchangeably in crop breeding and selection.

PCA is a technique which identifies plant traits contributing to the most observed variation among accessions (Afuape et al., 2011; Ahmadizadeh & Felenji, 2011; Felenji et al., 2011; Okii et al., 2014) which in turn assists selection of parent lines for breeding purposes. Moreover, Okii et al. (2014) reported that PCA-biplot demonstrates diagrammatically the genetic diversity pattern of germplasm based on their morphological characteristics. Yet, PCA scatter matrix can be used to identify the contribution of each morphological characteristic from the accessions to form group or cluster. This approach was used for grouping in various crops such as in wheat varieties (Malik, Sharma, Sharma, Kundu, Verma, Sheoran, Kumar, & Chatrath, 2014); in sweet potato (Afuape et al., 2011); and in common bean (Okii et al., 2014). The current PCA-biplot clustered the materials in to four groups. Accessions falling in the same quadrant are assumed to be the same or closely related (Malik et al., 2014). If the accession located in the plot at an angle of 90° then the two accessions have no close relationship. The fact that the accessions are clustered into four groups indicates that there is exploitable variability between the accessions for core planting material selection. Especially in GII, although some accessions like *Banba* and *Safira* showed close relationship, majority have very weak similarity as is shown from the plot. The PC bi-plot analysis, further, indicated that descriptors like ST, Emer_DAP, T_shape, T_set and most of the flowering pattern are placed at long vectors in the plot and explained most variation among the accessions compared to PH, T-wt and maturity. Descriptors that are in the same quadrant with <90° angle means they have very close positive relationship and *vice versa*. All the flowering patterns showed similar variation (<90°) with PH and T_set as well with maturity and T_wt. Similarly, accessions falling alongside of the trait are more explained by the respective character or trait. The plot, thus, will help to identify desired traits contained by each accession for further breeding activity.

5. Conclusion

The current study on characterization of potatoes using morphological traits is the first in its kind in Eritrea. The work has provided initial result how morphological descriptors employing multivariate analysis can be used to characterize and classify accessions. The PCA indicated that a total of 84.63% variation was explained by the first four components among the 21 accessions. The highest variation observed in PCA1 was mainly contributed by the flowering patterns while the PCA2 was attributed to the yield components. The morphological characters such as flowering patterns and yield components are thus found to be useful phenotypic markers for genetic diversity study. Four groups of accessions were identified using PC analysis. It was noted there was no sharp relationship between the accessions and geographic origin of the seeds. This is ascribed to the free movement of resources from one Zoba to other by farmers. The identified morphological characteristics contained by each accession will help to select accession based on their character for breeding program.

Acknowledgements

The authors wish to acknowledge the Japanese International Cooperation Agency (JICA) for funding the research work as well as Hamelmalo Agricultural College (HAC) and Zoba Maekel Agricultural Research Unit for providing graciously land for the research. We are also grateful to Biosciences eastern and central Africa – International Livestock Research Institute (BecA-ILRI) Hub for financial support through the Africa Biosciences Challenge Fund Program (ABCF). The ABCF is funded by the Australian Department for Foreign Affairs and Trade (DFAT) through the BecA-CSIRO partnership; the Syngenta Foundation for Sustainable Agriculture (SFSA); the Bill & Melinda Gates Foundation (BMGF); the UK Department for International Development (DFID) and; and the Swedish International Development Cooperation Agency (Sida). Last but not least, we sincerely thank Mr. Peter Willadsen, CSIRO for reviewing the paper and providing constructive comments.

References

Afuape, S. O., Okocha, P., & Njoku, D. (2011). Multivariate Assessment of the Agromorphological Variability and Yield Components among Sweet Potato (*Ipomoea batatas* (L.) Lam) Landraces. *African Journal of Plant Science, 5*(2), 123-132.

Ahmadizadeh, M., & Felenji, H. (2011). Evaluating Diversity among Potato Cultivars Using Agro-Morphological and Yield Components in Fall Cultivation of Jiroft Area. *American-Eurasian Journal of Agricultural & Environmental Sciences, 11*(5), 655-662.

Arslanoglu, F., Aytac, S., & Oner, E. K. (2011). Morphological Characterization of the Local Potato (*Solanum tuberosum* L.) Accessions Collected from the Eastern Black Sea Region of Turkey. *African Journal of Biotechnology, 10*(6), 922-932.

Biniam, M. G., Githiri, S. M., Tadesse, M., & Kasili, R. W. (2015). Analysis of Diversity among Potato Accessions Grown in Eritrea Using Single Linkage Clustering. *American Journal of Plant Sciences, 6*, 2122-2127. http://dx.doi.org/10.4236/ajps.2015.613213

Crawford, D. J. (2000). *Plant Molecular Systematics: Macromolecular Approaches*. Wiley, New York.

Del, E., Perez, C. M., Alvarez, O. C., Arrazate, C. H. A., Damian, M. A. T. M., & Lomeli, A. P. (2007). Morphological Variation in *Guajillo* Chili Pepper Plants (*Capsicum annuum* L.). *African Crop Science Conference Proceedings, 8*(99), 327-332.

Felenji, H., Aharizad, S., Afsharmanesh, G. R., & Ahmadizadeh, M. (2011). Evaluating Correlation and Factor Analysis of Morphological Traits in Potato Cultivars Fall Cultivation of Jiroft Area. *American-Eurasian Journal of Agricultural & Environmental Sciences, 11*(5), 679-684.

Fongod, A. G. N., Mih, A. M., & Nkwatoh, T. N. (2012). Morphological and Agronomical Characterization of Different Accessions of Sweet Potatoes (*Ipomoea batatas*) in Cameroon. *International Research Journal of Agricultural Science and Soil Science, 2*(6), 234-245.

Gana, A. S., Shaba, S. Z., & Tsado, E. K. (2013). Principal Component Analysis of Morphological Traits in Thirty-Nine Accessions of Rice (*Oryza sativa* L.) Grown in a Rainfed Lowland Ecology of Nigeria. *Journal of Plant Breeding and Crop Sciences, 5*, 120-126.

Huaman, Z., Williams, J. T., Salhuana, W., & Vincent, L. (1977). Descriptors for the Cultivated Potato and for the Maintenance and Distribution of Germplasm Collections. AGPE: IBPGR/77/32. International Board For Plant Genetic Resources Rome, Italy.

Malik, R., Sharma, H., Sharma, I., Kundu, S., Verma, A., Sheoran, S., Kumar, R., & Chatrath, R. (2014). Genetic Diversity of Agro-Morphological Characters in Indian Wheat Varieties Using GT biplot. *Australian Journal of Crop Science, 8*(9), 1266-1271.

Mohammadi, S. A., & Prasanna, B. M. (2003). Review and Interpretation: Analysis of Genetic Diversity in Crop Plants—Salient Statistical Tools and Considerations. *Crop Sciences, 43*, 1235-1248. http://dx.doi.org/10.2135/cropsci2003.1235

Nováková, A., Šimáčková, K., Bárta, J., & Čurn, V. (2010). Utilization of DNA Markers Based on Microsatellite Polymorphism for Identification of Potato Varieties Cultivated in the Czech Republic. *Journal of Central European Agriculture, 11*(4), 415-422. http://dx.doi.org/10.5513/JCEA01/11.4.854

Okii, D., Tukamuhabwa, P., Odong, T., Namayanja, A., Mukabaranga, J., Paparu, P., & Gepts, P. (2014). Morphological Diversity Of Tropical Common Bean Germplasm. *African Crop Science Journal, 22*(1), 59-67.

Rashidi, V. (2009). Effect of Plant Spacing and Seed Tuber Size on Yield and Yield Components of Potato Cultivars. *Journal of Modern Science in Agriculture, 5*(16), 19-26.

Rosa, P. M., de Campos, T., de Sousa, A. C. B., Sforça, D. A., Torres, G. A. M., & de Souza, A. P. (2010). *Potato Cultivar Identification Using Molecular Markers*. http://dx.doi.org/10.1590/S0100-204X2010000100015

Sinha, A. K., & Mishra, P. K. (2013). Morphology Based Multivariate Analysis of Phenotypic Diversity of Landraces of Rice (O*ryza sativa* L.) of Bankura District of West Bengal. *Journal of Crop and Weed, 9*(2), 115-121.

Tairo, F., Mneney, E., & Kullaya, A. (2008). Morphological and Agronomical Characterization of Sweet potato [*Ipomoea batatas* (L.) Lam.] Germplasm Collection from Tanzania. *African Journal of Plant Science, 2*(8), 077-085.

Upadhyaya, H. D., Ortiz, R., Bramel, P. J., & Singh, S. (2003). Development of a Groundnut Core Collection Using Taxonomical, Geographical and Morphological Descriptors. *Genetic Resources and Crop Evolution, 50*, 139-148. http://dx.doi.org/10.1023/A:1022945715628

Waniale, A., Wanyera, N., & Talwana, H. (2014). Morphological and Agronomic Traits Variations for Mungbean Variety Selection and Improvement in Uganda. *African Crop Science Journal, 22*(2), 123-136.

Appendix A

Table A-1. List and source of potato genotypes used in this study

Accession No.	Common Name	Village	Subzoba	Zoba	Source
1	Yeha	Geremi	Serejeka	Maekel	Farmers
2	Tsaeda_Embaba_I	Afdeyu	Serejeka	Maekel	Farmers
3	Keyh_Embaba_I	Afdeyu	Serejeka	Maekel	Farmers
4	Tsaeda_Embab_II	Adiregit	Serejeka	Maekel	Farmers
5	Carneshim	Adiregit	Serejeka	Maekel	Farmers
6	Shashemanie_I	Mendefera	Mendefera	Debub	Farmers
7	Zafira_I	SelaeDaero	Galanefhi	Maekel	Farmers
8	Round_Sudan	SelaeDaero	Galanefhi	Maekel	Farmers
9	Oval_Sudan	SelaeDaero	Galanefhi	Maekel	Farmers
10	Keyh_Embaba_II	Debarwa	Debarwa	Debub	Famers
11	Tsaeda_Embaba_III	Debarwa	Debarwa	Debub	Farmers
12	Banba	Halhale	Debarwa	Debub	NARI
13	Baren	Halhale	Debarwa	Debub	NARI
14	Orla	Halhale	Debarwa	Debub	NARI
15	Slaney	Halhale	Debarwa	Debub	NARI
16	Shashemanie_II	Adi-Blay	Emnihaili	Debub	Farmers
17	Keyh Embaba_III	Adi_mongonti	Mendefera	Debub	Farmers
18	Ajeba	Adi mongonti	Mendefera	Debub	Farmers
19	Zafir_II	Mendefera	Mendefera	Debub	Farmers
20	Safira	AdiInadi	Mendefera	Debub	Farmers
21	Grandnain	Adi-Mongonti	Mendefera	Debub	Farmers

9

Genetic Control of Beta-carotene, Iron and Zinc Content in Sweetpotato

Ernest Baafi[1], Kwadwo Ofori[2], Edward E. Carey[3], Vernon E. Gracen[2], Essie T. Blay[2] & Joe Manu-Aduening[1]

[1]CSIR-Crops Research Institute, P. O. Box 3785, Kumasi, Ghana

[2]West Africa Centre for Crop Improvement, University of Ghana, Legon

[3]International Potato Centre (CIP), Ghana

Correspondence: Ernest Baafi, CSIR-Crops Research Institute, P. O. Box 3785, Kumasi, Ghana. E-mail: e.baafi@gmail.com

Abstract

Micronutrients deficiency is a major contributor to poor health in developing countries. It can be alleviated by biofortification or enrichment of staple crops with micronutrients. Sweetpotato is a major staple crop in numerous tropical countries and is naturally biofortified. In spite of extensive promotion of orange-fleshed sweetpotato varieties (OFSPs), they are poorly utilized as staple food in most parts of West Africa because of their low dry matter and high sugar content. Beta-carotene is positively correlated with iron and zinc content in sweetpotato. Development of sweetpotato cultivars with end-user preferred traits and higher content of beta-carotene, iron and zinc will alleviate their deficiencies. Knowledge on the genetic control of these traits is critical for their improvement in sweetpotato. This study used diallel mating design to estimate general combining ability (GCA) and specific combining ability (SCA) of storage root beta-carotene, iron and zinc content to determine the genetic control of these traits for sweetpotato breeding. A general model for estimating genetic effect, Gardner and Eberhart analysis II (GEAN II), was used for data analysis. Genetic variability for the traits indicated that they were mostly controlled by additive gene effect. Significant heterosis was found indicating that levels of these micronutrients can be improved in sweetpotato through breeding.

Keywords: beta-carotene content, genetic control, iron content, sweetpotato, zinc content

1. Introduction

Micronutrients are vitamins and minerals required in small amounts that are essential to human health, development, and growth. They include folic acid, iodine, iron, vitamin A and zinc. They play a central role in metabolism and maintenance of tissue function (Shenkin, 2006). Micronutrient deficiencies are a chronic deprivation of these nutrients and constitute a huge public health problem, adversely affecting a third of the world's population (Darnton-Hill et al., 2005; Simon et al., 2013). Nearly 100 million preschool children in the world suffer from vitamin A deficiency (WHO, 2009). Thousands of death of children under five years worldwide are attributed to vitamin A deficiency (Feyrer, Politi, & Wei, 2013), while anemia, which is caused by inadequate intake of iron, affects 1.6 billion people around the globe (De Benoist, Erin, Ines, & Mary, 2008; WHO, 2013). Severe and extensive deficiencies also prevail for zinc (Black et al., 2008). Micronutrients deficiency can be alleviated in several ways including dietary diversification, fortification, supplementation and biofortification (Kumah, 2013). Among these different ways, biofortification which is the genetic enhancement of edible parts of plants with micronutrient concentration is the safest and the most cost effective. Biofortification can be achieved by breeding for farmer and consumer preferred cultivars that have adequate amount of micronutrients.

Sweetpotato (*Ipomoea batatas* (L) Lam) is an important root crop, particularly in the tropical countries such as Uganda, Rwanda and Burundi in Eastern Africa and in Papua New Guinea where annual per capita fresh roots consumption is over 150 kg (Lebot, 2009; Lebot, 2010; Warammboi, Dennien, Gidley, & Sopade, 2011). Sweetpotato is rich in minerals and vitamins (Ray, & Tomlins, 2010), making it a naturally biofortified crop. Improving beta-carotene, iron and zinc content in sweetpotato cultivars with traits preferred by farmers and consumers will go a long way to alleviate these micronutrients deficiency. Sweetpotato breeding objectives, until recently were exclusively based on heritability of traits (Jones, 1986; Jones, & Dukes, 1980; Jones, Schalk, & Dukes, 1979). But, High estimates of heritability indicate that superior parents tend to produce the best progenies (Rex, 2002), and only additive component of inheritance can be estimated accurately. Therefore, the expected amount of genetic gain realized in subsequent generations of breeding will be obtained only if all the genetic effects are additive (Miller, Williams, Robinson, & Comstock, 1958). In a heterozygous crop like the sweetpotato, non-additive effects are also important and heritability is a poor estimate of non-additive effect. Studies of inheritance in sweetpotato are complicated by its highly heterozygous hexaploid nature and complex segregation ratios. However, because of its ability to be vegetatively propagated, its genetic variation can be partitioned into general combining ability (GCA) and specific combining ability (SCA) using crosses of heterozygous varieties. For this reason, inference can be made on the gene action controlling a given trait based on the relative estimates of GCA and SCA.

North Carolina II (NCII) mating design has been used for trait inheritance studies in sweetpotato (Gasura, Mashingaidze, & Mukasa, 2008; Oduro, 2013; Sseruwu, 2012; Todd, 2013). The gene action controlling production and utilization constraints in sweetpotato has also been studied using diallel mating design (Chiona, 2009; Elisa, Humberto, & Luis, 2000; Mihovilovich, Mendoza, & Salazar, 2000; Mwanga, Yencho, & Moyer, 2002; Shumbusha et al., 2014). The diallel mating design provides more genetic information on a complex crop like sweetpotato. Diallel design in addition to estimating GCA and SCA variance components can also be used to determine cumulative gene effect of breeding populations (Griffing, 1956; Hayman, 1954a, 1954b; Hayman, 1957; Hayman, 1958). Furthermore, it provides information on heterosis effect, which provides a basis for the development of heterotic groups (Gardner, 1982). The objective of this study was to determine the inheritance of storage root beta-carotene, iron and zinc content in sweetpotato for breeding farmer and consumer preferred cultivars that have high amount of these micronutrients to curb their deficiencies.

2. Materials and Methods

2.1 Experimental Sites

Hybridization block was established at the Crops Research Institute of the Council for Scientific and Industrial Research (CSIR-CRI) at Fumesua in the Ashanti region of Ghana in the minor cropping season in 2012. The F_1 progenies produced were evaluated at three locations spanning over three ecozones of Ghana in the minor cropping season in 2013. These were the CSIR-CRI research station at Fumesua in the Ashanti region (forest ecozone), and the National Agricultural Research Stations at Wenchi in the Brong Ahafo region (transition ecozone) and Pokuase in the Greater Accra region (coastal savanna ecozone).

2.2 Genetic Materials Used

Crosses among four parental genotypes, two with low beta-carotene, relatively low iron and low zinc content (Histarch and Ogyefo), and the other two with high beta-carotene, relatively high iron and high zinc content (Apomuden and Beauregard) were made. The parents were three released varieties (Apomuden, Histarch and Ogyefo), and one breeding line (Beauregard) (Table 1). Histarch and Ogyefo are white-fleshed varieties while Apomuden and Beauregard are orange-fleshed. Progeny families and their respective number of seeds are presented in Table 2. The seeds were germinated on moist filter paper in a Petri dish after sand paper scarification, followed by transplanting unto prepared nursery pots in a screen house for the establishment of seedling nursery. Vine cuttings from each genotype were hardened and multiplied in the field after eight weeks in seedling nursery for the establishment of the trials.

2.3 Experimental Layout

The four parental genotypes were crossed using full diallel mating design. Sweetpotato is a highly heterozygous crop which makes each cross between two different parent plants genetically distinct cross and that the variation in the F_1 families produced is equivalent to an F_2 generation in homogenous crop. Twelve families comprising 234 F_1 progenies (123 crosses and 111 reciprocals) (Table 2) were raised in the seedling nursery but due to poor vigour of some genotypes, 196 F_1 progenies out of the 234 were evaluated alongside their parents at the three locations above using an alpha lattice design with two replications. All entries were planted in a single row on ridges at five plants per genotype at spacing of 0.3 m within row and 1m between rows. Four node vines from the middle portion to the tip were used for planting. Genotypes within family were randomised to adjacent plots.

Table 1. Genotypes used in the study and their storage root attributes

Parent	Origin/Status	β-carotene content (mg/100g)DW	Iron content (mg/100g)DW	Zinc content (mg/100g)DW	Storage root Yield (t/ha)	Storage root flesh colour
Apomuden	Release (2005)	33.67	2.56	1.53	19.7	Deep/dark orange
Beauregard	Introduction	24.31	2.12	1.13	37.5	Intermediate orange
Histarch	Release (2005)	9.85	1.49	0.86	24.4	white
Ogyefo	Release (2005)	6.83	1.69	1.01	15.4	White

Table 2. Progeny families and number of seeds per family used

Parents	Histarch	Ogyefo	Apomuden	Beauregard
Histarch		30	30	22
Ogyefo	20		22	6
Apomuden	13	30		13
Beauregard	13	4	31	

[S]F$_1$ families obtained from crosses above diagonal; F$_1$ progenies obtained from reciprocals below diagonal.

2.4 Data Collection

Harvesting was done at three and half months after planting on whole plot and one large, one medium and one small storage root were randomly selected for determination of beta-carotene, iron and zinc content. Storage roots selected were 3 cm or more in diameter and without cracks, insect damage or rotten parts (Ekanayake, Malagamba, & Midmore, 1990). The storage roots were washed, peeled and cut into four equal parts longitudinally. Two opposite quarters of the peeled storage roots were selected, sliced into pieces and 50g fresh sample weighed into a polythene envelope. The fresh samples were frozen using deep-freezer and freeze-dried for 72 hours using freeze-dryer. The freeze-dried samples were milled and the milled samples used for the determination of beta-carotene, iron and zinc content using the near-infrared reflectance spectroscopy (NIRS) (Tumwegamire et al., 2011). About 3g of milled sample was placed in a cuvette and placed in a NIRS monochromator model XDS (NIRSystems, Inc., Silver Springs, MD, USA) for scanning and their spectra collected between 400 and 2498nm. Each sample was scanned twice. The average spectrum of each sample was stored and the NIRS results (beta-carotene, iron and zinc content) in an Excel format obtained from computer connected to the NIRS.

2.5 Data Analysis

F$_1$ progenies with missing data were eliminated from the analysis. Data for 156 F$_1$ progenies (80 crosses and 76 reciprocals) out of the 196 and their four parents were used for the analyses. Analysis of Variance (ANOVA) was first performed on data of all parents and their F$_1$ derived individuals using the approach of Buerstmayr, Nicola, Uwe, Heinrich and Elisabeth (2007) to determine the mean performance of parents and F$_1$ progenies. The average cross performance was used to elucidate GCA and SCA using the approach of Gardner and Eberhart (1966) analysis II (GEAN II). Average cross or family performance was obtained as average value from the respective F$_1$ progenies mean performance. GEAN II is a method of analysing diallel crosses data from heterogeneous parents/populations ("varieties"). This method assumes parents and crosses performance to be fixed effect and environments random effect (Harold, Hugo, Kevin, & Magnie, 2001). The method fit parents and parent cross means, X_{ij} to the linear model $X_{ij} = \mu_v + \frac{1}{2}(V_i + V_j) + \sigma h_{ij}$; where, μ_v = mean effect of parents; V_i and V_j = estimates of variety effect for the ith and the jth parents, respectively; and h = estimate of heterosis effect when parent i is crossed to parent j ($\sigma = 0$ when i = j, and 1 when i ≠ j). Heterosis effect is further partitioned as $H_{ij} = h + h_i + h_j + s_{ij}$; where, h = estimate of average heterosis; h_i and h_j = estimates of variety heterosis (expressed as deviation from h) and indicates general combining ability (GCA); and S_{ij} = estimate of specific heterosis from crossing parents i and j which indicates specific combining ability (SCA). The GEAN II analysis was carried out using SAS 9.2 computer software (SAS, 2002), based on the macros in DIALLEL-SAS05 (Zhang, Kang, & Kendall, 2005). Contrary to the Griffing's Model, GEAN II works with condition if I>J, delete, so data for crosses and reciprocals were not analysed simultaneously (full diallel with parents) but separately (Half diallel with parents). Mid-parent and better parent heterosis were calculated following Fonseca and Patterson (1968) as shown in equation 1 and 2.

$$Ht \, (\%) = ((F_1 - MP) / MP) \times 100) \tag{1}$$

$$Hbt \, (\%) = ((F_1 - BP) / BP) \times 100) \tag{2}$$

Where, Ht = mid-parent heterosis; Hbt = better parent heterosis; MP = mid-parent value; BP = better parent value; F$_1$ = F$_1$ progeny value.

3. Results

3.1 Performance of Parents and F_1 Progenies across Three Environments

Genotype by environment interaction (G x E) was significant (P<0.05) for beta-carotene, iron and zinc content across the reciprocals and only zinc content for the crosses (Table 3). Highly significant differences (P<0.01) were found between the genotypes for beta-carotene, iron and zinc content across the crosses and the reciprocals. There was significant (P<0.05) effect of environment on all the traits except for beta-carotene content for the crosses. While overall heterosis was significant (P<0.05) for beta-carotene and iron content for the crosses and all the traits for the reciprocals, average heterosis was significant (P<0.05) for beta-carotene, iron and zinc content across the crosses and reciprocals. Variety heterosis (which indicates GCA) and SCA were significant (P<0.01) only for beta-carotene content for the crosses but were significant for beta-carotene, iron and zinc content across the reciprocals.

Apomuden performed best as parent for beta-carotene content (37.19mg/100gDW) but, its overall cross performance (14.66 mg/100gDW) was not significantly different from Beauregard (14.86mg/100gDW) and Histarch (10.13mg/100gD) (Table 4). Ogyefo and Histarch had the lowest iron content among the parents with means 1.68 mg/100gDW and 1.49 mg/100gDW. Their means in the overall crosses were 1.64 mg/100gDW and 1.56 mg/100gDW, respectively. The iron content for Apomuden (2.56 mg/100gDW) was the highest followed by Beauregard (2.12 mg/100gDW) but, there was no significant difference in their cross performance with respective means of 1.73 mg/110DW and 1.79 mg/100gDW. The performance of the parents for zinc content was in the same trend as the iron content with Apomuden (1.53 mg/100gDW) obtaining the highest value followed by Beauregard (1.12 mg/100gDW). No significant difference was observed between the cross performance. Significant differences were observed between some crosses and their reciprocals for beta-carotene content. These were Ogyefo x Beauregard and Beauregard x Ogyefo, and Beauregard x Histarch and Histarch x Beauregard (Table 4)

Table 3. Mean squares for the four parents and all their crosses across three environments

Source of variation	Df	Crosses			Reciprocals		
		Beta-carotene content	Iron content	Zinc content	Beta-carotene content	Iron content	Zinc content
Environment (Env.)	2	23.974^{ns}	1.18**	0.73**	26.08*	1.43**	0.90**
Rep. (Env.)	3	4.696^{ns}	0.05^{ns}	0.02^{ns}	2.87^{ns}	0.03^{ns}	0.02^{ns}
Entry	9	801.825**	0.64**	0.21**	719.56**	0.72**	0.23**
Env. x Entry	18	13.436^{ns}	0.05^{ns}	0.02**	11.24*	0.05*	0.02*
Overall heterosis (h_{ij})	5	305.932**	0.360**	0.1056^{ns}	213.01**	0.36**	0.129*
Average heterosis (h)	1	533.216**	1.141**	0.2310*	686.72**	1.03**	0.210*
Variety heterosis (h_j) (GCA)	3	264.286**	0.233^{ns}	0.1992^{ns}	90.03**	0.30**	0.145*
SCA	2	104.175**	0.001^{ns}	0.0003^{ns}	80.17**	0.02*	0.003*

*Significant at P< 0.05; **Significant at P<0.01; ^{ns} Not significant.

Table 4. Beta-carotene, iron and zinc content of the four parents and their crosses over three environments

Parents	F_1 means				Mean Performance	
	Apomuden	Ogyefo	Beauregard	Histarch	Overall crosses	Parents
Beta-carotene content (mg/100g)DW						
Apomuden		7.9	23.84	13.29	14.66	37.19
Ogyefo	8.79		3.98	4.04	6.82	5.95
Beauregard	22.56	11.55		19.1	14.86	25.64
Histarch	11.59	4.62	8.13		10.13	3.67
Lsd (5%)					5.39	5.39
Iron content (mg/100g)DW						
Apomuden		1.65	1.88	1.61	1.73	2.56
Ogyefo	1.78		1.77	1.47	1.64	1.68
Beauregard	1.91	1.96		1.72	1.79	2.12
Histarch	1.52	1.51	1.51		1.56	1.49
Lsd (5%)					0.35	0.35
Zinc content (mg/100g)DW						
Apomuden		1.05	1.09	0.94	1.03	1.53
Ogyefo	1.1		1.08	0.92	1.05	1.01
Beauregard	1.08	1.16		0.96	1.04	1.12
Histarch	0.9	0.94	0.89		0.93	0.86
Lsd (5%)					0.23	0.23

$^S F_1$ means for crosses above diagonal; F_1 means for reciprocals below diagonal

3.2 Estimates of Variety Effect, Variety Heterosis and Average Heterosis for Beta-Carotene, Iron and Zinc Content

Variety effect (v_j) was significant for beta-carotene, iron and zinc content (Table 5). Variety effect (v_j) for beta-carotene ranged from -14.44 mg/100gDW (Histarch) to 19.08 mg/100gDW (Apomuden). Those for iron and zinc content ranged from -0.47 mg/100gDW to 0.59 mg/100gDW and -0.269 mg/100gDW to 0.394 mg/100gDW, respectively. Histarch produced the lowest values and Apomuden the highest values. All the parents showed significant (P<0.01) variety effect for beta-carotene content. Beauregard did not show significant (P>0.05) variety effect for iron and zinc content and Ogyefo for zinc content. Variety heterosis (h_j) which indicates GCA was significant for all the traits, but not all the parents indicated significant values across the traits (Table 5). Values for beta-carotene content ranged from -5.06 (Apomuden) to 7.40 (Histarch). Those for iron content ranged from -0.25 (Apomuden) to 0.22 (Ogyefo). Values for zinc content ranged from -0.175 to 0.145 and these values were given by Apomuden and Ogyefo. Average heterosis was significant (P<0.01) for beta-carotene, iron and zinc content (Table 5).

Table 5. Estimates of variety effect (v_j), average heterosis (h) and variety heterosis (hj) for beta-carotene, iron, and zinc content over three environments

Parents	Traits					
	Beta-carotene (mg/100g)DM		Iron (mg/100g)DW		Zinc (mg/100g)DW	
	Variety effects (V_j)	Variety heterosis (h_j)	Variety effects (V_j)	Variety heterosis (h_j)	Variety effects (V_j)	Variety heterosis (h_j)
Crosses						
Apomuden	19.08**	-5.06**	0.59**	-0.25**	0.394**	-0.166*
Ogyefo	-12.16**	-4.00**	-0.28*	0.06^{ns}	-0.121^{ns}	0.072^{ns}
Beauregard	7.53**	1.66^{ns}	0.16^{ns}	0.08^{ns}	-0.004^{ns}	0.055^{ns}
Histarch	-14.44**	7.40**	-0.47**	0.11^{ns}	-0.269**	0.039^{ns}
Std. error	1.19	1.03	0.10	0.09	0.08	0.07
Average Heterosis	-6.09±0.88**		-0.28±0.08**		-0.127±0.06*	
Reciprocals						
Apomuden	19.08**	-4.87**	0.59**	-0.24**	0.394**	-0.175*
Ogyefo	-12.16**	1.75*	-0.28*	0.22*	-0.121^{ns}	0.145*
Beauregard	7.53**	0.55^{ns}	0.16^{ns}	0.07^{ns}	-0.004^{ns}	0.048^{ns}
Histarch	-14.44**	2.58**	-0.47**	-0.04^{ns}	-0.269**	-0.019^{ns}
Std. error	1.01	0.87	0.11	0.09	0.08	0.07
Average Heterosis	-6.91±0.75**		-0.27±0.08**		-0.121±0.06*	

*Significant at P<0.05; **Significant at P<0.01; ^{ns} Not significant.

3.3 Better Parent, Mid-Parent and Specific Heterosis for Beta-Carotene, Iron and Zinc Content Over Three Environments

Better parent heterosis (h_{ij}) for beta-carotene ranged from -84% for crosses Ogyefo x Beauregard to -22% for crosses Histarch x Ogyefo (Table 6). Mid-parent heterosis ($\hat{h}_{ij}$) ranged from -75% for crosses Ogyefo x Beauregard to 30% for crosses Beauregard x Histarch. Specific heterosis (s_{ij}) was significant (P<0.05) for all the crosses except for crosses Apomuden x Ogyefo, Histarch x Apomuden, Beauregard x Ogyefo, and Beauregard x Histarch (Table 6). Specific heterosis (s_{ij}) was not significant for any of the crosses for iron and zinc content (Table 6). For iron content, better parent heterosis (h_{ij}) ranged from -41% for crosses Histarch x Apomuden to -1% for crosses Beauregard x Histarch. Mid-parent heterosis ($\hat{h}_{ij}$) also ranged from -25% for crosses Histarch x Apomuden to 3% for crosses Beauregard x Ogyefo. Better parent heterosis (h_{ij}) for zinc content had values which ranged from -41% for crosses Histarch x Apomuden to 4% for crosses Beauregard x Ogyefo while mid-parent heterosis ($\hat{h}ij$) ranged from -25% for crosses Histarch x Apomuden to 8% for crosses Beauregard x Ogyefo. F_1 progenies with superior performance over the parents for beta-carotene, iron and zinc content are presented in Table 7. Their beta-carotene content ranged from 13.55 mg/100gDW to 41.71 mg/100gDW. Their iron content ranged from 1.44 to 2.12 mg/100gDW while their zinc content ranged from 0.82 mg/100gDW to 1.26 mg/100gDW.

Table 6. Estimates of heterosis effect for beta-carotene, iron, and zinc content over three environments

Cross	Trait								
	Beta-carotene			Iron			Zinc		
	Better parent (h_{ij}) (%)	Mid-parent $(\hat{h}_{ij})$	Specific heterosis (s_{ij})	Better Parent (h_{ij}) (%)	Mid-parent $(\hat{h}_{ij})$	Specific Heterosis (s_{ij})	Better Parent (h_{ij}) (%)	Mid-parent $(\hat{h}_{ij})$	Specific heterosis (s_{ij})
Apomuden x Ogyefo	-79	-63	1.48^{ns}	-36	-22	0.004^{ns}	-31	-19	0.000^{ns}
$^{\$}$Ogyefo x Apomuden	-76	-59	-2.74^{**}	-30	-16	-0.054^{ns}	-28	-15	-0.015^{ns}
Apomuden x Beauregard	-36	-24	1.92^{*}	-27	-20	-0.011^{ns}	-29	-4	-0.005^{ns}
$^{\$}$Beauregard x Apomuden	-39	-28	2.38^{**}	-25	-18	0.012^{ns}	-29	3	0.000^{ns}
Apomuden x Histarch	-64	-35	-3.39^{**}	-37	-21	0.007^{ns}	-39	-22	0.005^{ns}
$^{\$}$Histarch x Apomuden	-69	-43	0.36^{ns}	-41	-25	0.042^{ns}	-41	-25	0.016^{ns}
Ogyefo x Beauregard	-84	-75	-3.39^{**}	-17	-7	0.007^{ns}	-4	1	0.005^{ns}
$^{\$}$Beauregard x Ogyefo	-55	-27	2.38^{ns}	-8	3	0.042^{ns}	4	8	0.016^{ns}
Ogyefo x Histarch	-32	-16	1.92^{*}	-13	-8	-0.011^{ns}	-9	-2	0.005^{ns}
$^{\$}$Histarch x Ogyefo	-22	-4	2.38^{**}	-10	-5	0.012^{ns}	-7	0	0.000^{ns}
Beauregard x Histarch	-26	30	1.48^{ns}	-1	-5	0.004^{ns}	-14	-4	0.000^{ns}
$^{\$}$Histarch x Beauregard	-68	-45	-2.75^{**}	-29	-17	-0.055^{ns}	-21	-6	-0.016^{ns}
Std. error (crosses)			0.79			0.70			0.51
Std. error (reciprocal)			0.67			0.07			0.05

$^{\$}$Reciprocals; *Significant at P<0.05; **Significant at P<0.01; ^{ns}Not significant

Table 7. List of F_1 progenies that showed superior performance over the parents across three environments

F_1 progenies	Beta-carotene (mg/100g)DW	Iron (mg/100g)DW	Zinc (mg/100g)DW
Apomuden x Beauregard-4	41.71	2.09	1.20
Apomuden	37.19	2.56	1.53
Beauregard x Apomuden-23	33.73	2.01	1.12
Beauregard x Apomuden-4	33.47	2.12	1.26
Beauregard x Histarch-9	32.49	2.07	1.09
Beauregard x Apomuden-19	25.97	1.96	1.13
Apomuden x Ogyefo-18	25.72	2.03	1.21
Apomuden x Histarch-15	25.65	1.77	0.99
Beauregard	25.64	2.12	1.13
Beauregard x Apomuden-17	25.47	1.85	0.96
Beauregard x Histarch-5	24.98	1.59	0.89
Beauregard x Histarch-7	23.20	1.82	1.07
Beauregard x Histarch-4	23.00	1.67	0.91
Beauregard x Apomuden-30	22.49	1.90	1.06
Histarch x Beauregard-21	19.55	1.71	0.90
Histarch x Apomuden-9	19.03	1.44	0.82
Ogyefo x Apomuden-19	18.62	1.96	1.14
Apomuden x Histarch-14	18.16	1.72	0.95
Beauregard x Histarch-6	14.86	1.48	0.86
Beauregard x Apomuden-11	13.55	1.60	0.97
Histarch	3.67	1.49	0.86
Ogyefo	5.95	1.69	1.01
*SEM (P<0.05)	2.63	0.11	0.07
CV (%)	41.1	11.1	11.5

*SEM=Standard error of mean

4. Discussion

The significant mean squares for both variety heterosis (GCA) and specific combining ability (SCA) for beta-carotene, iron and zinc content means that additive and non-additive effects were involved in the expression of beta-carotene, iron and zinc content. However, the substantially greater amounts of the GCA sum of squares for the traits compared with the sum of squares for SCA suggests that additive effects were more important than non-additive effects for beta-carotene, iron and zinc content. This indicates that most of the genetic variation found were additive in nature and majority of the total sum of squares of the traits due to differences among generation performance may be explained by variety effects (vj) and variety heterosis. Consequently, variety effects of the parents were important predictors of the cross performance. This shows predominance effect of additive gene action in beta-carotene, iron and zinc content and suggests that there would be no complications in breeding sweetpotato varieties with high content of beta-carotene, iron and zinc through selection. Oduro (2013), also found similar result for beta-carotene content on different sweetpotato genotypes. Similar results have also been reported on other traits in sweetpotato (Gasura et al., 2008; Mwanga et al., 2002; Shumbusha et al., 2014; Sseruwu, 2012). Kumah (2013), reported significant effect of GCA and SCA for iron and zinc content in sorghum but found additive gene action conditioning grain zinc content while both additive and non-additive gene effects control grain iron content.

Significant (P<0.01) differences between the genotypes show significant genetic differences and indicates that meaningful selection and improvement on beta-carotene, iron and zinc content in sweetpotato is possible. G x E interaction is key in evaluating genotype adaptation and development of genotypes with improved end-product quality (Ames, Clarke, Marchylo, Dexter, & Woods, 1999). Presence of G x E suggest that progress from selection may be complicated since it may be difficult to separate genotypic effects from environmental effects. Significant differences were observed between some crosses and their reciprocals for beta-carotene content. These were Ogyefo x Beauregard and Beauregard x Ogyefo, and Beauregard x Histarch and Histarch x Beauregard. In addition, G x E was not significant for any of the traits for the crosses except zinc content, but G x E was significant for all the traits across the reciprocals. Moreover, overall heterosis was significant for all the traits except zinc content for the crosses, while variety heterosis and SCA were significant (P<0.05) for all the traits across the reciprocals and only beta-carotene content for the crosses. These differences between the crosses and their reciprocals may be attributed to maternal or cytoplasmic effects. Maternal or cytoplasmic effects are influences of parents on offspring phenotype occurring through pathways other than inherited DNA. If existing, maternal effect could have inflated the GCA mean squares at the expense of SCA. This has consequences for the interpretation of the results and perhaps for several others that concluded that additive gene action was predominant over non-additive effects. Maternal effects have been reported to influence a number of traits in sweetpotato (Chiona, 2009; Lin et al., 2007; Oduro, 2013).

Significance of overall heterosis indicates opportunity for exploiting heterosis for increase beta-carotene, iron and zinc content in sweetpotato storage roots. Other studies in sweetpotato have shown that there is exploitable heterosis in sweetpotato (Baafi et al., 2016; Grüneberg, Mwanga, Andrade, & Dapaah, 2009). However, among the three kinds of heterosis, average heterosis (h) was the most important. Average heterosis (h) contributed by a particular set of parents used in crosses is the differences between the mean of all crosses and the mean of all parents (Gardner, 1967). The high values of variety heterosis (h_j) of Histarch and Beauregard for beta-carotene and iron content, and Ogyefo and Beauregard for zinc content indicates that these parents have good general combining ability for the respective traits. High variety heterosis (h_j) show differences in occurrence of dominant alleles between parents (Crossa, Gardner, & Mumm, 1987). Negative values may be attributed to unrealized performance expectation of the parents in the progenies. This is because negative values of variety heterosis for breeding varieties/population seem to represent an unfulfilling performance expectation from high variety effect(v_j) and a high average heterosis effect (h) (Harold et al., 2001). Apomuden and Beauregard had the highest beta-carotene, iron and zinc content among the four parents. Parents with higher beta-carotene content had higher iron and zinc content due to the strong positive genotypic correlation among the traits (Baafi, 2014). This indicates that breeding sweetpotato genotypes with high beta-carotene, iron and zinc content will not be too difficult due to the strong positive genotypic association between the traits if suitable parents are used. Apomuden and Beauregard were good parents for the traits and can be inter-crossed to develop elite genotypes with sufficient genetic variability for improvement on beta-carotene, iron and zinc content in sweetpotato.

5. Conclusion

Genetic variability exists for beta-carotene, iron and zinc content in sweetpotato, and much of this genetic variation is additive in nature. This means that beta-carotene, iron and zinc content in sweetpotato are mostly controlled by additive gene effects rather than dominance and epistasis. Significance of overall heterosis indicates some opportunity for exploitation of heterosis for increasing beta-carotene, iron and zinc content in sweetpotato storage roots.These indicates that the parents used can be inter-crossed to develop elite genotypes with sufficient genetic variability for breeding sweetpotato varieties that combined higher amounts of these traits to alleviate malnutrition in Ghana and beyond.

Acknowledgements

Many thanks to the Alliance for a Green Revolution in Africa (AGRA) for sponsoring this study through West Africa Centre for Crop Improvement (WACCI), University of Ghana. We are also grateful to the International Potato Centre for supporting this study through the SASHA Project.

References

Ames, N. P., Clarke, J. M., Marchylo, B. A., Dexter, J. E., & Woods, S. M. (1999). Effect of Environment and

Genotype on Drurum Wheat Gluten Strenght and Paster Viscoelasticity. *Cereal Chem., 76*, 582-586. http://dx.doi.org/10.1094/CCHEM.1999.76.4.582

Baafi, E. (2014). Development of End-User Preferred Sweetpotato Varieties in Ghana. Phd Thesis. West Africa Centre for Crop Improvement (WACCI), University of Ghana.

Baafi, E., Manu-Aduening, J., Gracen, V. E., Ofori, K., Carey, E. E., & Blay, E. T. (2016). Development of End-User Preferred Sweetpotato Varieties. *Journal of Agricultural Science, 8*(2), 57-73. ISSN 1916-9752. E-ISSN 1916-9760. doi:10.5539/Jas.V8n2p57. URL:http://dx.doi.org/10.5539/Jas.V8n2p57.

Black, R. E., Allen, L. H., Bhutta, Z. A., Caulfield, L. E., De Onis, M., Ezzati, M., … Rivera, J. (2008). Maternal and Child Undernutrition: Global and Regional Exposures and Health Consequences. *Lancet, 371*, 243-260. http://dx.doi.org/10.1016/S0140-6736(07)61690-0

Buerstmayr, H., Nicola, K., Uwe, S., Heinrich, G., & Elisabeth, Z. (2007). Agronomic Performance and Quality of Oat (*Avena Sativa* L.) Genotypes of Worldwide Origin Produced under Central European Growing Conditions. *Field Crops Res., 101*, 92-97. http://dx.doi.org/10.1016/j.fcr.2006.12.011

Chiona, M. (2009). Towards Enhancement of B-Carotene Content of High Dry Mass Sweetpotato Genotypes in Zambia. Phd Thesis. University of KwaZulu-Natal, Pietermaritzburg, Republic of South Africa.

Crossa, J., Gardner, C. O., & Mumm, R. H. (1987). Heterosis among Populations of Maize (*Zea Mays* L.) with Different Levels of Exotic Germplasm. *Theor Appl Genet., 73*, 445-450. http://dx.doi.org/10.1007/BF00262514

Darnton-Hill, I., Webb, P., Harvey, P. W. J., Hunt, J. M., Dalmiya, N., Chopra, M., … De Benoist, B. (2005). Micronutrient Deficiencies and Gender: Social and Economic Costs. *American Society for Clinical Nutrition, 81*, 1198S-1205S.

De Benoist, B., Erin, M., Ines, E., & Mary, C. (2008). Worldwide Prevalence of Anemia 1993-2005: WHO Global Database on Anemia, Geneva, Switzerland: World Health Organization, 2008.

Ekanayake, I. J., Malagamba, P., & Midmore, D. J. (1990). Effect of Water Stress on Yield Indices of Sweetpotatoes. In: Howeler, R.H.(Ed.). *Proceedings of 8th Symposium of the International Society for Tropical Root Crops. Bangkok, Thailand. 724pp.*

Elisa, M., Humberto, A. M., & Luis, F. S. (2000). Combining Ability for Resistance to Sweetpotato Feathery Mottle Virus. *Hortscience, 35*(7), 1319-1320.

Feyrer, J., Politi, D., & Wei, N. D. (2013). The Cognitive Effects of Micronutrient Deficiency: Evidence from Salt Iodization in the United States. *Working Paper 19233.* Retrieved from http://www.nber.org/papers/w19233. [Accessed on 15th October 2015]

Fonseca, S., & Patterson, F. L. (1968). Hybrid Vigour in Seven Parental Diallel Cross in Common Wheat (*Triticum Aestivum* L.). *Crop Sci., 8*, 85-88. http://dx.doi.org/10.2135/cropsci1968.0011183X000800010025x

Gardner, C. O. (1967). Simplified Methods for Estimating Constants and Computing Sum of Squares for Diallel Cross Analysis. *Fitotec Latinoam, 4*,1-12.

Gardner, C. O. (1982). Genetic Information from the Gardner-Eberhart Model for Generation Means. Somefi Saltiuo, Coahuila, Mexico.

Gardner, C. O., & Eberhart, S. A. (1966). Analysis and Interpretation of the Variety Cross Diallel and Related Populations. *Biometrics, 22*, 439-452. http://dx.doi.org/10.2307/2528181

Gasura, E., Mashingaidze, A. B., & Mukasa, S. B. (2008). Genetic Variability for Tuber Yield, Quality, and Virus Disease Complex in Uganda Sweetpotato Germplasm. *Afr Crop Sci J., 16*(2), 147-160.

Griffing, B. (1956). Concept of General and Specific Combining Ability in Relation to Diallel Crossing Systems. *Aust J Biol Sci., 9*, 463-493. http://dx.doi.org/10.1071/BI9560463

Grüneberg, W., Mwanga, R., Andrade, M., & Dapaah, H. (2009). Challenge Theme Paper 1: Sweetpotato Breeding. In: Andrade, M. Barker, I. Dapaah, H., Elliot, H., Fuentes, S., Grüneberg, W., Kapinga, R., Kroschel, J., Labarta, R., Lemaga, B., Loechl, C., Low, J., Lynam, J. Mwanga, R., Ortiz, O., Oswald, A. and Thiele, G. 2009. Unleashing the Potential of Sweetpotato in Sub-Saharan Africa: Current Challenges and Way Forward. International Potato Center (CIP), Lima, Peru. Working Paper 2009-1, 1-42.

Harold, R. M., Hugo, C., Kevin, V. P., & Magnie, S. B. (2001). Heterotic Relationships among Nine Temperate

and Subtropical Maize Populations. *Crop Sci., 41*, 1012-1020.
http://dx.doi.org/10.2135/cropsci2001.4141012x

Hayman, B. I. (1954a). The Theory and Analysis of the Diallel Crosses. *Genetics, 39*, 798-809.

Hayman, B. I. (1954b). The Analysis of Variance of Diallel Tables. *Biometrics, 10*, 235-244.
http://dx.doi.org/10.2307/3001877

Hayman, B. I. (1957). Interaction, Heterosis and Diallel Crosses *Genetics, 42*, 336-355.

Hayman, B. I. (1958). The Theory and Analysis of Diallel Crosses, 2. *Genetics, 43*, 63-85.

Jones, A. (1986). Sweetpotato Heritability Estimates and Their Use in Breeding. *Horticultural Science, 21*, 14-17.

Jones, A., & Dukes, P. D. (1980). Heritability of Sweetpotato Resistance to Root Knot Nematodes Caused by *Meloidogyne Incognita* and *M. Javanica. Journal of American Society of Horticultural Science, 105*, 154-156.

Jones, A., Schalk, J. M., & Dukes, P. D. (1979). Heritability Estimates for Resistance in Sweetpotato Soil Insects. *Journal of American Society of Horticultural Science, 104*, 424-426.

Kumah, A. A. (2013). Options for Enhancing Grain Iron and Zinc Concentrations in Sorghum. Research Program on Dryland Cereals. International Crops Resaerch Institute for the Semi-Arid Tropics (ICRISAT). Powerpoint 2nd May 2013.

Lebot, V. (2009). Tropical Root and Tuber Crops – Cassava, Sweetpotato, Yams and Aroids. Wallingford, UK: Cabi Publishing Group.

Lebot, V. (2010). Sweet Potato. In: Bradshaw J. E.(Ed.), Root and Tuber Crops, *Handbook of Plant Breeding, 7*, 97. Doi 10.1007/978-0-387-92765-7_3, (C) Springer Science+Business Media, Llc 2010.
http://dx.doi.org/10.1007/978-0-387-92765-7_3

Lin, K. H., Lai, Y. C., Chang, K. Y., Chen, Y. F., Hwang, S. Y., & Lo, H. F. (2007). Improving Breeding Efficiency for Quality and Yield of Sweetpotato. *Botanical Studies, 48*, 283-292.

Mihovilovich, E., Mendoza, H. A., & Salazar, L. F. (2000). Combining Ability for Resistance to Sweetpotato Feathery Mottle Virus. *Horticultural Science, 35*, 1319-1320.

Miller, P. A., Williams, J. C., Robinson, H. F., & Comstock, R. E. (1958). Estimates of Genotypic and Environmental Variances and Covariances in Upland Cotton and Their Implications in Selection. *Agron. J., 50*, 126-131. http://dx.doi.org/10.2134/agronj1958.00021962005000030004x

Mwanga, R. O. M., Yencho, G. C., & Moyer, J. W. (2002). Diallel Analysis of Sweetpotatoes for Resistance to Sweetpotato Virus Disease. *Euphytica, 128*, 237-248. http://dx.doi.org/10.1023/A:1020828421757

Oduro, V. (2013). Genetic Control of Sugars, Dry Matter and Beta-carotene in Sweetpotato *(Ipomoea Batatas [L.] Lam).* Phd Thesis. West Africa Centre for Crop Improvement (WACCI), University of Ghana.

Ray, R. C., & Tomlins, K. I. (2010). Sweetpotato: Post Harvest Aspects in Food, Feed and Industry, New York: Nova Science Publishers Inc.

Rex, B. (2002). Breeding for Quantitative Traits in Plants. Stemma Press, Minnesota, USA.

SAS. (2002). SAS Institute. 2002. SAS/Stat 9 User's Guide. Vol. 1, 2, and 3. SAS Inst., Cary, NC.

Shenkin, A. (2006). Micronutrients in Health and Disease. *Postgraduate Medical Journal, 82*(971), 559-567.
http://dx.doi.org/10.1136/Pgmj.2006.047670.

Shumbusha, D., Tusiime, G., Edema, R., Gibson, P., Adipala, E., & Mwanga, R. O. M. (2014). Inheritance of Root Dry Matter Content in Sweetpotato. *Afr Crop Sci J., 22*(1), 69-78.

Simon, W., Rafael, P., Klaus, E., Olivia, M., Mario, V. C., Imelda, A., & Urs, B. (2013). Burden of Micronutrient Deficiencies by Socio-Economic Strata in Children Aged 6 Months to 5 Years in the Philippines. *BMC Public Health,* 1167. http://www.biomedcentral.com/1471-2458/13/1167

Sseruwu, G. (2012). Breeding of Sweetpotato *(Ipomoea Batatas (L.) Lam.) f*or Storage Root Yield and Resistance to Alternaria Leaf Petiole and Stem Blight *(Alternaria Spp.)* in Uganda. PhD Thesis. University of KwaZulu-Natal, Pietermaritzburg, Republic of South Africa.

Todd, M. S. (2013). Application of near-Infrared Spectroscopy to Study Inheritance of Sweetpotato Composition Traits. PhD Thesis. North Carolina State University, Raleigh, North Carolina.

Tumwegamire, S., Rubaihayo, P. R., Labonte, D. R., Diaz, F., Kapinga, R., Mwanga, R. O., & Gruneberg, W. J. (2011). Genetic Diversity in White- and Orange-Fleshed Sweetpotato Farmer Varieties from East Africa Evaluated by Simple Sequence Repeat Markers. *Crop Sci., 51*, 1132-1142.

Warammboi, J. G., Dennien, S., Gidley, M. J., & Sopade, P. (2011). Characterization of Sweetpotato from Papua New Guinea and Australia: Physicochemical, Pasting and Gelatinisation Properties. *Food Chem., 126*, 1759-1770. http://dx.doi.org/10.1016/j.foodchem.2010.12.077

WHO. (2009). Global Prevalence of Vitamin A Deficiency in Populations at Risk 1995–2005. WHO Global Database on Vitamin A Deficiency. Geneva: WHO; 2009.

WHO. (2013). "Is It True That Lack of Iodine Really Causes Brain Damage?,"Online Q&A, May 2013. http://www.who.int/features/qa/17/en/index.html.

Zhang, Y., Kang, S. M., & Kendall, R. L. (2005). Diallel-Sas05: A Comprehensive Program for Griffing's and Gardner–Eberhart. American Society of Agronomy Analyses. *Agron J., 97*, 1097-1106. http://dx.doi.org/10.2134/agronj2004.0260

10

Ecology and Diversity of Diatoms in *Kuttanadu* Paddy Fields in Relation to Soil Regions, Seasons and Paddy-Growth-Stages

Dhanya Vijayan[1] & J. G. Ray[2]

[1] Research Scholar, Environment Science Research Lab, St. Berchmans College, Changanacherry, Kerala, India

[2] Professor, School of Biosciences, Mahatma Gandhi University, Kottayam, Kerala, India

Correspondence: J. G. Ray, School of Biosciences, Mahatma Gandhi University, Kottayam, Kerala, India. E-mail: jgray@mgu.ac.in

Abstract

This assessment of ecology and diversity of Diatoms in *Kuttanadu*, is continuation of the previous publications of Green-algae and Blue-green-algae of the same region; all the three investigations were carried out simultaneously. The unique *Kuttanadu* wetlands (90°17'N to 90°40'N; 760°19'E to 760°33'E), a well known 'Rice bowl' of Kerala, spread over 53,639 hectares, is located in Alapuzha District of Kerala, India. This wetland was once known for its high biodiversity, but currently severely affected by intensive green-revolution activities of past few decades. Main objective of the present report is to explain the diversity and ecological status of Diatoms in relation to the environment conditions currently existing in the region. Duration of the study was from December 2009 to October 2010. Soil samples of three different soil-types in two different crop-seasons, at two diverse crop-growth-stages are analyzed and compared. Altogether 120 composite soil samples randomly collected from the broad wetlands are analyzed. A rich community of Diatoms, of 40 species is found out. The order Naviculales is observed as the dominant Diatom (40%) in these paddy wetlands. Ecological parameters of Diatoms were found highest in the Lower *Kuttanadu* soil region, during *Virippu* season, at the seedling- stage of the crop, whereas the lowest value for most of the parameters were observed in *Kayal* soils during *Puncha* season at the seedling-stage. Apart from the specific soil factors influencing Diatom population characteristics in the region, crop seasons and soil phosphorus are found to have significant influence on Diatoms in this wetland paddy soils.

Keywords: diatoms, paddy-wetland, crop-seasons, soil-regions, paddy-growth-stages

1. Introduction

This is the third in the series of reports on ecology and diversity of algae of *Kuttanadu* wetlands, the first two parts, one on green-algae and the other on blue-green-algae are already published (Vijayan & Ray, 2015a, 2015b).

Diatoms are one of the most diverse groups of algae, estimated to be more than 100,000 species, with cosmopolitan distribution (Demirbas & Demirbas, 2011). Uniqueness of Diatoms is that they are not just universally abundant group of algae, but are highly ecologically diverse (Jena, Ratha, & Adhikary, 2006). Moreover, they are one of the most successful groups of photosynthetic eukaryotes (Muruganantham, Gopalakrishnan, Chandrasekharan, & Jayachandran, 2012). Seasonal change of diatom community in flooded paddy soil environments is a widely discussed topic worldwide.

Another major significance of the study of Diatoms is that they are well known bio-indicators, capable of providing specific information on varieties of specific environment characteristics (Seirieyssol, Chatelard, & Cubizolle, 2010) of specific habitats. Some of them are highly tolerant to harsh environmental conditions, while some others are highly sensitive to soil conditions such as desiccation, freezing and abrupt heating (Souffrean, 2010). Such characteristics make the group ideal ecological indicators (Mazumder, 2012).

Like all other algae, diatoms also contribute significantly to carbon sequestration; but the role of diatoms in the recycling of silica is highly specific (Kesic, Tuney, Zerben, Guden & Sukatar, 2013). Diatoms constitute a fundamental link between primary and secondary production in aquatic systems. Therefore, the major ecological importance of diatoms is as a component in the food chains of specific aquatic species. Influence of diatoms on the mechanical, hydraulic, and physico-chemical properties of soil is also well known.

Many species of Diatoms are now well known for the bio-oil content (McGinnis, Dempster, & Sommerfeld, 1997) in the cells. Some species of diatoms are already commercially cultivated as oil yielding resources and other agents of bio-production (Wanga et al., 2014; Xue et al., 2015).

Therefore, assessment of the diversity of diatoms of any specific ecosystem, especially the ecological details of them in relation to specific environment conditions, has universal significance. Such investigations may also be highly useful in the identification of economically valuable species of diatoms. Discovery of the ecological details of such valuable diatoms have applications in the commercial cultivation of them as bio-resources.

It is true that Diatom diversity in many parts of India have been well documented by various investigators (Gandhi, 1956; Pandey & Pandey, 1980; Mohan, 1987; Nautiyal & Verma, 1997; Venkataraman, 2005; Jena et al., 2006; Seirieyssol et al., 2010; Pareek, Singh, & Singh, 2011; Muruganantham et al., 2012). However, reports of diatom community in Indian paddy fields are quite rare. This is the first report of diatom community of Kuttanadu wetland paddy fields; the uniqueness of the present report is that it provides detailed ecological description of the wet soil diatom community in relation to three different types of wetland soils, two different crop-seasons and at two different paddy-growth-stages.

2. Material and Methods

Study area, sampling sites and methodology of sampling were the same as that of the previous reports on green algae and blue-green algae (Vijayan & Ray, 2015a, 2015b) as all these three reports are based on the same soil sampling, and the studies were carried out simultaneously.

2.1 Study Area

Kuttanadu (9^0 17'- 9^0 40' N and 76^0 19'- 6^0 33' E) is the delta of five eastward flowing rivers from the Western Ghats to the Arabian Sea. This delta is seen on the south eastern boundary of the Vembanadu Lake, a 'Ramsar site', which opens to the sea at Kochi. It is the well known unique 'Rice bowl' of Kerala spread over 53,639 hectares (Sudhikumar, Mathew, Sunish, & Sebastian, 2005). The Kuttanadu paddy fields are below sea-level

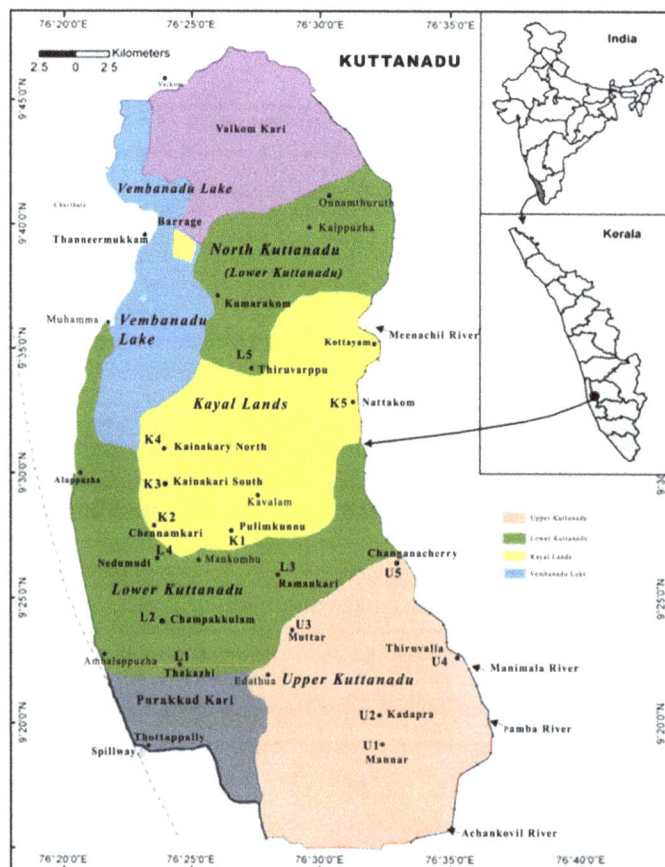

Figure 1. Map of Kuttanadu showing the three different kinds of soils in the broad region and the sampling sites

wetlands where cultivation is carried out after pumping out the excess water of the broad fields, which during the season of paddy cultivation remain separated from the surrounding backwaters by artificially reclaimed protective 'barrages'. Majority of the fields are cultivated twice a year. The non-monsoon crop, '*Puncha*', is the main crop, which is sown in November or December and harvested by the end of March. The additional monsoon crop, '*Virippu*' is sown in May to the end of June and harvested in September or October. Upper *Kuttanadu*, Lower *Kuttanadu* and *Kayal* lands are the three soil regions of *Kuttanadu*. Three kinds of soils reported from the area are '*Karappadom*' (river-borne alluvial soils found mostly in the Upper *Kuttanadu*), '*Kari*' (deep black in colour with high proportion of organic matter, characterized by acidity and salinity found mainly in the Lower *Kuttanadu*and, and rarely in some parts of the upper *Kuttanadu* zones) and '*Kayal*' (the reclaimed shallow beds of *Vembanadu* Lake, mainly silt-loam to silt-clay-loam) soils (Pillai, Ponniah, Vincent, & David, 1983). All the geographic and other details of the three soils regions are well described in the previous reports (Ray, Dhanya, & Binoy, 2014). The *Kuttanadu* is a unique wetland ecological zone with high biodiversity, especially of indigenous fishes, but quite destroyed by intensive green-revolution efforts of past few decades in the area. It is believed that once up on a time, the brackish water environments of *Kuttanadu* promoted a rich diversity of mangroves and provided ideal habitats for diverse flora and fauna.

2.2 Soil Sampling Procedure

Soil samples were systematically collected from all the three soil regions in two different paddy cultivating seasons of a year; in each season, samples of two different paddy growth stages were collected. At each site two broad specific paddy fields located at 1-2 km intervals were randomly selected for sampling. In each specific field, several plots (5-10) of about 10m² areas were randomly identified for sampling. Several soil samples (5-10 from a plot) from the upper 1-5 cm soil layer from all these plots were collected and put together in a sterile cotton bag using aseptic implement. Repetitive samples put in each cotton bag from different plots of a single paddy field on air drying were mixed thoroughly to a composite sample. There were five sites in a soil region. Since there were five sites, two sampling fields for each site and two crop seasons (the *Puncha* - summer crop of December 2009 to March 2010; the *Virippu* - the additional crop of June 2010 to October 2010) to sample and two crop-growth stages (seedling stage and panicle stage) at each season to sample, altogether, 40 composite soil samples (5 x 2 x 2 x 2) were collected from each soil region. Altogether, there were 120 composite soil samples (3 x 40) representing the three soil regions of *Kuttanad* - the 'Upper *Kuttanadu*' or '*Karappadom*', the 'Lower *Kuttanadu* or '*Kari*' ' and the '*Kayal*'zones for the soil study and for diatom studies.

2.3 Physico-chemical Analyses of Soil Samples

Details of physico-chemical analyses of soils are already reported (Ray, Dhanya, & Binoy, 2014). Physico chemical analyses of soils were completed within two weeks after the collection. Determination of pH (1water: 2.5 soil), was done using pH meter (Systronics 324) organic carbon (Walkley & Black, 1934) total nitrogen (Microkjeldahl method), and available Phosphorus (Bray & Kurtz, 1945), Potassium (Flame photometric method), Calcium and Magnesium (Atomic Absorption Spectrometry, Perkin Elmer A Analyst-400) were done by standard analytical procedures (Jackson, 1973).

2.4 Study of Diatom Flora

Methods of Van de Vijver and Beyens (1998) and Jena et al. (2006) were used in the microscopic examination of soil samples for diatoms. Microphotograph of each specimen was taken using Olympus digital Camera attached to light microscope. Taxonomical descriptions of Gandhi (1956), Sarode and Kamat (1984), Taylor, Archibald and Harding (2007) were followed in the identification of species.

2.5 Scanning Electron Microscopic (SEM) Analyses of Diatoms

Soil samples for diatom analyses were placed in air tight plastic bags to prevent moisture loss from the algae and the samples were repeatedly sprayed with iodine to preserve them. The samples were then sieved through a 75 micron nylon mesh. The SEM studies of cleaned specimens were carried out after ion coating in high vacuum mode Jeol make model- JSM 6390 LA machine. The specimen were fixed to carbon tapes (1 x15 cm) placed on aluminium stubs, coated with gold and underwent magnetron sputtering by auto fine coater - Jeol JFC-1600. The accelerating voltage was 20KV. SEM analyses were carried out in the laboratory of the sophisticated test and Instrumentation Centre (STIC), University of Science and Technology, Cochin.

2.6 Environment Relationships and Population Dynamics of Diatoms

Correlation of diatoms to environmental factors such as region, season, and physiochemical soil parameters are described. Method of Dey, Tayung and Bastia (2010) is followed in the study of relative abundance of species in the community. Diversity index (Shannon Wiener Index) and species richness of all species of diatoms in different seasons were worked out as per Shannon & Wiener, (1949). Species richness was calculated following the method of Whittaker, (1977). Classification of taxa was done as per the classification of Guiry & Guiry, (2012).

3. Results

Altogether 40 species of diatoms (Table 1; Figures 2 & 3) belonging to 19 genera have been recorded from *Kuttanadu* paddy fields during the entire study period (2009-10). Diatoms collected belonged to nine orders, which are Melosirales (1genus, 2 species), Thalassiorales (1 genus, 1 species), Fragilariales (3genera, 4 species), Thalassiophysales (1 genus, 1species), Naviculaes (8 genera, 16 species), Bacillariales (2 genera, 12 species), Cymbellales (2 genera, 1 species), Eunotiales (1 genera, 1 species) and Achanthales (1 genus, 1 species).

Table 1. Occurrence of diatoms in *Kuttanadu* paddy fields in different regions, seasons and paddy-growth-stages: (R1-Upper *Kuttanadu*, R2-Lower *Kuttanadu* and R3-*Kayal* Lands; S1-Pre-monsoon (*Puncha*) crop and S2-Monsoon (*Virippu*) crop; G1-Seedling-growth-stage and G2-Mature-growth-stage); Symbols (+) presence of the species, (–) absence of the species

No	Diatom species	R1				R2				R3			
		S1		S2		S1		S2		S1		S2	
		G1	G2	G1	G2	G1	G2	G1	G2	G1	G2	G1	G2
	Order: Melosirales												
1	*Melosira variansAgardh*	+	+	+	+	+	+	+	+	+	+	+	+
2	*Melosira italica* (Ehrenberg) Kutzing	+	+	+	+	+	+	+	+	+	+	+	+
	Order: Thalassiosirales												
3	*Cyclotella meneghiniana*Kutzing	+	+	+	+	+	+	+	+	+	+	+	+
	Order: Fragilariales												
4	*Fragilaria capucina* Desmazieres	+	+	+	+	+	+	+	+	+	+	+	+
5	*Staurosira elliptica* (Schumann) Williams & Round	+	+	+	+	+	+	+	+	+	+	+	+
6	*Synedraulna* (Nitzsch) Ehrenberg	+	+	+	+	+	+	+	+	+	+	+	+
7	*Synedratabulate* (C.Agardh) Kutzing	+	+	+	+	+	+	+	+	+	+	+	+
	Order: Thalassiophysales												
8	*Amphora pediculus*(Kutzing) Grunowex. A. Schmidt	+	+	+	+	+	+	+	+	+	+	+	+
	Order: Naviculales												
9	*Navicula capitatoradiata* Germain	+	+	+	+	+	+	+	+	+	+	+	+
10	*Navicula cuspidata*(Kutzing) Kutzing	+	+	+	+	+	+	+	+	+	+	+	+
11	*Navicula ranomafenensis* Kutzing	+	+	+	+	+	+	+	+	+	+	+	+
12	*Navicula reinhardtii* (Grnow) Grunow	+	+	+	+	+	+	+	+	+	+	+	+
13	*Navicula salinarum* Grunow	+	+	+	+	-	-	+	+	+	+	+	+
14	*Navicula pupula* Kutzing	+	+	+	+	-	+	+	+	-	-	+	+
15	*Navicula viridula*(Kutzing) Ehrenberg	+	+	+	+	-	-	+	-	-	-	+	+
16	*Navicula viridis*(Nitzsch) Kutzing	+	+	+	+	-	-	+	-	-	-	+	+
17	*Navicula lanceolata* Ehrenberg	+	+	+	+	-	-	+	+	-	-	+	+
18	*Capartogramma crucicula*	+	-	+	+	+	+	+	+	-	+	+	+

No	Diatom species	R1				R2				R3			
		S1		S2		S1		S2		S1		S2	
		G1	G2	G1	G2	G1	G2	G1	G2	G1	G2	G1	G2
	Grunow (R.Ross)												
19	*Craticula cuspidate* (Kutzing) D.G.Mann	+	+	+	+	-	-	+	+	-	-	+	+
20	*Craticula halophila* (Kutzing) D.G.Mann	+	+	+	+	+	+	+	+	+	-	-	-
21	*Frustulia vulgaris* (Thwaites) De Toni	+	+	+	+	+	+	+	+	+	+	+	+
22	*Nedium affine* (Ehrn) Pfitzer	+	-	+	+	+	+	+	+	+	+	+	+
23	*Diadesmus confervaceae* (Kutzing) DG Mann	+	+	+	+	+	+	+	+	+	+	+	+
24	*Sellaphora ulna* (Nitzsch) Ehrenberg	+	-	+	+	+	+	+	+	+	+	+	+
	Order: Bacillariales												
25	*Pinnularia acrosphaeria* W Smith	+	+	+	+	+	+	+	+	+	+	+	+
26	*Pinnularia biceps* W. Gregory	+	+	+	+	+	+	+	+	-	-	+	+
27	*Pinnularia brauni* (Grunow) Cleve	+	+	+	+	+	+	+	+	-	-	+	+
28	*Pinnularia gibba* Ehrenberg	+	+	+	+	+	+	+	+	+	+	+	+
29	*Pinnularia viridiformis* Krammer	+	+	+	+	+	+	+	+	+	+	+	+
30	*Nitzschia reversa* W Smith	+	+	+	+	+	+	+	+	+	+	+	+
31	*Nitzschia sigmoidea* (Nitzsch) W.Smith	+	+	+	+	+	+	+	+	+	+	+	+
32	*Nitzschia clausii* Hantzsch	+	+	+	+	+	+	+	+	+	+	+	+
33	*Nitzschia filiformis* (W Smith) Vanurk	+	+	+	+	+	+	+	+	+	+	+	+
34	*Nitzschia linearis* W Smith	+	+	+	+	+	+	+	+	+	+	+	+
35	*Nitzschia nana* Grunow	+	+	+	+	+	+	+	+	+	+	+	+
36	*Nitzschia obtusa* Rabenhorst	+	+	+	+	+	+	+	+	+	+	+	+
	Order: Cymbellales												
37	*Cymbella kolbei* Hustedt	+	+	+	+	+	+	+	+	+	+	+	+
38	*Gomphonema acuminatum* Ehr	-	+	+	+	+	+	+	+	-	+	+	+
	Order: Eunotiales												
39	*Eunotia bilunaris* (Ehrenberg) Mills	+	+	+	+	+	+	+	+	+	+	+	+
	Order:Achnanthales												
40	*Achnanthidium minutissimum* (Kutzing) Czarnecki	-	-	-	-	+	+	+	+	-	-	-	-
	Total species	35	36	39	39	33	35	40	38	29	30	38	38

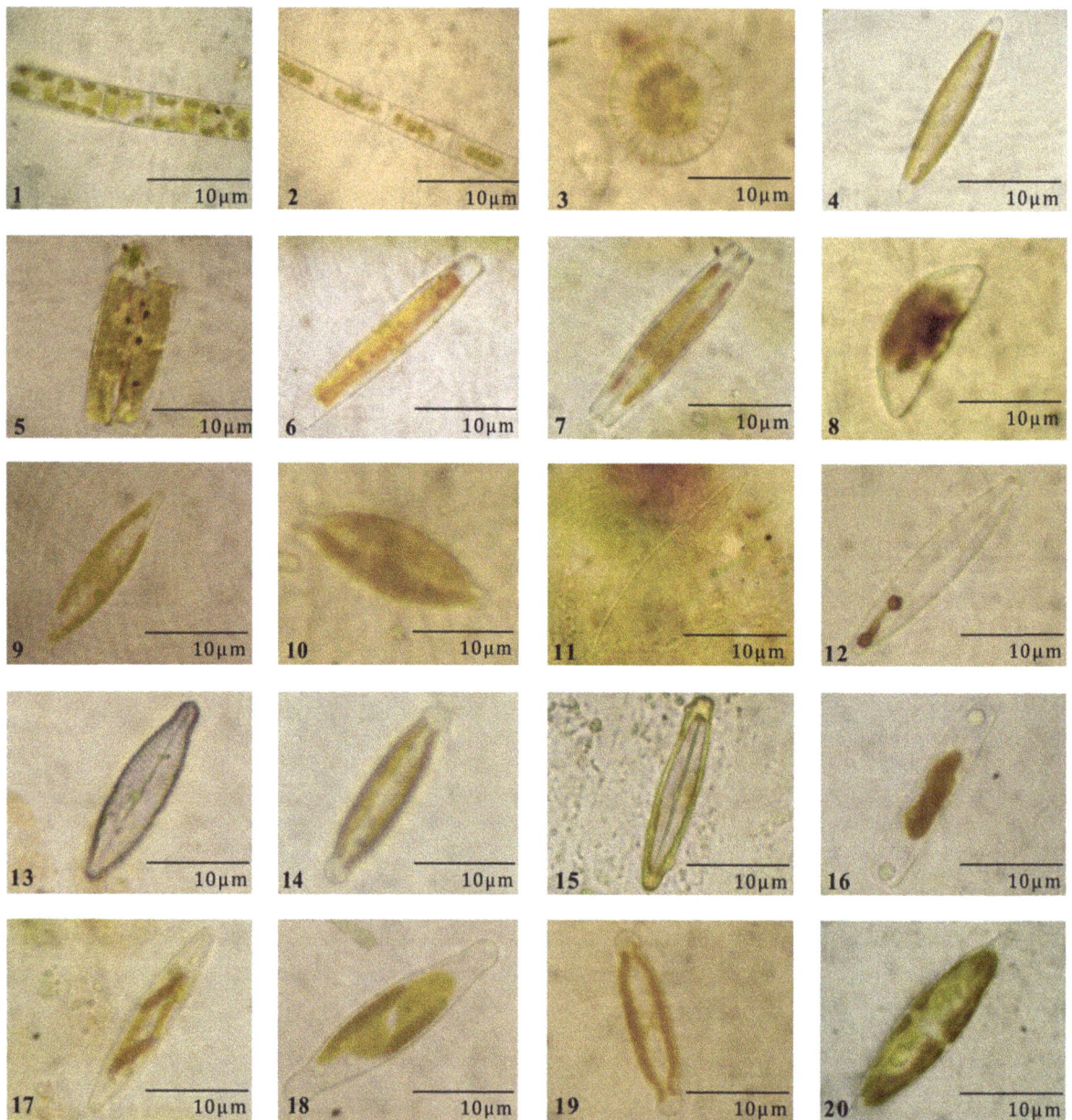

Figure 2. Diatoms 1 – 20

1. *Melosira varians,* 2. *Melosira italica,* 3. *Cyclotella meneghiniana,* 4. *Fragilaria capucina,* 5. *Staurosira construens,* 6. *Synedra ulna,* 7. *Synedra tabulate,* 8. *Amphora pediculus,* 9. *Navicula capitatoradiata,* 10. *Navicula cuspidate,* 11. *Navicula ranomafenensis,* 12. *Navicula reinhardtii,* 13. *Navicula salinarum,* 14. Navicula *pupula,* 15. *Navicula viridula,* 16. *Navicula viridis,* 17. *Navicula lanceolata,* 18. *Capartogramma crucicula,* 19. *Craticula cuspidata,* 20. *Craticula crucicula.*

Figure 3. Diatoms 21 – 40

21. *Frustulia vulgaris*, 22. *Nedium affine*, 23. *Diadesmis confervacea*, 24. *Sellaphora ulna*, 25. *Pinnularia acrosphaeria*, 26. *Pinnularia biceps*, 27. *Pinnularia brauni*, 28. *Pinnularia gibba*, 29. *Pinnularia viridiformis*, 30. *Nitzchia reversa*, 31. *Nitzchia sigmoidea*, 32. *Nitzschia clausii*, 33. *Nitzschia filiformis*, 34. *Nitzschia liebertruthii*, 35. *Nitzschia nana*, 36. *Nitzschia obtusa*, 37. *Cymbella kolbei*, 38. *Gomphonema acuminatum*, 39. *Eunotia bilunaris*, 40. *Achanthidium minutissimum*

Relative abundance of diatoms (Table 2; Figure 4) in percentage was; 5% from the order Melosirales, 10% from Fragilariales, 2% from Thalassiosirales, 2% from Thalassiophysales, 40% from Naviculales, 30% from Bacillariales, 5% to from Cymbellales, 3% from Eunotiales and 1% from Achanthales. Navicula and Nitzschia were the dominant genera with 9 and 7 species respectively, followed by the genus Pinnularia with 5 species.

Table 2. Relative abundance (%) of occurrence of diatoms in *Kuttanadu* paddy fields in different regions, seasons and crop stages (R1-Upper *Kuttanadu*, R2-Lower *Kuttanadu* and R3-*Kayal* Lands; S1-Pre-monsoon (*Puncha*) crop and S2-Monsoon (*Virippu*) crop; G1-Seedling-growth-stage and G2-Mature-growth-stage)

No	Diatom species	R1				R2				R3				Abundance (%)
		S1		S2		S1		S2		S1		S2		
		G1	G2	G1	G2	G1	G2	G1	G2	G1	G2	G1	G2	
	Order: Melosirales													
1	*Melosira varians* Agardh	1	1	4	6	1	1	6	4	1	1	3	3	26.67
2	*Melosira italica* (Ehrenberg) Kutzing	1	1	5	4	1	3	2	1	6	5	1	1	25.83
	Order: Thalassiosirales													
3	*Cyclotella meneghiniana* Kutzing	2	2	2	5	1	2	4	5	2	4	6	6	34.17
	Order: Fragilariales													
4	*Fragilaria capucina* Desmazieres	2	2	4	4	1	1	4	4	1	1	5	5	28.33
5	*Staurosira elliptica* (Schumann) Williams & Round	1	1	5	5	1	1	2	2	3	5	1	4	25.83
6	*Synedra ulna* (Nitzsch) Ehrenberg		3	3	3	1	2	4	4			2	2	20.00
7	*Synedra tabulata* (C.Agardh) Kutzing		1	2	4	1	1	5	5	3	1	4	5	26.67
	Order: Thalassiophysales													
8	*Amphora pediculus* (Kutzing) Grunow ex. A. Schmidt		1	3	4		1	4	4	1	2	4	3	22.50
	Order: Naviculales													
9	*Navicula capitatoradiata* Germain	1	1	1	1	1	1	6	7	1	1	1	1	19.17
10	*Navicula cuspidata* (Kutzing) Kutzing	1	1	1	1	1	1	1	1	1	1	1	1	10.00
11	*Navicula ranomafenensis* Kutzing	1	1	1	4	1	1	2	2	1	1	2	4	17.50
12	*Navicula reinhardtii* (Grnow) Grunow	1	1	7	6	1	1	3	3	1	1	5	2	26.67
13	*Navicula salinarum* Grunow	1	1	3	4			3	5	1	1	1	1	17.50
14	*Navicula pupula* Kutzing	1	1	1	1		1	4	3			6	4	18.33
15	*Navicula viridula* (Kutzing) Ehrenberg	1	1	4	5		6					5	3	20.83
16	*Navicula viridis* (Nitzsch) Kutzing	1	1	1	1		5					4	4	14.17
17	*Navicula lanceolata* Ehrenberg	2	1	1	1			4	5			7	8	24.17
18	*Capartogramma crucicula* Grunow (R.Ross)	1		1	1	2	2	2	2		3	3	2	15.83
19	*Craticula cuspidata* (Kutzing) D.G.Mann	1	1	1	1			3	5			3	3	15.00
20	*Craticula halophila* (Kutzing) D.G.Mann	1	1	3	1	1	1	4	4	1				14.17
21	*Frustulia vulgaris* (Thwaites) De Toni	2	1	5	5	2	1	6	6	3	3	6	6	38.33
22	*Nedium affine* (Ehrn) Pfitzer	1		4	4	1	1	4	3	1	1	5	5	25.00
23	Diadesmus confervaceae (Kutzing) DG Mann	2	1	5	5	1	1	6	6	1	1	8	5	35.00
24	*Sellaphora ulna* (Nitzsch) Ehrenberg	1		3	5	2	2	3	4	2	2	4	1	24.17
	Order: Bacillariales													
25	*Pinnularia acrosphaeria* W Smith	2	3	4	4	1	2	2	2	1	1	3	3	23.33
26	*Pinnularia biceps* W. Gregory	3	2	2	4	1	1	4	4			2	2	20.83
27	*Pinnularia brauni* (Grunow) Cleve	2	2	5	5	1	1	5	6			6	6	32.50
28	*Pinnularia gibba* Ehrenberg	2	1	4	4	1	1	4	2	1	2	2	1	20.83
29	*Pinnularia viridiformis* Krammer	2	1	3	3	1	1	2	2	1	1	2	2	17.50
30	*Nitzschia reversa* W Smith	3	3	4	6	1	1	6	7	1	2	7	8	40.83
31	*Nitzschia sigmoidea* (Nitzsch) W. Smith	1	1	1	4	1	1	4	5	1	1	6	7	27.50
32	*Nitzschia clausii* Hantzsch	1	2	8	6	1	1	6	7	1	1	7	8	40.83
33	*Nitzschia filiformis* (W Smith) Vanurk	1	1	3	5	1	2	2	1	1	1	1	1	16.67
34	*Nitzschia linearis* W Smith	1	1	1	1	1	1	1	1	1	1	1	3	11.67
35	*Nitzschia nana* Grunow	2	2	5	4	1	1	7	7	2	1	4	6	35.00
36	*Nitzschia obtusa* Rabenhorst	2	2	1	1	1	1	2	2	2	2	5	5	21.67

N o	Diatom species	R1				R2				R3				Abundance (%)
		S1		S2		S1		S2		S1		S2		
		G 1	G 2	G 1	G 2	G 1	G 2	G 1	G 2	G 1	G 2	G 1	G 2	
	Order: Cymbellales													
37	*Cymbella kolbei* Hustedt	1	1	1	1	1	1	1	1	1	1	1	1	10.00
38	*Gomphonema acuminatum*Ehr		1	1	1	1	1	1	1		1	2	2	10.00
	Order: Eunotiales													
39	*Eunotia bilunaris* (Ehrenberg) Mills	2	2	3	4	2	1	1	1	1	2	2	2	19.17
	Order: Achnanthales													
40	*Achnanthidium minutissimum*(Kutzing) Czarnecki					1	1	1	1					3.33
	Total No. of species	35	36	39	39	33	35	40	38	29	30	38	38	

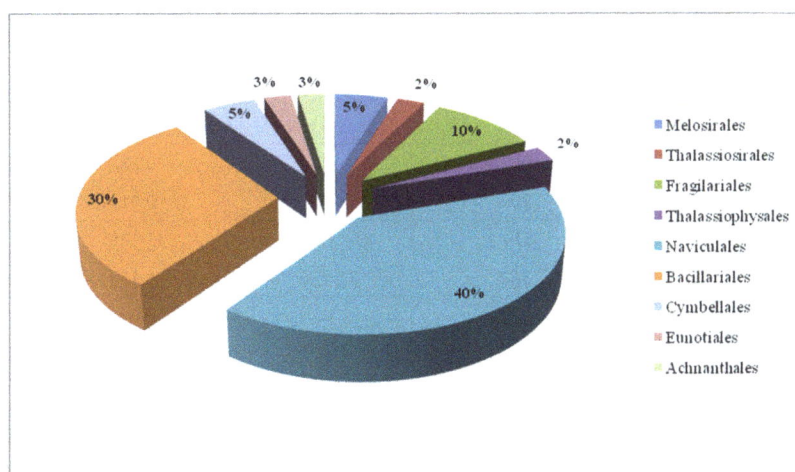

Figure 4. Relative abundance of various orders of Diatoms in *Kuttanadu*

3.1 Seasonal Variations in the Number of Species and Total Isolates

The highest number of species of diatoms (40 species) and the highest number of total isolates (355) were observed at panicle-growth-stage in the *Puncha* cultivation season of *Kayal* lands, and the lowest number of species (25 species) and the lowest total isolates (69) was also found at the panicle-growth-stage of in the '*Virippu*' cultivation of Kayal lands.

The most abundant species in Upper *Kuttanadu* and Lower *Kuttanadu* was found to be *Pinnularia brauni*, but that of '*Kayal*' lands was *Nedium affine*. In the '*Puncha*' season *Pinnularia braunii* was found to be the most abundant species in *Kuttanadu* but that in the '*Virippu*' season was *Frustulia vulgaris*.

3.2 Ecological Characteristics of the diatom community in Kuttanadu

The ecological characteristics included in the present study were relative abundance, species richness, species evenness and diversity index. The relative abundance of all the 40 algal species calculated from 120 soil samples belonging to the three different regions of *Kuttanadu* was found out; *Nitzchia clausii* (74.17%), showed highest relative abundance followed by *Frustulia vulgaris* (72.5%) and *Nedium affine* (70%). The least abundant algae in the whole *Kuttanadu* were *Staurosira construens* (20.83%). The most abundant species in 'Upper *Kuttanadu*' and 'Lower *Kuttanadu*' was found to be *Pinnularia brauni* with relative abundances 85% and 82.5% respectively and that in '*Kayal*' lands was *Nedium affine* with relative abundance 82.5%.

Species richness and Diversity Index of diatoms (Table 3) varied from 0.884-1.158 in different regions of *Kuttanadu* during the two crop seasons and crop stages. The species richness was found maximum (1.158) in 'Lower *Kuttanadu*', during '*Virippu*' season at panicle-growth-stage. Minimum species richness (0.884) was noticed in '*Kayal*' lands during '*Puncha*' cultivation at seedling stage. The species diversity index of diatoms in *Kuttandu* was found varying from 1.06 - 1.17 in different regions of *Kuttanadu* during the two crop seasons and crop stages. The diversity index was found maximum (1.17) in 'Lower *Kuttanadu*', during '*Virippu*' season at seedling stage. Minimum diversity index (1.06) was noticed in *'Kayal'* lands during *'Puncha'* cultivation season at seedling-growth-stage of the crop. The species evenness showed very slight variation from 0.315 - 0.317 in different regions of *Kuttanadu* during the two crop seasons and crop-growth-stages.

Table 3. Species richness and diversity index of diatoms in different seasons and vegetations (UK – upper Kuttanadu; LK – lower Kuttanadu; KL – Kayal lands; R1-Upper *Kuttanadu*, R2-Lower *Kuttanadu* and R3-*Kayal* Lands; S1-Pre-monsoon (*Puncha*) crop and S2-Monsoon (*Virippu*) crop; G1-Seedling-growth-stage and G2-Mature-growth-stage)

Region	Season	Growth Stage	No. of Species	Total Isolates	Sps rich	Div.Ind
UK	S1	G1	35	51	1.066	1.12
		G2	36	50	1.097	1.13
	S2	G1	39	116	1.188	1.16
		G2	39	134	1.188	1.16
LK	S1	G1	33	37	1.006	1.10
		G2	35	43	1.066	1.12
	S2	G1	40	142	1.219	1.17
		G2	38	135	1.158	1.15
KL	S1	G1	29	44	0.884	1.06
		G2	30	51	0.914	1.07
	S2	G1	38	138	1.158	1.15
		G2	38	136	1.158	1.15

Sps. Rich – species richness; Div.Ind – Diversity Index.

3.3 Physico-Chemical Characteristics of the Soils as per Ray et al. (2014)

Variations in pH, total organic carbon (C), total/ Kjeldal nitrogen (N), plant available phosphorus (P), potassium (K), calcium (Ca) and magnesium (Mg) studied in relation to crop seasons and different stages of paddy growth in the three different regions of *Kuttanadu* are given in Table 4.

Table 4. Average soil chemical characteristics of wetland paddy fields of *Kuttanadu* (Ray et al., 2014)

No	Location	Crop season	Growth stage	Average p^H	Average TN (%)	Average OC (%)	Average available nutrients (kg/ha)			
							P	K	Ca	Mg
1	UK	*Puncha*	*Seedling*	3.94	0.429	3. 07	10.86	998.6	3858.8	788.8
2	UK	*Puncha*	*Panicle*	3.88	0.392	4.06	18.26	656.9	2378.6	512.2
3	UK	*Virippu*	*Seedling*	3.40	0.289	0.94	129.00	298.7	441.4	177.7
4	UK	*Virippu*	*Panicle*	3.65	0.459	2.63	110.00	496.5	758.2	205.5
5	LK	*Puncha*	*Seedling*	4.00	0.413	3.97	7.14	855.0	1842.4	785.8
6	LK	*Puncha*	*Panicle*	4.29	0.374	3.54	25.34	389.2	2841.7	840.2
7	LK	*Virippu*	*Seedling*	3.74	0.370	1.68	109.00	316.7	1078.0	321.9
8	LK	*Virippu*	*Panicle*	3.70	0.520	2.16	95.00	408.9	1437.4	377.4
9	KL	*Puncha*	*Seedling*	4.03	0.390	3.28	9.60	391.2	1249.6	748.5
10	KL	*Puncha*	*Panicle*	4.61	0.435	3.70	9.46	396.0	2331.8	992.1
11	KL	*Virippu*	*Seedling*	4.05	0.400	2.49	129.00	440.9	1466.6	407.3
12	KL	*Virippu*	*Panicle*	4.23	0.431	2.68	58.00	408.3	1618.7	399.9

TN – Total nitrogen; OC – Total organic carbon.

3.4 SEM Analyses of Soil Samples for Diatoms

SEM analyses of the *Kuttanadu* soil samples revealed the presence of seven species of diatoms (Figure 5). Randomly selected six soil samples from the three regions of Kuttanadu were analyzed for diatoms. *Cyclotella meneghiniana* Kutzing, and *Achnanthidium atomus* (Hustedt) Monnier, Lange- Bertalot and Ector were reported from 'Upper *Kuttanadu*' samples. *Aulocoseira granulata* (Ehrenberg) Simonsen, and *Diadesmisconfervaceae*

Kutzing and *Eunotiasp* were reported from 'Lower *Kuttanadu*' soil samples. *Fragilariformavirescens* Ralfs (Williams) and Rounds and *Pinnularia sp.* were observed from 'Kayal' land samples.

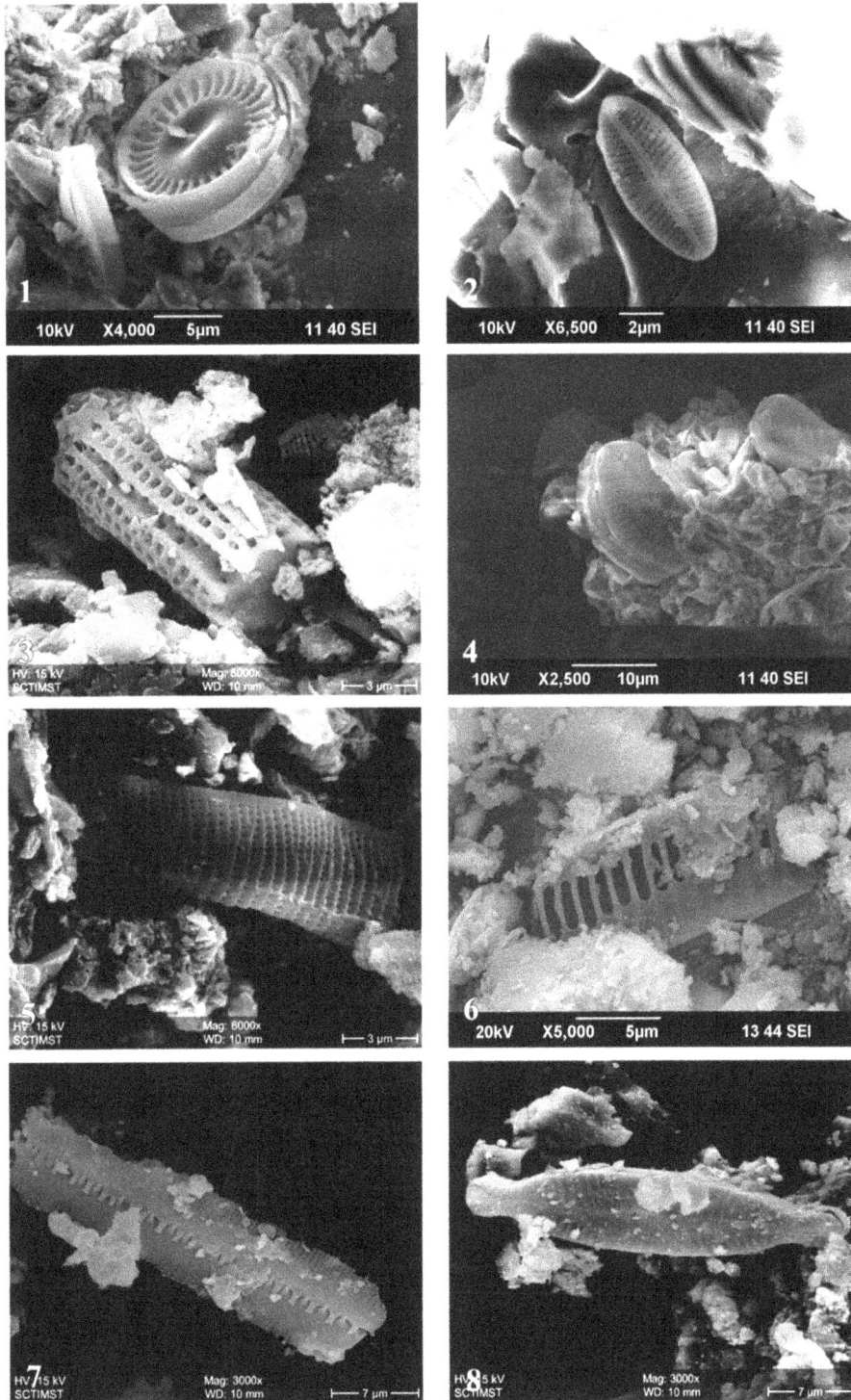

Figure 5. SEM pictures of Diatoms 1 - 8

1. *Cyclotella meneghiniana* Kutzing, 2. *Achnanthidium atomus* (Hustedt) Monnier, Lange -Bertalot and Ector, 3. *Aulocoseira granulata* (Ehrenberg) Simonsen, 4. *Diadesmis confervaceae* Kutzing, 5. *Eunotia macroglossa* R. C. Furey, R. L. Lowe and J. R. Johansen, 6. *Pinnularia* – showing Chambered striae, 7. Fragilariforma virescens Ralfs (Williams) and Rounds 8. Pinnularia *sp.*

4. Discussion

Soil is a complex heterogeneous environment (El-Gamal, 2008). Human influences affect its physical, chemical, and biological characteristics. In general, high diatom diversity is used as an indicator to anthropogenic disturbance and environmental contaminations (Vacht, Puusepp, Koff, & Reitalu, 2014). Among algae, diatom communities are considered as indicators of both organic and anthropogenic pollutions. In the present study, diatom diversity was reported to be high in almost all regions, seasons and growth stages, which indicated that the disturbed nature of this agro-ecosystem. In the current investigation, the highest number of diatom species, total isolates, species richness and diversity index were observed in the Lower *Kuttanadu* paddy soils in the seedling stage (early monsoon or pre-monsoon) of *Virippu* cultivation, which gradually decreased towards panicle-growth-stage (the monsoon season).

Comparatively lower diatom diversity is observed in non-monsoon crop season (*Puncha* cultivation) than in the monsoon crop (*Virippu*), which showed the influence of climate in determining the diatom density of *Kuttanadu* wetland paddy soils. Ecological conditions have prominent role in determining the algal diversity (Lin, Chou & Wu, 2013). Redekar and Wagh (2000), Pareek et al. (2011), Muralidhar and Murthy (2014), and Jayabhaye (2010) noticed that diatom diversity reaches maximum during monsoon or pre-monsoon months and minimum during summer. Contradictory to these findings, Muruganantham et al. (2012), observed maximum diversity of diatoms in summer.

According to Metting (1981), Neustupa and Skaloud (2008) and Lames –Da-Silva, Branco & Necchi- Junior, (2010) light play a significant role in the growth and differentiation of algae. In the present study, highest diatom diversity and abundance was observed in seedling stage of second cultivation (*Virippu*), when light is abundant due to less canopy of the crop plant. Roger and Reynaud (1982) also observed that diatom community increases in paddy fields during seedling stage of the crop. Lin et al. (2013), Fujita and Ohtsuka (2005), also supported this view that diatom cell density reaches the highest number just after transplanting of rice and thereafter decreases.

In the current investigation, the Diatom species richness and diversity index were found to be correlated (Pearson's correlation coefficient) to soil parameters such as pH, N, P, K, Ca and Mg of different regions of *Kuttanadu* at different crop-growth-stages and crop-seasons. A variety of physical and biological factors influence the distribution and diversity of diatoms. Comerton and Houghton (1978) observed that inorganic nutrients stimulates or suppress algal growth in the soil. Nutrient concentration is the most important factor that influences the distribution and abundance of Diatom species (Broady, 1979; Mohan, Shukla, Patil, Shetye, & Kerkar, 2011). In the present study, highest Diatom diversity was observed in 'Lower *Kuttanadu*' soil. Soil factors, vegetation and topography affect the algal biomass (Patrick, Groffman, Driscoll, Fahy, & Siccama, 2001), especially diatoms (Venkitachalapathy & Karthikeyan, 2013). Soil of 'Lower *Kuttanadu*' region is clay loams characterized by a fairly high amount of nitrogen. Diatom diversity, community behaviour and composition are influenced by nitrates (Muralidhar & Murthy, 2014). But in the present study, nitrates in Kuttanadu soils are not found to be correlated to any diversity parameters of Diatoms there. On the other hand, species richness and diversity index showed a significant positive correlation to phosphorus. This is in agreement with Comerton and Houghton (1978) that phosphorus increases the algal count, especially that of Diatoms in natural environment.

In the present study, species richness and diversity index showed significant negative correlation with pH. This is in agreement with the observations of Muralidhar and Murthy (2014) that Diatoms are sensitive to pH and salinity. Lukesova (2001) observed that Diatoms prefer neutral to alkaline pH. But, Diatoms showed high diversity in *Kuttanadu* soils, where pH ranges from 3.4 to 4.6, which is in agreement with the observations of Muralidhar and Murthy (2014) that pH has no role in the periodicity and density of Diatoms in soils.

5. Conclusion

Overall, the present analysis of the ecological characteristics of Diatoms of *Kuttanadu* wetland paddy fields in relation to soil regions, crop-seasons and paddy-growth-stages appears quite useful to the exploration of the economic potentials of many individual species of Diatoms. Since many of the diatoms observed in this investigation are found quite specific to soils or seasons or crop growth stages, further investigations on the environment factors contributing to limited occurrence of them in specific locations alone, would enable identification of them as bio-indicators of diverse environmental conditions.

Acknowledgements

Authors wish to acknowledge facilities provided by Laboratory of the Sophisticated Test and Instrumentation Centre (STIC), University of Science and Technology, Cochin for SEM analyses of diatoms in soils.

References

Bray, R. H., & Kurtz, L. T. (1945). Determination of total, organic, and available forms of phosphorus in soils. *Soil Science, 59*, 39-45. http://dx.doi.org/10.1097/00010694-194501000-00006

Broady, P. A. (1979). Quantitative studies on the terrestrial algae of Signy Island, South Orkney Islands. *British Antartic Survey Bulletin, 47*, 31-41.

Comerton, M., & Houghton, J. A. (1978). The effects of fertilizers on the algal flora of peat. *Proceedings of the Royal Irish Academy, 8*, 233-245.

Demirbas, A., & Demirbas, M. F. (2011). Importance of algae oil as a source of biodiesel. *Energy Conservation and Management, 52*, 163-170. http://dx.doi.org/10.1016/j.enconman.2010.06.055

Dey, H. S, Tayung, K., & Bastia, A. K. (2010). Occurence of nitrogen fixing Cyanobacteria in local rice fields of Orissa, India. *Ecoprint, 17*, 77-85.

El-Gamal, A. D., Nady, A., Ghanem, E., Eisha, Y. E. l- Ayouty., & Shehatha, E. F. (2008). Studies on soil algal flora in Kafrel- sheikh Governnorate Egypt. *Egyptian Journal of Phycology, 9*, 1-23.

Fujitha, Y., & Ohtsuka, T. (2005). Diatoms from paddy fields in northern Laos. *Diatom, 21*, 71-89.

Gandhi, H. P. (1956). A Preliminary account of the soil diatom flora of Kolhapur. *The Journal of Indian Botanical Society, 35*, 402-408.

Guiry, M. D., & Guiry, G. M. (2012). *Worldwide electronic publication, National University of Ireland, Galway.* Retrieved from http://www.algaebase.org

Jackson, M. L. (1973). *Soil Chemical analysis* (pp. 487-498). Prentice Hall of India, Private limited, NewDelhi, India.

Jayabhaye,U. M. (2010). Studies on Phytoplankton diversity in Sawana Dam, Maharashtra India, Shodh Samiksha aur Mulyankan. *International Research Journal, 11*, 11-12.

Jena, M., Ratha, S. K., & Adhikary, S. P. (2006). Diatoms (Bacillariophyceae) from Orissa state and Neighbouring regions, India. *Algae, 21*(4), 377-392. http://dx.doi.org/10.4490/ALGAE.2006.21.4.377

Kesic, K., Tuney, I., Zerben, D., Guden, M., & Sukatar, A. (2013). Morphological and molecular identification of pinnate diatoms isolated from Urla, Izmil, Coast of the Aegean Sea. *Turkish Journal of Biology, 37*, 530-537. http://dx.doi.org/10.3906/biy-1205-40

Lames–Da-Silva, N. M., Branco, L. H. Z., & Necchi- Junior, O. (2010). Corticolous green algae from tropical forest remnants in the north west region of Sao Paulo State, Brazil. *Revista Brasil Botany, 33*, 215-226.

Lin, C., Chou, T., & Wu, J. (2013). Biodiversity of soil algae in the farmlands of mid-Taiwan. *Biological Studies, 54*, 4. http://dx.doi.org/10.1186/1999-3110-54-41

Lukesova, A. (2001). Soil Algae in Brown coal and Lignite Post mining areas in Central Europe (Czech Republic and Germany). *Restoration Ecology, 9*, 341-350. http://dx.doi.org/10.1046/j.1526-100X.2001.94002.x

Mazumder, A., Govili., Ghosh, A. K., & Ravindra, R. (2012).Significant research on diatom in Antartica Lake during last decade. *Journal of Algal Biomass Utilization, 3*(4), 74-79.

McGinnis, K. M., Dempster, T. A., & Sommerfeld, M. R. (1997). Characterization of the growth and lipid content of the diatom *Chaetoceros muelleri*. *Journal of Applied Phycology, 9*(1), 19-24. http://dx.doi.org/10.1023/A:1007972214462

Metting, B. (1981). The Systematics and Ecology of Soil Algae. *The Botanical Review, 147*(2), 195-312. http://dx.doi.org/10.1007/BF02868854

Mohan, R. R., Shukla, S. K.., Patil, S. M., Shetye, S. S., & Kerkar, K. K. (2011). Diatoms from surface sediments of Enderby basin of Indian Sector of Southern Ocean. *Journal Geological Society of India, 78*, 36-44. http://dx.doi.org/10.1007/s12594-011-0065-9

Mohan, S. K. (1987). Bacillariophyceae of two tropical South Indian lakes of Hyderabad. *Botanical Bulletin, Academia Sinica, 28*, 13-24.

Muruganantham, P., Gopalakrishnan, T., Chandrasekharan, R., & Jayachandran, S. (2012). Seasonal variations and diversity of Planktonic diatoms of Muthupet and Aarukattuthura, South east coast of India. *Advances in Applied Science and Research, 3*(2), 919-929.

Nautiyal, P., & Verma, J. (2009). Taxonomic Richness and diversity of the epilithic diatom flora of the two biogeographic regions of Indian sub-continent. *Bulletin of the National Institute of Ecology, 19*, 1-4.

Neustupa, J., & Skaloud, P. (2008). Diversity of sub–aerial algae and Cyanobacteria on tree bark in tropical mountain habitats. *Biologia, 63,* 806-812. http://dx.doi.org/10.2478/s11756-008-0102-3

Pandey, U. C., & Pandey, D. C. (1980). Diatom flora of Allahabad (India)-1. *Proc. Indian Natn. Sci. Acad. B46, 3,* 350-355.

Pareek, R., Singh, G. P., & Singh, R. (2011). Some fresh water diatoms of Galta Kund, Jaipur,India. *Journal of Soil Science and Environmental Management, 2*(4), 110-116.

Patrick, J. B., Groffman, P. M., Driscoll, C. T., Fahy, T. J., & Siccama, T. G. (2001). Plant Soil Microbial interactions in a northern hard wood forest. *Ecology, 82*(4), 965-978.

Pillai, V. K., Ponniah, A. G., Vincent, D., & David Raj, I. (1983). Acidity in Vembanadu Lake causes fish mortality, Marine Fisheries Information Series, T & E Series No 53.

Ray, J. G., Dhanya, V., & Binoy, T. T. (2014). Globally unique *Kuttanadu* wetland paddy soil of South India: Soil fertility in relation to seasons and different stages of crop. *International journal of agriculture photon, 125,* 296-304.

Redekar, P. D., & Wagh, A. B. (2000). Planktonic diatoms of the Zuari estuary, Goa (West Coast of India). *Seaweed Research Utilization, 22*(1 & 2), 107-112.

Roger, P. A., & Reynaud, P. A. (1982). Free living blue green algae in tropical soils. In Y. Dommergues & H. Diem (eds) Microbiology of tropical soils and plant productivity. *Martinus Nijhoff Publisher La Hague,* 147- 168. http://dx.doi.org/10.1007/978-94-009-7529-3_5

Sarode, P. T., & Kamat, N. D. (1984). *Fresh water diatoms of Maharashtra* (p. 338).

Serieyssol, K., Chatelard, S., & Cubizolle, H. (2010). Diatom fossils in mires: a protocol for extraction, preparation and analysis in palaeoenvironmental studies, *Mires and Peat, 7*(12), 1-11.

Shannon, C. E., & Wiener, W. (1949). *The Mathematical Theory of Communication.* University of Illinois, press Urbana.

Souffreau, C., Vanormelingen, P., Verleyen, E., Sabbe, K., & Vyverman, W. (2010). Tolerance of benthic diatoms from temperate aquatic and terrestrial habitats to experimental desiccation and temperature stress. *Phycologia, 49*(4), 309-324. http://dx.doi.org/10.2216/09-30.1

Sudhikumar, A. V., Mathew, M. J., Sunish, E., & Sebastian, P. A. (2005). Seasonal variation in spider abundance in *Kuttanadu* rice agroecosystem, Kerala, India (Araneae). *European Arachnology, Acta zoologica bulgarica, l*(1), 181-190.

Taylor, J. C., Archibald, C. G. M., & Harding, W. R. (2007). *An illustrated guide to some common diatom species from South Africa* (WRC Report TT 282/07).

Vacht, P., Puusepp, L., Koff, T., & Reitalu, T. (2014). Variability of riparian soil diatom communities and their potential as indicators of anthropogenic disturbances. *Estonian Journal of Ecology, 63*(3), 168-184. http://dx.doi.org/10.3176/eco.2014.3.04

Van de Vijver, & Beyens, L. (1998). A preliminary study on the soil diatom assemblages from Ile de la Possession (Crozet, Subantartica). *Polar Biology, 25,* 721-729.

Venkatachalapathy, R., & Karthikeyan , P. (2013). Benthic diatoms in river influenced by Urban pollution, Bhavani region, Cauvery river, South India. *International Journal of Innovative technology and exploring engineering, 2*(3), 206-210.

Venkataraman, K. (2005). Coastal and marine biodiversity of India. *Indian Journal of Marine Sciences, 34*(1), 57-75.

Vijayan, D., & Ray, J. G. (2015b). Ecology and diversity of cyanobacteria in Kuttanadu. Paddy wetlands, Kerala, India. *American Journal of Plant Sciences, 6,* 2924-2938. http://dx.doi.org/10.4236/ajps.2015.618288

Vijayan, D., & Ray, J. G. (2015a). Green algae of a unique tropical wetland, Kuttanadu, Kerala, India, in relation to soil regions, seasons and paddy growth stages. *International Journal of Science, Environment and Technology, 4*(3), 770-803.

Walkley, A., & Black, I. A. (1934). An examination of Degtjareff Method for determining soil organic matter and a proposed modification of the chronic acid titration method. *Soil Science, 37,* 29-37. http://dx.doi.org/10.1097/00010694-193401000-00003

Whittaker, R. H. (1977). Evolution of species diversity in land community. In M. K. Heeht, W. C. Stee & B. Wallace (Eds.), *Evolutionary Biology.* Plenum, New York. http://dx.doi.org/10.1007/978-1-4615-6953-4_1

Wanga, X. W., Jun-Rong Lianga, Chun-Shan Luoa, Chang-Ping Chena, b, & Ya-Hui Gao (2014). Biomass, total lipid production, and fatty acid composition of the marine diatom *Chaetoceros muelleri* in response to different CO2 levels. *Bioresource Technology, 161,* 124-130. http://dx.doi.org/10.1016/j.biortech.2014.03.012

Xue, J., Ying-Fang Niu, Tan Huang, Wei-Dong Yang, Jie-Sheng Liu, & Hong-Ye, Li (2015) Genetic improvement of the microalga *Phaeodactylum tricornutum* for boosting neutral lipid accumulation. *Metabolic Engineering, 27,* 1-9. http://dx.doi.org/10.1016/j.ymben.2014.10.002

11

Response of Potato Varieties to Potassium Levels in Hamelmalo Area, Eritrea

Daniel Zeru Zelelew[1] & Biniam Mesfin Ghebreslassie[1,2]

[1] Department of Horticulture, Hamelmalo Agricultural College, Eritrea

[2] Deprtment of Horticulture, Jomo Kenyatta University of Agriculture and Technology, Kenya

Correspondence: Biniam Mesfin Ghebreslassie, Department of Horticulture, Hamelmalo Agricultural College, Eritrea; Deprtment of Horticulture, Jomo Kenyatta University of Agriculture and Technology, Kenya. E-mail: bm95913@yahoo.com

Abstract

Poor soil fertility and lack of high yielding certified varieties are of the major potato production tribulations in Eritrea. Top soils are continually removed due to water run-off and thus soil fertility and productivity has declined as a result. An experiment was designed to assess the response of potato varieties to different levels of potassium application at Hamelmalo Agricultural College, Eritrea. Three varieties (Ajiba, Zafira and Picasso) and five potassium levels (0, 75, 150, 225 and 300 kg K_2O/ha) along with all possible interactions were used. Experimental design following factorial Randomized Complete Block Design (RCBD) in three replications was employed. Data was collected on yield and tuber quality parameters. The result of the study indicated that there were significant variations in the performances of varieties in terms of yield and quality parameters in which Ajiba was found to be more responsive and high yielding. Tuber number, tuber diameter, tuber weight per plant, total yield, total soluble solids, specific gravity and tuber moisture content showed significant differences due to the application of potassium. As a result, the highest tuber weight (1.14 kg/plant) and yield (49.38 tones/ha) were recorded from Ajiba treated with 300 kg K_2O/ha. The result further revealed that there is a promising profit return by investing more on potassium application upto 300 kg K_2O/ha. It is, thus, recommended that potassium fertilizers should be introduced to optimize productivity in Hamelmalo area, Eritrea.

Keywords: Eritrea, potassium fertilizer, potato, varieties, tuber yield and quality

1. Introduction

Potato (*Solanum tuberosum* L.) is the fourth major food crop, next to wheat, rice and maize (Rana, 2008). It is highly recommended food security crop that can safe guard low-income countries from the risks posed by rising international food prices (FAO, 2009). It is a source of both food and income in many of the densely populated highlands of Sub-Saharan Africa. Taking into consideration the prospect for growth in the market for fresh potatoes and the current international market conditions characterized by high prices of cereals, potato can be taken as a good benchmark for rural development in sub-Saharan Africa (Gildemacher et al., 2009). In the last few years potato is becoming one of the priority crops in the highland and midland of Eritrea more particularly in Zoba Maekel and Zoba Debub (Biniam, Githiri, Tadesse, & Kasili, 2014). It is widely grown by small-scale farmers, contributing to food security as a direct food source and cash crop (MoA, 2010), with low input and low output practice, in an estimated total area of 2, 000 ha (Biniam et al., 2014). Land holding and yield of potato varies among farmers and sites but it was estimated to be 12 tones/ha on average (MoA, 2010) which is very low as compared with international and regional standards.

Potato is a short duration, high yielding and exhaustive crop. Balanced use of nutrients is essential for sustainable productivity of crops. In many potato producing areas nitrogen (N) and phosphorus (P) fertilizers are being used while potassium (K) application is ignored which causes serious decrease in the status of potassium in soils of potato growing areas (Pervez, Ayyub, Shabeen, & Noor, 2013). This is particularly true to Eritrean case due to the assumption that the soil is developed from K rich parent material and contains sufficient amount of K to support crop growth. However, this assumption is based on the work done before forty-seven years by Murphy (1968), which indicated that the available K content of most Ethiopian (including Eritrean) soils is high. The most commonly used fertilizer types currently in Eritrea are Di-ammonium Phosphate (DAP), urea and farmyard manure (Biniam et al., 2014). Potassium has a crucial role in higher productivity of potato tubers because it plays an important role in photosynthesis, regulation of opening and closing of stomata, favours high energy status which helps in timely and appropriate nutrient translocation and water uptake in plants (Bergmann, 1992). Besides, adequate supply of K can help to reduce internal blackening and mechanical damage, and has been associated with increased stress tolerance (FAO, 2009). Potassium is an essential nutrient for all plants and has a major effect upon yield and quality of potatoes as well as the general health and vigor of the crop (Abd El-Latif, Osman, Abdullah, & Abdelkader, 2011). Abay and Sheleme (2011) found that the highest tuber yield (53.33 t/ha) which is 11.4% yield advantage over the control from application of 280 kg K/ha.

Potato seeds used in Eritrea consist of a mixture of many cultivars that were originally imported and seed tubers that have been saved from these over many generations by local farmers (MoA, 2010). Reports of the Ministry of Agriculture, the State of Eritrea (2012) had indicated that the high cost and shortage of inputs especially seeds; fungicides and fertilizers have severely limited their application. Potatoes grow in Eritrea both under rain fed and irrigated condition often in small parcels of lands with limited resources. To improve soil fertility, farmers apply manures and fertilizers only when they are available and affordable (Biniam et al., 2014). The dose, time and methods of application are so different among the farmers from different area of production, the authors added. Several recent research activities to improve and optimize productivity and quality of potato through the application of K have been reported in other countries and promising results have been obtained and recommended by the authors (Abay & Sheleme, 2011; Abubaker, AbuRayyan, Amre, Alzubil, & Haidi, 2011; Noor, 2010; Wassie, 2009). However, no similar research was conducted on the response of potato varieties to K levels in Eritrea. An experiment was, therefore, designed to carry a field experiment at Hamelmalo area. Promising results were obtained and their application will immensely boost up potato productivity in Eritrea.

2. Materials and Methods

The research project was carried out in the Research Farm, of Hamelmalo Agricultural College; Hamelmalo, Eritrea in 2014. It is located 12 km North of Keren at 38°27'42" East of longitude and 15°52'21" North of the latitude. It is situated at an altitude of 1285meters above the sea level. The climate of the area is semi-arid with 429.1 mm annual rain fall for the growing season. Prior to planting of the potato crop, nine soil samples from different areas of the experimental blocks were taken with the help of an auger at 0-30 cm depth. Composite and representative soil sample was prepared thorough mixing of samples taken. It was air dried and then passed through 2 mm sieve, the soil sample was analyzed and its physico-chemical properties are shown in Table 1.

The field was thoroughly ploughed two times and harrowed once to a fine texture before clearing and leveling. Farmyard manure (FYM) of 20 t/ha was applied as recommended by (Rana, 2008). Urea (N) and DAP (P_2O_5) were applied at 225 and 135 kg/ha, respectively as per the recommendation of Hochmuth and Hanlon (2000). All doses of FYM and P and half dose of N were applied uniformly to all the treatments during field preparation. The remaining half dose of N was applied 30 days after planting. The experiment was laid out in factorial arrangement Randomized Complete Block Design (RCBD) with three replications. The treatments consisted of three varieties of potato (Ajiba, Zafira and Picasso) and five K levels (0, 75, 150, 225 and 300).

Table 1. Physico-Chemical properties of the experimental field soil at 30 cm depth

Chemical analysis		Chemical analysis	
Sand %	61.9	N%	0.03
Clay %	13.2	P (ppm)	1.07
Silt %	24.9	Ca^{++} (meq/100g)	16
Soil texture Class	Sandy loam	Mg^{++} (meq/100g)	5
pH	8.01	K^+ (meq/100g)	0.09
EC (ms/cm)	0.13	Na^+ (meq/100g)	0.29
OM%	0.11	CEC (meq/100g soil)	22.8

Full doses of potassium (K_2O) were applied through band method at the time of planting as per the treatments to each experimental plot (3x3m). Healthy pre-sprouted potato seed tubers were planted at a spacing of 75 x 30 cm on 15th September, 2014. Irrigation was applied immediately after planting and subsequent irrigations were given as per the requirement of the crop. To avoid any over lapping of inputs, plots were 80 cm apart from each other. All other cultural practices were done uniformly. Ajiba and Zafira were harvested on 25th December 2014 whereas Picasso was harvested on 15th of January 2015. Data was collected on number of tubers per plant, tuber diameter, tubers weight per plant and tuber yield per hectare at the time of harvest. Specific gravity was determined from the raw tubers by adopting weight-in-air/weight-in-water method as prescribed by CIP (2006). Total Soluble Solids (TSS) was determined using Refract meter. Tuber moisture content was determined using an oven and balance method as described by Kabir and Lemaga (2003) and CIP (2006). The data obtained were subjected to statistical analysis using the analysis of variance by GENSTAT software (4th ed) and IBM SPSS statistical package version 20 at 5% level of significance (95% confidence limit) for the analysis of variance.

3. Results and Discussions

3.1 Yield and Yield Components

All yield and yield components studied in the present investigation were significantly influenced by the applied levels of K and varieties. However, only tuber number per plant was found to be significantly affected by the interaction treatments of K and varieties.

3.1.1 Tuber Number per Plant

Number of tubers per plant had shown gradual and significant (p<0.001) increase within creasing K levels (Table 2). The highest tuber number per plant (9.08) was obtained from the application of 300 kg K_2O/ha while the lowest (6.92) was obtained from control. In a similar study Adhikary and Karki (2006) and Wibowo, Wijaya, Sumartono, and Pawelzik (2014) found that addition of K_2SO_4 fertilizer increased numbers of tubers produced. This could be due to the significant role of K on photosynthesis, favors high energy status which helps the crop for timely and appropriate nutrients translocation and water absorption by roots resulting in more availability of photo synthates to produce more number of tubers per plant (Bergmann, 1992). There was significant (p<0.001) difference among potato varieties as well in their production of tubers per plant. The highest number of tubers per plant (11.11) was produced by Ajiba followed by Zafira (8.58). This is ascribed to the existence of differences among genotypes in their adaptability to the specific environment and nutrient use efficiency (Wassie, 2009). The interaction effect of K level and potato varieties on number of tubers per plant was statistically significant. Ajiba responded more to K up to 150 kg K_2O/ha. Zafira showed slow increase in tuber number produced in response to K in the range of 0 to 225 kg K_2O/ha but showed higher response as K level increased beyond 225 kg K_2O/ha. Picasso showed steady response and number of tubers produced increased gradually with every increase of K levels (Table 2). Both Zafira and Picasso produced their higher number of tubers 10.57and 5.61, respectively when treated with 300 kg K_2O/ha (Figure 1). All the varieties produced minimum number of tubers per plant at zero application (control). The results speculates that application of more than 150 kg K_2O/ha for Ajiba could be excessive dose that caused reduction in tuber production while it is too low for Zafira and Picasso, as K requirement of potato variety varies (Rana, 2008). Similar results were reported by Ali, Costa, Abedin, Sayed, and Basak (2009) for sweet potato variety.

3.1.2 Tuber Diameter

The results shown in Table 2 revealed that K level increased tuber diameter significantly (p<0.001). The highest mean value (5.34 cm) was obtained from application of 225 kg K_2O/ha and the smallest tuber diameter (4.77 cm) was recorded from control. Potassium requirement of potatoes increased within creasing tuber size as the functions of K is related with translocation of carbohydrates from source (leaves) to sink (tuber) resulting in increased tuber size (Adhikary & Karki, 2006). The fact that the experimental field was poor in K content (Table 1) has led to better response of the crop to K treatment. This is in accordance with the findings of Bansal and Trehan (2011). The authors concluded that K application increases the size of tubers, especially, in low to medium soil types. Tindall and Westermann (1994) also noted that insufficient K can result in reduced yields and produced smaller-sized tubers. The varieties had significant variations (p<0.001) with regard to their response. Picasso which produced larger sized tubers (5.31 cm) was found to be superior to the other varieties studied. Table 2 indicates that Ajiba which had highest tuber number produced smallest tuber size (4.84 cm). Similarly, Abong, Okoth, Imungi and Kabira (2010) found that tuber diameter varied significantly among potato cultivars. This could be due to the peculiar genetic characteristics of the varieties: Ajiba, Zafira and Picasso produce large, large oval to long-oval, and very large to large with red eyes tubers, respectively (NIVAP, 2011). However, the varieties did not show any significant difference in response to K on tuber diameter (Figure 1).

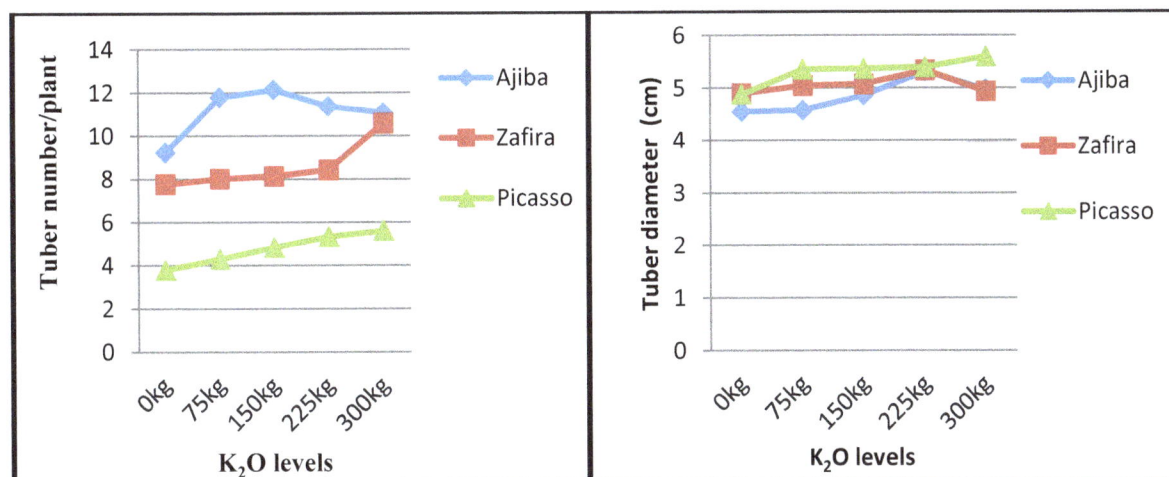

Figure 1. Interaction effects of K and potato varieties on number of tubers per plant (left) and tuber diameter (right)

3.1.3 Tuber Weight (kg/plant)

Increasing K application from 0 to 300 kg/ha had increased tuber weight per plant significantly (p<0.001). Maximum and minimum mean values of 0.91 and 0.54 kg per plant were produced from the application of 300 kg K_2O/ha and control, respectively .This is because, higher application of K facilitates the crop to have better nutrients and water absorption that improve growth and development of the crop and ultimately tuber weight (Bergmann, 1992). Adhikary and Karki (2006) found sharp response of potato to K_2O application on tuber weight. Table 2 further illustrates that tuber weight per plant had significantly (p<0.001) influenced by varietal treatment. Ajiba was superior of the three varieties evaluated followed by Zafira. In agreement with this finding, Abubaker et al. (2011) reported significant differences between varieties in their seasonal yield and tuber production per plant. Such variation is ascribed to the fact that nutrient usage of potato varies among cultivars and different environmental conditions (Peirce, 1987). Nevertheless, all the varieties exhibited steady increase in their tuber weight with the increase of K levels (Figure 2). Slight tuber weight decrease was noted at 150 kg K_2O/ha for Ajiba as a result of sudden site specific late blight occurrence few days before the crop maturity.

3.1.4 Total Yield (t/ha)

Results in Table 2 revealed that potato tuber yield increased with increasing K levels and had significant (p<0.001) differences to all varieties. Highest tuber yield (40.25 t/ha) which is 39.20 % yield advantage over the control was obtained from the application of 300 kg K_2O/ha. Consistent to the current results, Abd El-Latif et al. (2011) reported that gradual and significant increase of total tuber yield as a result of increased K level. Highest yield was obtained from the application of 285 kg K_2O/ha, the authors added. This is attributed to the importance of K in carbohydrate formation and transformation (Van der Zaag, 1981). Moreover, according to Wassie (2009), appreciable increase in yield was noted in response to the application of K fertilizers, especially from soils with very low or below critical K level. The varieties investigated also showed significant (p<0.001) differences (Table 2). Ajiba produced highest tuber yield with 8.67 and 52.09 % yield advantage over Zafira and Picasso, respectively. Contradictory to the current finding Vaezzadeh and Naderidarbaghshahi (2012) noted that the effect of cultivar on tuber yield per unit area was not significant. The interaction between potassium and variety was not found to have any significant influence on the total yield of potato, although there was a trend of increase among the treatments. On contrary Wassie (2009) reported that potato varieties showed significant difference in response to K application. Maximum yield (49.38 t/ha) was produced from Ajiba treated with 300 kg/ha followed by Zafira (45.20 t/ha) (Figure 2).

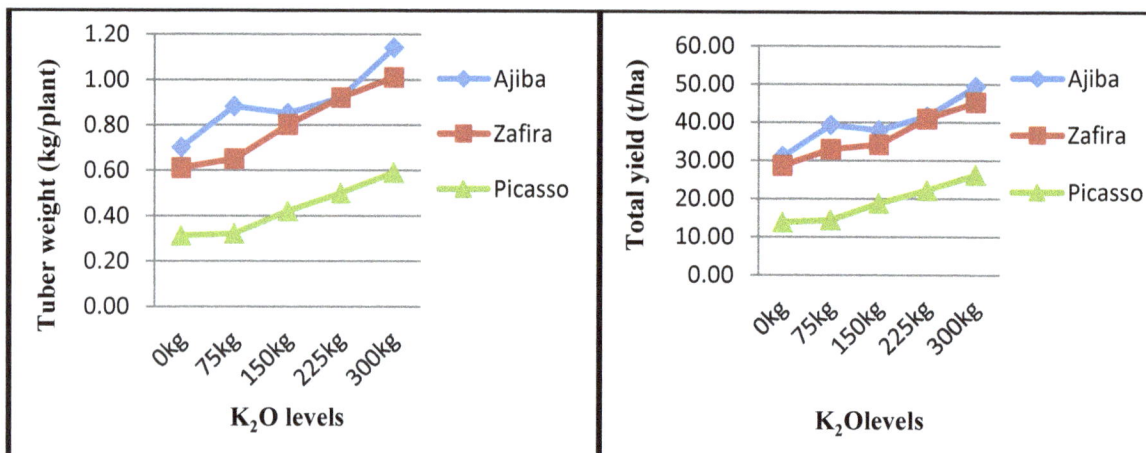

Figure 2. Interaction effects of K and varieties on tubers weight (left) and tuber yield (right)

Table 2. The effect of potassium and variety on yield and yield components of potato

K_2O (kg/ha)	Tuber number/plant	Tuber diameter (cm)	Tuber weight (kg/plant)	Yield (t/ha)
0	6.92	4.77	0.54	24.47
75	8.02	4.99	0.62	28.85
150	8.37	5.09	0.69	30.22
225	8.37	5.34	0.78	34.85
300	9.08	5.16	0.91	40.25
LSD	**0.876**	**0.214**	**0.127**	**3.463**
CV %	**11.10**	**4.40**	**18.60**	**11.30**
Varieties				
Ajiba	11.11	4.85	9.00	39.79
Zafira	8.58	5.05	0.80	36.34
Picasso	4.77	5.31	0.43	19.06
LSD	**0.678**	**0.166**	**0.098**	**2.683**
CV %	**11.10**	**4.40**	**18.60**	**11.30**

3.2 Tuber Quality Parameters

In the present study, variety and K application had significant effect in all the tuber quality aspect studied. However, interaction treatments of K by variety had significant influence only on TSS and moisture percentage without any influence on specific gravity.

3.2.1 Total Soluble Solids (TSS) (^{o}Brix)

The K level treatment exhibited significant differences (p<0.001) in the TSS level of varieties. TSS level tends to decrease with increasing potassium doses (Table 3). The highest TSS (5.58^{o}Brix) was obtained from control. Minimum TSS (5.00 ^{o}Brix) was recorded from plots treated with 300 kg K_2O/ha. K application has a potential to decrease reducing sugar content of potato tubers by activating starch synthesis (Marschner, 1995). The result is in agreement with the finding of Pervez et al. (2013) where the TSS level was significantly affected by the application of K. However, Abd El-Latif et al. (2011) reported that application of K fertilizers had no any significant effect on tuber TSS level. There was also significant differences (p<0.001) of TSS level among the varieties (Table 3). The highest (5.44^{o}Brix) was recorded from Ajiba followed by Zafira. This could be due to the inherited characteristic differences among the varieties used. This is fully supported by Baloch in (2010) who reported that potato varieties differ markedly in various plant characters including TSS content. Similarly, potato tuber quality characteristics are governed by both the variety of potato and the conditions under which it is grown (Van der Zaag, 1992). Interaction effects of K by variety were significant (p<0.001) for TSS. The TSS level decreased for Ajiba with increasing K_2O application, while it had increased for Zafira as the dose increases from 0 to 150 K_2O beyond which starts to decrease. On the other hand, Picasso had shown very low response of TSS to K_2O application. Accordingly, highest TSS (6.40°Brix) was produced by Ajiba in control (Figure 3). Previously, it was reported by Havlin, Beaton, Tisdale and Nelson (2005) that varieties or hybrids of crops response to production and quality varies depending on the inputs used.

3.2.2 Specific Gravity

Potassium application had significant (p<0.001) effect on specific gravity of potato tubers. As it is shown in Table 3 specific gravity had increased with increasing K levels from 0-150 kg K_2O/ha. Application of 150 kg K_2O/ha produced highest (1.11) specific gravity. Increasing K application above this level had decreased specific gravity. In agreement to this finding Berger, Potterton and Hobson (1961) reported that K fertilization generally reduces specific gravity if applied in excess. Specific gravity is closely related to the tuber starch content or total solids and thus dry matter content of tubers (Dampney, Wale, & Sinclair, 2011). Potassium application has the potential to decrease reducing sugar content of potato tubers by activating starch synthesis (Marschner, 1995) that has a tendency to increase tuber specific gravity. However, Noor (2010) reported that K application did not have significant effect on specific gravity. Based on varietal effect specific gravity was found to have significant (p<0.001) differences among the varieties evaluated (Table 3). Highest specific gravity (1.12) was recorded from Picasso. Zafira produced the lowest (1.05) mean value. In a similar study Abong et al. (2010) found that specific gravity and dry matter contents had significant difference among potato varieties. The interaction of K and varieties had no significant influence on specific gravity. All the varieties evaluated responded to K application in a similar way and their tuber specific gravity had increased with increasing K application only in the range of 0-150 K_2O kg/ha (Figure 3). Maximum and minimum specific gravity of these varieties was recorded at 150 K_2O kg/ha and control, respectively. This implies that application of 150 K_2O kg/ha was optimum for higher specific gravity. However, Havlin et al. (2005) reported that with a given environment, one variety may have greater response to applied nutrient than the other.

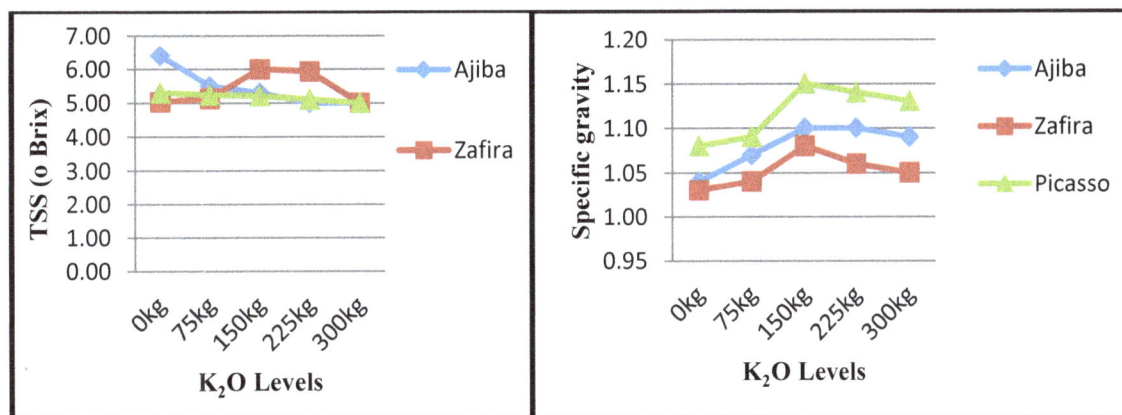

Figure 3. Interaction effects of K and potato varieties on tuber TSS (left) and specific gravity (right)

3.2.3 Tuber Moisture Content (%)

Statistically K treatment was found to have significant effect on tuber moisture content. Tuber moisture content had increased in the range of 150-300 kg K_2O/ha. Maximum tuber moisture content (83.83 %) was obtained from the application of 300 kg K_2O/ha. Potassium plays vital role in water use efficiency of crops, maintaining the turgidity of plant cells and amount of water in plant cells so as to increase the amount of water in plant organ like tubers (Abd El-Latif et al., 2011). The result of current experiment is supported by the findings of (Bergmann, 1992) whom reported that with the application of K, the water contents of the plasma volume were influenced increasing the water contents of the storage tissues and reducing the dry matter content. Based on varietal treatments, statistically tuber moisture content showed significant differences (Table 3). Ajiba exhibited low moisture content (82.40%), while Zafira (83.30%). Interaction of K and varieties had significant influences on moisture content of potato tuber. The varieties had great variation in their responses to K levels. Picasso shown decreased moisture content as the K level increased from 0-225 kg/ha after which starts to increase. On the other hand, Ajiba and Zafira showed reduction in moisture content up to the level of 150 kg K_2O/ha then increased as level increased. This could be attributed to genetic make-up of varieties in which one variety may have greater or lesser response to applied nutrient than the other (Havlin et al., 2005).

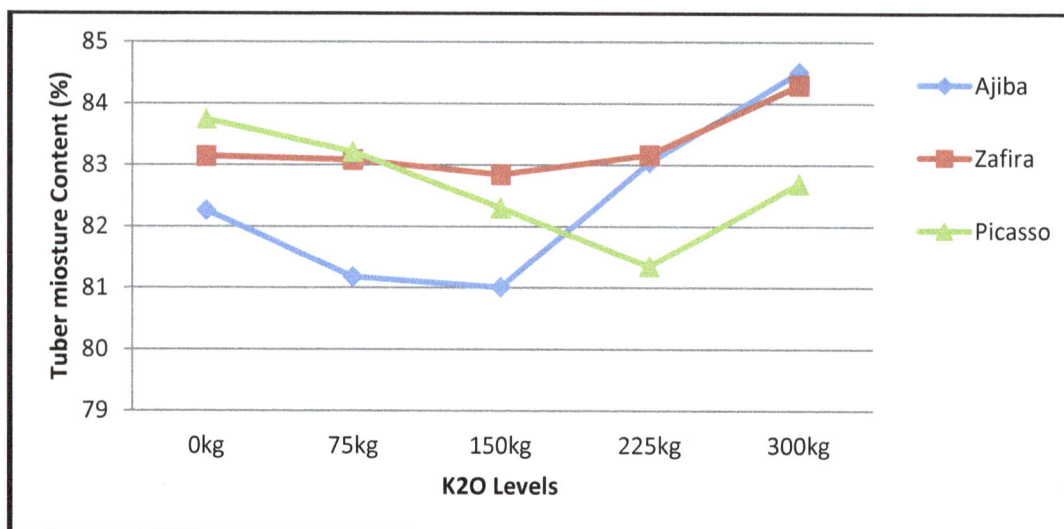

Figure 4. Interaction effects of K and potato varieties on Tuber moisture content (%)

3.3 Correlation among Yield and Quality Parameters

The correlation analysis shown that, tuber number per plant had positive and significant correlation with tuber weight per plant (r = 0.801) and total tuber yield (r = 0.855). This indicates that as number of tubers produced by a single plant increases, total tuber weight per plant and overall tuber yield can be increased. However, it had negative and significant correlations with tuber diameter (r = -0.416) and tuber specific gravity (r = -0.306). As the number of tubers per plant increased the tubers produced will be of smaller size and lower specific gravity. This is in line with the findings of Van der Zaag (1992) who stated that tuber size of the harvested product depends on the total tuber yield and the number of tubers per m². Total tuber yield had positive and significant correlation (0.462) with K application. This is because of the importance of K in carbohydrate formation and transformation and movement of starch from potato leaves to tubers (Van der Zaag, 1981). Tuber specific gravity and K were positively correlated (r = 0.362) while, TSS and K application were negatively correlated (r = -0.325). This is ascribed to the activating role of K to starch synthesis through decreasing the reducing sugar content of potato tubers (Marschner, 1995).

Table 3. The effect of potassium and variety on the tuber components of potato

K₂O (kg/ha)	TSS (° Brix))	Specific Gravity	Moisture content (%)
0	5.58	1.05	83.05
75	5.29	1.07	82.49
150	5.5	1.11	82.05
225	5.34	1.1	82.52
300	5.00	1.09	83.83
LSD (p=0.05)	0.183	0.027	0.866
CV %	3.50	2.60	1.10
Varieties			
Ajiba	5.44	1.08	82.40
Zafira	5.42	1.05	83.30
Picasso	5.17	1.12	82.66
LSD (p=0.05)	0.141	0.021	0.671
CV %	3.50	2.60	1.10

4. Conclusion

Potassium had significant effect on all yield and quality parameters of potato varieties studied. Application of K had consistent and positive effect on the yield of potato tubers. Maximum tuber yield (40.25 t/ha) was recorded from the application of 300 kg K_2O/ha. Potato varieties had significant differences in their performance, and also shown significant differential response to K application. Ajiba was superior and high yielding in the study area. The highest yield (49.38 t/ha) was obtained from the variety of Ajiba in response to 300 kg K_2O/ha. Despite K being one of the most vital macro nutrients of crops, it has never been introduced and used by potato growers in Eritrea. It is, therefore, recommended to carryout extensive experiments on screening for K need and dosage in major potato growing areas of the country and K levels above 300 kg/ha need to be tested. It is, also recommended that to improve yield and quality of crops produced, K fertilizers should be introduced to Hamelmalo and other similar areas of the country.

Acknowledgement

The authors wish to thank the Eritrean National Board for Higher Education in collaboration with Hamelmalo Agricultural College (HAC) for funding the research project.

References

Abay, A., & Sheleme, B. (2011). The influence of Potassium Fertilizer on the Production of Potato (*Solanum tuberosu L.*) at Kembata in Southern Ethiopia. *Journal of Biology, Agriculture and Healthcare, 1*(1), 1-12.

Abong, G. O., Okoth, M. W., Imungi, J. K., & Kabira, J. N. (2010). Evaluation of Selected Kenyan Potato Cultivars for Processing into Potato Crisps. *Agriculture and Biology Journal of North America, 1*(5), 886-893. http://dx.doi.org/10.5251/ abjna.2010.1.5.886.893

Abd El-latif, K. M., Osman E. A. M., Abdullah, R., & Abdelkader, N. (2011). Response of Potato Plants to Potassium Fertilizer Rates and Soil Moisture Deficit. *Advances in Applied Science Research, 2*(2), 388-397.

Abubaker, S., AbuRayyan, A., Amre A., Alzubil, Y., & Hadidi, N. (2011). Impact of Cultivar and Growing Season on Potato under Center Pivot Irrigation System. *World Journal of Agricultural Sciences, 7*(6), 718-721.

Adhikary, B. H., & Karki K. B. (2006). Effect of Potassium on Potato Tuber Production in Acid Soils of Malepatan, Pokhara. *Nepal Agricultural Research Journal, 7*, 42-48.

Ali, M. R, Costa, D. J., Abedin, M. J., Sayed, M. A., & Basak, N. C. (2009). Effect of Fertilizer and Variety on the Yield of Sweet Potato. *Bangladesh Journal of Agricultural Research, 34*(30), 473-480. http://dx.doi.org/10.3329/bjar.v34i3.3974

Baloch, M. N. (2010). Vegetable crops. In M. N. Malik. *Horticulture* (pp. 490-491). Biotec Book. New Delhi, India.

Bansal, S. K., & Trehan S. P. (2011). Effect of Potassium on Yield and Processing Quality Attributes of Potato. *Karnataka Journal of Agricultural Sciences, 24*(1), 48-54.

Berger, K. C., Potterton, P. E., & Hobson, E. I. (1961). Yield Quality and Phosphorus Uptake of Potatoes as Influenced by Placement and Composition of Potassium Fertilizers. *American Potato Journal, 38*, 272-285. http://dx.doi.org/10.1007/BF02862172

Bergmann, W. (1992). Nutritional disorders of plants development, visual and analytic diagnosis. *Gustav Fischer, Jena.*

Biniam, M. G., Githiri S. M., Tadesse, M., & Kasili, W. R. (2014). Diagnostic Survey on Potato Production Practices in Eritrea. *Journal of Agricultural and Biological Science, 9*(12), 444-453.

CIP. (2006). *Procedures for standard evaluation trials of advanced potato clones*. An International Cooperators' Guide.

Dampney, P., Wale, S., & Sinclair, A. (2011). *Review Potash Requirements of Potatoes*. Report of Agriculture & Horticulture Development Board 2011.

FAO. (2009). *Potato and food price inflation*. International Year of the Potato 2008.

Gildemacher, P. R., Kaguongo, W., Ortiz, O., Tesfaye, A., Woldegiorgis, G., Wagoire, W. W., … Struik P. C. (2009). Improving Potato Production in Kenya, Uganda and Ethiopia: A System Diagnosis. *Potato Research, 52*, 173-205. http://dx.doi.org/10.1007/s11540-009-9127-4

Havlin, J. L., Beaton, J. D., Tisdale, S. L., & Nelson W. L. (2005). Soil fertility and fertilizers: *An introduction to nutrient management* (7th ed.). Pearson Educational, Inc, NJ, USA.

Hochmuth, G. J., & Hanlon, E. A. (2000). *IFAS standardized fertilization recommendations for vegetable crops.* Circular 1152. University of Florida, Institute of Food and Agricultural Sciences, Gainesville, FL.

Kabir, J. N., & Lemaga, B. (2003). Potato processing; Quality evaluation procedures for research and food industry applications in East and Central Africa. *Kenya Agricultural Research Institute,* Nairobi, Kenya.

Marschner, H. (1995). *Mineral nutrition of higher plants* (2nd ed.). Academic press limited, San Diego, USA.

Ministry of Agriculture. (2010). *Vegetable crops research program.* Annual report of National Agricultural Research Institute. Halhale, Eritrea.

Murphy, H. F. (1968). *A report on the Fertility status and other data on some soils of Ethiopia,* Bulletin, College of Agriculture, Haile Sellasie I University, Experiment station, Dire Dawa, Ethiopia, No 44.

NIVAP. (2011). *Netherland catalogue of potato variety.* AC Den Haag, the Netherlands. Retrieved November 20, 2014, from www.nivap.nl

Noor, M. A. (2010). *Physiomorphological Determination of Potato Crop Regulated by Potassium Management.* (Doctoral thesis) submitted to Institute of Horticultural Sciences University of Agriculture, Faislabad, Pakistan.

Peirce, L. C. (1987). *Vegetables: Characteristics, Production and Marketing.* John Wiley and Sons, Inc.

Pervez, M. A., Ayyub, M. I. C. M., Shabeen, M. R., & Noor, M. A. (2013). Determination of Physiomorphological Characteristics of Potato Crop Regulated by Potassium Management. *Pakistan Journal of Agricultural Sciences, 50*(4), 611-615.

Rana, M. K. (2008). *Olericulture in India.* Kalyani Publishers, Ludiana; India.

Tindall, T., & Westermann, D. T. (1994). *Potassium fertility management of potatoes.* University of Idaho Potato School (Mimeo). Idaho State Univ., Pocatello, ID.

Vaezzadeh, M., & Naderidarbaghshahi, M. (2012). The Effect of Various Nitrogen Fertilizer Amounts on Yield and Nitrate Accumulation in Tubers of Two Potato Cultivars in Cold Regions of Isfahan (Iran). *International Journal of Agriculture and Crop Sciences, 4*(22), 1688-1691.

Van der Zaag, P. (1981). *Soil fertility requirements for potato production.* Technical information bulletin, December 1981, 14, CIP, Lima, Peru.

Van der Zaag, W. D. E. (1992). *Potatoes and their production in the Netherlands* (3rd ed.). Netherlands Potato Consultative Institute, CH the Hague, the Netherlands. Retrieved December 23, 2014, from www.potato.nl

Wassie, H. (2009). On Farm Verification of Potassium Fertilize Effect on the Yield of Irish Potato Grown on Acidic Soils of HagereSelam, Southern Ethiopia. *Ethiopian Journal of Natural Resources, 11*(2), 207-22.

Wibowo, C., Wijaya, K., Sumartono, G. H., & Pawelzik, E. (2014). Effect of Potassium Level on Quality traits of Indonesian Potato Tubers. *Asia Pacific Journal of Sustainable Agriculture Food and Energy, 2*(1), 11-16.

12

Compartmentalization of Metabolites and Enzymatic Mediation in Nutritive Cells of Cecidomyiidae Galls on *Piper Arboreum* Aubl. (Piperaceae)

Gracielle Pereira Bragança[1], Denis Coelho de Oliveira[2] & Rosy Mary dos Santos Isaias[1]

[1]Universidade Federal de Minas Gerais, Instituto de Ciências Biológicas, Departamento de Botânica, Laboratório de Anatomia Vegetal. Caixa postal 486.31270-901, Belo Horizonte, Minas Gerais, Brazil

[2]Universidade Federal de Uberlândia, Instituto de Biologia, Laboratório de Anatomia e Desenvolvimento Vegetal. Caixa postal 593, Uberlândia, Minas Gerais, Brazil

Correspondence: Rosy Mary dos Santos Isaias, Universidade Federal de Minas Gerais, Instituto de Ciências Biológicas, Departamento de Botânica, Laboratório de Anatomia Vegetal. Caixa postal 486.31270-901, Belo Horizonte, Minas Gerais, Brazil. E-mail: rosy@icb.ufmg.br

Abstract

Galling insects commonly change the chemical profile of their host plant tissues during gall induction and establishment. As a consequence, galls accumulate a wide range of metabolites in specialized cells, which may be organized in a nutritive tissue and in outer storage cells. The nutrients compartmentalized in nutritive cells may be directly assessed or metabolized via enzymatic mediation, while the gall outer cortex may accumulate secondary metabolites. These secondary metabolites may configure a specialized chemical barrier against the attack of natural enemies. Either the nutritive inner cells or the outer cortical cells, with their specific metabolic apparatus, should differentiate under the chemical constraints of each host plant-galling herbivore interaction. This premise is herein addressed by the investigation of the histochemical profile of the non-galled leaves and galls induced by Diptera: Cecidomyiidae on *Piper arboreum*. The spatial compartmentalization of the nutritive and defensive metabolites indicates the new functions assumed during the redifferentiation of the host plant cells. The enzymatic mediation of the primary metabolites by sucrose synthase and invertases favors the nutritive requirements of the galling Cecidomyiidae or the structural maintenance of the gall. The accumulation of secondary metabolites is restrict to the tissue layers not involved in nutrition, and may act in the chemical protection against predators or parasitoids. Current results systematically document metabolites compartmentalization, evidence the impairment of toxic compounds storage in cells surrounding the larval chamber, as well as, detect the redirection of nutritive substances to the site of the Cecidomyiidae feeding. The activity of sucrose synthase is restrict to the nutritive tissue in the galls on *Piper arboreum*, and reinforces previous detection of this enzyme mediation in carbohydrate metabolism in Cecidomyiidae galls.

Keywords: chemical profile, cecidomyiidae, galling insect diet, gall metabolism

1. Introduction

Galling insects alter the morphogenetical patterns of their host plant organs by inducing cell redifferentiation (sensu Lev-Yadun, 2003), division and growth (Oliveira & Isaias, 2010a; Isaias, Oliveira, & Carneiro, 2011; Isaias, Oliveira, Carneiro, & Kraus, 2014; Magalhães, Oliveira, Suzuki, & Isaias, 2014). The new morphogenetical patterns generate specialized cells and tissues at the gall site, whose chemical and functional features are distinct from those of the host organs (Oliveira, Carneiro, Magalhães, & Isaias, 2011; Castro, Oliveira, Moreira, Lemos-Filho, & Isaias, 2012). The gall metabolic requirements generate a sink of primary and secondary metabolites, which may compartmentalize in protective and/or nutritive tissues (Carneiro, Castro, & Isaias, 2014). The carbohydrates drained to the gall developmental site are responsible for plant cell machinery support (Oliveira, Christiano, Soares, & Isaias, 2006; Castro et al. 2012), and can also act as potential signaling molecules for cell division and growth (Koch, 2004; Wind, Smeenkens, & Hanson, 2010; Isaias, Oliveira, Moreira, Soares, & Carneiro, 2015). The signaling function in gall sites has also been attributed to reactive oxygen species (ROS) (Oliveira & Isaias 2010a; Oliveira, Moreira, & Isaias, 2014; Isaias et al., 2014). ROS accumulation demands phenolics scavenging (Isaias et al., 2015), which, in turn, influence IAA regulation (Bedetti, Modolo, & Isaias, 2014).

Proteins, lipids, reducing sugars or starch accumulate in the inner layers of the gall, i. e., the nutritive tissue, and can provide resources for the nutrition of the galling insect (Price, Waring, & Fernandes, 1986, 1987; Bronner, 1992). Nevertheless, in some systems, the availability of these metabolites to the galling insect or to the maintenance of cell metabolism depends on enzymatic activities (Bronner, 1992; Oliveira Magalhães, Carneiro, Alvim, & Isaias, 2010; Oliveira & Isaias, 2010b; Carneiro et al., 2014). The patterns of enzymatic activities depend on the feeding behavior of the galling insects (Bronner, 1992), and on the chemical profile of the host plants (Oliveira et al. 2010, 2014; Oliveira & Isaias, 2010b).

The enzymatic mediation of carbohydrates metabolism has been documented just for four galling herbivore-host plant systems in the Neotropics (Oliveira et al., 2010, 2011; Oliveira & Isaias, 2010b; Carneiro et al., 2014). These four systems involve gall inducing Cecidomyiidae and Psylloidea with their divergent peculiarities, and demonstrated the non-exclusiveness of the carbohydrates accumulation to Cecidomyiidae galls. Among the investigated carbohydrates, sucrose is synthesized in the cytosol from photosynthetically fixed carbon, starch reserves or lipid metabolism (Wind et al., 2010), and is transported via phloem to other plant parts. The main enzymes responsible for sucrose metabolism are sucrose synthase (SuSy) and invertases, which catalyse the conversion of sucrose into glucose and fructose (Koch, 2004). In general, SuSy activity is associated to sink tissues and starch accumulation, while invertases mediate cell respiration, tissue growth and development (Koch, 2004; Wind et al., 2010). In galls, the metabolism of these enzymes was also related to the formation of histochemical gradients and maintenance of the gall tissues in *Aspidosperma australe* (Oliveira & Isaias, 2010b). The sites of the galling herbivore's nutrition seems to be crucial for the redifferentiation of the nutritive cells, as documented for *Nothotrioza cattleiani* galls on *Psidium cattleianum* (Carneiro, Pacheco, & Isaias, 2015).

Outside the nutritive cells, Cecidomyiidae galls usually have sclerenchymatic cells, forming a mechanical protective layer, and an outer parenchymatic cortex. These outer cell layers usually accumulate secondary metabolites, such as alkaloids, flavonoids, phenolics and tannins (Nyman & Julkunen-Titto, 2000; Oliveira et al., 2006; Formiga, Soares, & Isaias, 2011; Isaias et al., 2014), and can protect the galling herbivores against the attack of parasitoids and predators, and the gall structure against cecidophagous (Price, Waring, & Fernandes, 1987).

Cell redifferentiation and metabolism in gall outer and inner tissue layers are herein revisited in a Cecidomyiidae - *Piper arboreum* system. We assume that primary and secondary metabolites accumulate in distinct tissue compartments, as expected, but some metabolic steps of this gall should be dependent on the host plant carbohydrates metabolism rather than on the Cecidomyiidae feeding mode. If this is true, some similarities between the compartmentalization and metabolism of Cecidomyiidae and Psylloidea galls should be found.

The intralaminar lenticular galls on *P. arboreum* have nutritive cells limiting the larval chamber, where carbohydrates should accumulate, as previously observed in other two Neotropical Cecidomyiidae galls (Oliveira et al., 2010, 2011). Similarly, the accumulation of cabohydrates in nutritive like cells has been documented in a Psyllloidea gall (Oliveira et al., 2011; Carneiro et al., 2014). Both gall inducers should not come into contact with non-palatable secondary metabolites, and therefore, enzymes mediation and a spatial compartmentalization of nutritive and defensive metabolites in gall developmental site are expected. Current analyses focus on the following questions: (I) does the accumulation of primary and secondary metabolites follow the expected compartmentalization? (II) Is there any disruption for metabolites accumulation in response to the galling stimuli of the Cecidomyiidae on *P. arboreum*? (III) Are there any similarities between the enzymatic mediation of Cecidomyiidae and Psylloidea galls? And (IV) should ROS signaling involve sugar mediation during gall development on the Cecidomyiidae intralaminar leaf galls on *P. arboreum*?

2. Methods

Samples of non galled leaves (NGL) and mature leaf galls (MG) induced by an unidentified species of Diptera: Cecidomyiidae on *Piper arboreum* were accompanied and collected from September 2013 to March 2014 at the ecological station of Universidade Federal de Minas Gerais in Belo Horizonte, Minas Gerais state, Brazil.

2.1 Histochemical Assays

Fresh samples (n ≥ 5) were free-hand sectioned and submitted to histochemical tests with the following reagents: saturated solution of sudan Red B in 70°GL ethanol during 5 min to detect lipids (Brundett, Kendrick, & Peterson, 1991); Fehling's reagent (Solution "A" - 7.9% copper sulfate, and solution "B" - 34.6% sodium potassium tartrate and 1% sodium hydroxide) heated to pre-boiling temperature for detecting reducing sugars

(Sass, 1951); Lugol's reagent (1% potassium iodine-iodide solution) during 5 min for starch detection (Johansen, 1940); 0.1% bromophenol blue in a saturated solution of magnesium chloride in ethanol during 15 min, and later washed in acetic acid and water, for the detection of proteins (Baker, 1958); 1% ferric chloride during 5 min, for phenolic compounds detection (Johansen, 1940); Dragendorff's reagent (Solution "A" - 12.5% bismuth nitrate in 25% acetic acid, and solution "B" 40% potassium iodide) during 5 min for the detection of alkaloids (Johansen, 1940); Wiesner's reagent (2% phloroglucinol in acidified solution) during 5 min for lignins detection (Johansen, 1940); fixation in 0.5% caffeine sodium benzoate in 90% butanol, followed by incubation in 1% p-dimethylaminocinnamaldehyde (DMACA) during 30 min for the detection of flavonoids (Feucht, Schmid, & Christ, 1986); 1% α-naftol and 1% dimethyl-p-phenylenediamine in phosphate buffer (pH 7.2) (NADI) during 30 min for the detection of terpenoids (David & Carde, 1964), and Lieberman-Buchard's reagent (concentrated solution of sulfuric acid and acetic acid, 1:1, v/v) during 1 min for the detection of triterpenes (Wagner, Bladt, & Zgainski, 1984). The sections were washed in water and photographed under an optical microscope (Zeiss Primo Star®) with a digital camera (Canon Power Shot A 630®). Blank sections were used for the comparison of results.

2.2 Enzymatic Activity

For the detection of the activity of acid phosphatase, the sections were incubated in 0.012% lead nitrate and 0.1M potassium sodium glicerophosphate in 0.5M acetate buffer (pH 4.5) for 24 hours, at room temperature. The sections were washed in distilled water, and incubated in 1% ammonium sulfate for 5 min. As a control, the samples were not submitted to potassium sodium glicerophosphate (Gomori, 1956). For the detection of phosphorylase activity, the sections were incubated for two hours in 1% glucose-1-phosphate in 0.1M acetate buffer (pH 6.0), at room temperature, and subsequently subjected to Lugol's reagent for 5 min. For control, the samples were not incubated in glucose-1-phosphate (Jensen, 1962). For observation of invertase activity, the sections were incubated for 3 hours at room temperature in a solution containing 0.024% tetrazolium blue (NBT), 0.014% phenazin methosulfate, 30U of glucose oxidase and 30 mM of sucrose, 0.38mM sodium phosphate buffer (pH 7.5). The control was subjected to the reaction media without sucrose (Zrenner, Salanoubat, Willmitzer, & Sonnewald, 1995; Doehlert & Felker 1987). For detection of sucrose synthase activity, the sections were fixed in 2% paraformaldehyde with 2% polyvinylpyrrolidone and 0.005M of dithiothreitol (pH 7.0) for 1 hour at 4°C. Later, they were incubated in a solution containing 5μL of 150 mM NADH, 5μl (1U) of phosphoglucomutase, 5μl of 3mM glucose 1,6-bisphosphate, 5μl (1U) of glucose-6-phosphate dehydrogenase, 5μl (1U) of UDPG pyrophosphorylase, 280μL of 0.07% aqueous solution of blue tetrazolium (NTB), 350 μl of buffer and 50μL substrate during 30 minutes. The buffer contained 10 mM MgCl2, 2 mM EDTA, 100 mM HEPES, 0.2% BSA and 2mM EGTA (pH 7.4). The substrate consisted of 15 mM UDP, 0.75 mM sucrose, 15 mM pyrophosphate. Two controls were used. In the first control, the glucose 1,6-bisphosphate and pyrophosphate were not added, and for the second control, the sucrose was suppressed (Wittich & Vreugdenhil 1998).

2.3 Histochemical Test for Reactive Oxygen Species (ROS)

ROS were detected by immersion of the sections in 0.5% 3.3'-diaminobenzidine (DAB) during 15 - 60min, in the dark (Rossetti & Bonatti 2001). The sections were washed in water and photographed under an optical microscope (Zeiss Primo Star®) with a digital camera (Canon Power Shot A 630®). Blank sections were used for the comparison of results.

3. Results

3.1 General Features

Leaf galls on *Piper arboreum* are intralaminar and lenticular. They project to both leaf surfaces, and has a uniseriate epidermis, an outer parenchymatic cortex, an inner cortex composed of sclerenchymatic cells, and a nutritive zone involving the larval chamber, which houses the galling Cecidomyiidae (Figure 1 A-C).

Figure 1. Leaf galls on *Piper arboreum* Aubl (Piperaceae)

A: Leaf with galls viewed by the adaxial surface. B: Hemisection of a mature gall evidencing the adaxial outer cortex, abaxial outer cortex, inner cortex, larval chamber, and nutritive tissue. C: Cecidomyiidae larva. ab, abaxial outer cortex; ad, adaxial outer cortex ic, inner cortex; lc, larval chamber; nt, nutritive tissue. Scale bars = 1cm (A), 0.5 mm (B), 2mm (C).

Table 1. Histochemical detection of metabolites and enzymes in non-galled leaves of *Piper arboreum* Aubl (Piperaceae)

Metabolites	Tissues of non-galled leaves						
	ADEP	ADHP	PP	SP	VB	ABHP	ABEP
Phenolic compounds	-	+	-	-	-	+	-
Flavonoids	-	+	-	-	-	+	-
Alkaloids	-	+	-	-	-	+	-
Triterpenes	-	+	-	-	-	+	-
Lignins	-	-	-	-	+	-	-
Lipids	-	-	-	-	-	+	-
Reducing sugar	-	+	-	-	-	+	-
Starch	-	-	+	+	-	-	-
Proteins	-	-	+	+	-	-	-
Invertase	-	+	-	-	-	+	-
Sucrose Sintase	-	-	-	-	+	-	-
Acid phosphatase	-	-	-	-	-	-	-
Phosphorylase	-	-	-	-	-	-	-
Reactive oxygen species	+	+	+	+	+	+	+

Note. ABEP = abaxial epidermis; ABHP = abaxial hypodermis; ADEP = adaxial epidermis; ADHP = adaxial hypodermis; PP = palisade parenchyma; SP = spongy parenchyma; VB = vascular bundles. (+) positive reaction, (-) negative reaction.

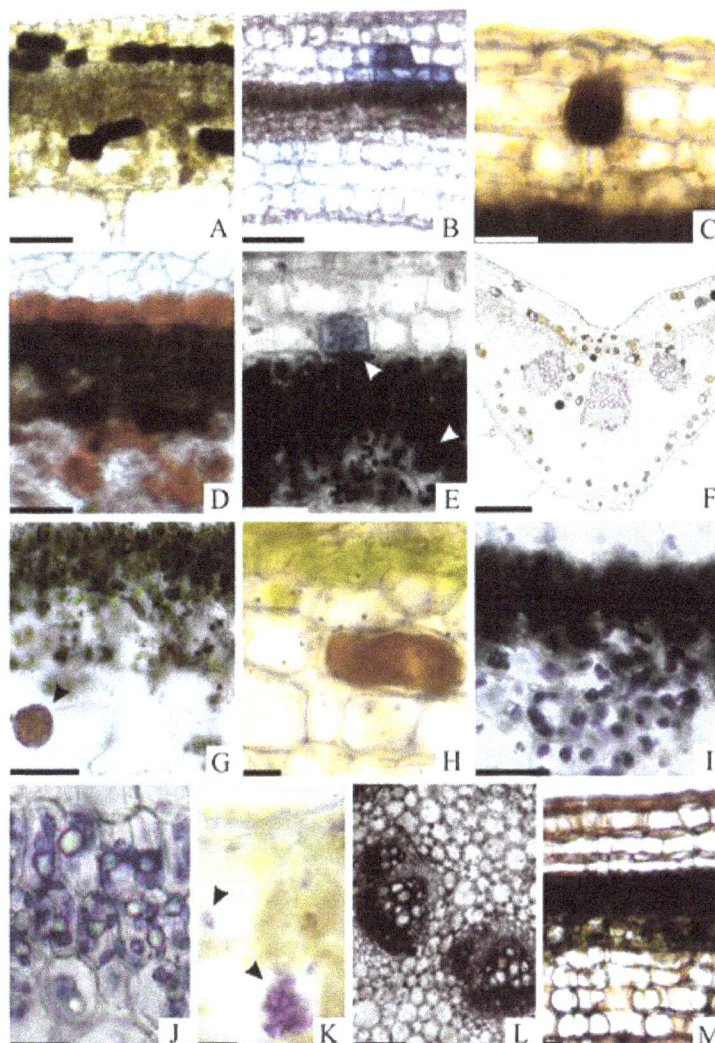

Figure 2. Histochemical tests in non-galled leaves of *Piper arboreum* Aubl (Piperaceae)

A: Reaction with ferric chloride evidencing the presence of phenolics in the hypodermis. B: DMACA revealing flavonoids in adaxial hypodermis. C: Alkaloids in the adaxial hypodermis. D: Reaction with Lieberman-Buchard's reagent showing triterpenes in the adaxial and abaxial hypodermis. E: NADI detecting terpenoids in adaxial hypodermis and idioblasts between the palisade and spongy parenchyma. F: Lignins in the cell walls of the xylem and pericyclic fibers. G: Sudan red demonstrating lipid droplets (arrows) in the abaxial hypodermis. H: Reducing sugars detected by Fehling`s reagent in the abaxial outer cortex. I: Lugol's reagent detecting starch in the palisade parenchyma and spongy parenchyma. J: Bromophenol blue indicating total proteins in the palisade parenchyma and spongy parenchyma. K: Activity of invertases in the adaxial hypodermis (arrow). L: Activity of sucrose synthase in the vascular bundles. M: Accumulation of ROS in the outer and inner gall cortices. Scale bars = 200µm (A-D, L), 50µm (E-I, K, M), 30µm (J).

3.2 Histochemical Profile of Non-Galled Leaves (NGL).

Phenolic compounds, flavonoids, alkaloids, and triterpenes were detected in the cells of hypodermis in black, dark blue, brown and red, respectively (Figure 2A-D, Table 1). Terpenoids were detected in blue in the cells of the hypodermis, and also in the idioblasts located between the palisade and spongy parenchymas (Figure 2E, Table 1). Lignins were evidenced in red in the cell walls of the xylem and of the pericyclic fibers (Figure 2F, Table 1). Lipids were detected as red droplets in the cells of the abaxial hypodermis, while reducing sugars, forming red precipitates, were observed in the adaxial and abaxial hypodermis (Figure 2G-H, Table 1). Starch grains and proteins were detected as dark blue grains or precipitates in the palisade and spongy parenchyma (Figure 2I-J, Table 1). The activity of invertases was evidenced as a dark blue precipitate in the hypoderm and parenchyma cells of the veins (Figure 2K, Table 1). The activity of sucrose synthase (SuSy) was detected as a purple precipitate in the vascular bundles (Figure2L, Table 1). The activity of phosphorylase and acid phosphatase was not observed. The ROS were detected in the epidermis and chlorophyllous parenchyma of the NGL (Figure 2M, Table 1).

3.3 Metabolites Compartmentalization in Cecidomyiidae Induced Galls on Piper Arboreum

The outer cortical cells of the mature galls (MG) developed from the hypodermis of the NGL, and accumulated phenolic compounds, flavonoids, alkaloids and terpenoids (Figure 3A-C, Table 2). Terpenoids occurred in idioblasts located between the palisade and spongy parenchyma of NGL, but they were not observed in the gall inner cortex (Figure3D, Table 2). Triterpenes were not evidenced in MG (Table 2). Lignins were detected in the cell walls of the xylem and of the pericyclic fibers both in NGL and MG. In MG, lignins were also observed in the cell walls of the sclerenchyma surrounding the nutritive tissue (Figure 3E, Table 2). Lipids were observed in the outer region, and also in the nutritive tissue of the MG (Figure 3F, Table 2). Reducing sugars were revealed in the cells of the adaxial and abaxial outer cortices (Figure 3G, Table 2). The detection of starch was more intense in the cells of the inner cortex and next to the larval chamber, increasing laterally towards the non-galled region. Starch was also detected in the lignified-walled cells (Figure 3H, Table 2). Proteins were detected in the nutritive tissue (Figure 3I, Table 2).

3.4 Enzymatic Activity

The activity of invertases was detected in the cells adjacent to the larval chamber, and formed a centrifugal gradient towards the non-galled region (Figure 3J, Table 2). The activity of SuSy was detected homogeneously throughout the nutritive tissue and vascular bundles (Figure 3K, Table 2). The activity of phosphorylase and acid phosphatase was not observed either in NGL or MG.

3.5 ROS Detection. In MG, the ROS accumulated in a centrifugal gradient, decreasing towards the outer cortical tissue layers (Figure 3L, Table 2).

Table 2. Histochemical detection of metabolites and enzymes in Cecidomyiidae galls on *Piper arboreum* Aubl (Piperaceae)

Metabolites	Tissues of mature galls						
	ADEP	AD	IC	NT	VB	AB	ABEP
Phenolic compounds	-	+	-	-	-	+	-
Flavonoids	-	+	-	-	-	+	-
Alkaloids	-	+	-	-	-	+	-
Triterpenes	-	-	-	-	-	-	-
Lignins	-	-	+	-	+	-	-
Lipids	-	+	-	+	-	+	-
Reducing sugar	-	+	-	-	-	+	-
Starch	-	-	+	+	-	-	-
Proteins	-	-	-	+	-	-	-
Invertase	-	-	-	+	-	-	-
Sucrose Sintase	-	-	-	+	+	-	-
Acid phosphatase	-	-	-	-	-	-	-
Phosphorylase	-	-	-	-	-	-	-
Reactive oxygen species	+	+	+	+	+	+	+

Note. AB = abaxial outer cortex; AD = adaxial outer cortex; ABEP = abaxial epidermis; ADEP = adaxial epidemis; IC = inner cortex; NT = nutritive tissue; VB = vascular bundles. (+) positive reaction, (-) negative reaction.

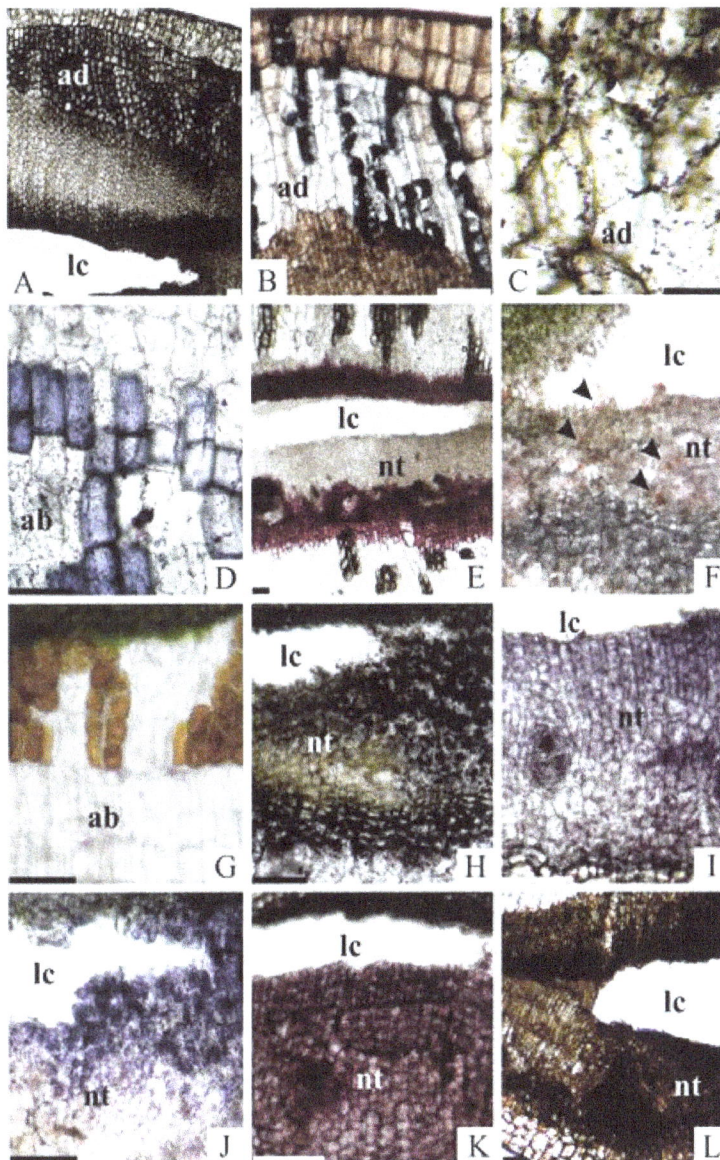

Figure 3. Histochemical tests in Cecidomyiidae galls on *Piper arboreum* Aubl (Piperaceae)

A: Ferric chloride evidencing phenolics in the outer cortex. B: DMACA revealing flavonoids in the outer cortex. C: Alkaloids detected by Draggendorff's reagent in the outer cortex (arrow). D: NADI detecting terpenoids in the outer cortex. E: Lignins in cell walls of the sclerenchymatic sheath. F: Sudan red demonstrating lipid droplets (arrows) in the nutritive tissue. G: Reducing sugars evidenced by Fehling`s reagent in the abaxial outer cortex. H: Lugol's reagent detecting starch in the nutritive tissue and lignified cells. I: Bromophenol blue indicating total proteins in the nutritive tissue. J: Activity of invertases in the nutritive tissue. K: Activity of sucrose synthase in the nutritive tissue. L: Accumulation of ROS in the outer and inner cortices of the gall. ab, abaxial outer cortex; ad, adaxial outer cortex; ic, inner cortex; lc, larval chamber; nt, nutritive tissue. Scale bars = 200μm (A-L), 50μm (F).

4. Discussion

4.1 Compartimentalization of Metabolites

Two distinct compartments with accumulation of primary and secondary metabolites were observed in the leaf galls on *P. arboreum*. Primary metabolites have been especially detected in the inner tissues, while secondary metabolites accumulated only in the outer cortical parenchyma, which corroborates the expected spatial functional division of gall tissues. The outer cells were converted from photosynthetic and respiratory compartments towards defensive tissues, while the inner cells assumed a specialized nutritional role in redifferentiated gall tissues.

The defensive compartment, i.e. the outer cortex of *P. arboreum* leaf gall, accumulated alkaloids, terpenes, and phenolics, which have been considered waste products of plant metabolism (Roberts & Wink, 1998). Nevertheless, they have been contemporarily evaluated as sources of nitrogen or energetic lipids, and ROS scavenging molecules (Isaias et al., 2015), all of them necessary for the maintenance both of the host plant and gall metabolism. In spite of their involvement in antioxidant mechanisms (Blokhina, Virolainen, & Fagerstedt, 2003; Detoni, Vasconcelos, Rust, Isaias, & Soares, 2011), alkaloids, terpenes, and phenolics may secondarily deter or discourage the attack of predators, due to their toxicity (Rhodes, 1994; Róstas, Maag, Ikegami, & Inbar, 2013). In galls, because of the high oxidative stress, the role of phenolics and flavonoid derivatives has been discussed as ROS dissipation, an efficient strategy to recover the redox-potential homeostasis (Isaias et al., 2015). Also, the phenolics are involved in IAA metabolism and consequently influence cell hypertrophy at gall site (Bedetti et al., 2014).

Even though the terpenes accumulated all over leaf mesophyll, they are detected exclusively in the outer compartment of the galls on *P. arboreum*, reinforcing the chemical protectiveness of the gall outer tissue layers. The impairment of the terpenic idioblasts differentiation in the nutritive tissue of the galls on *P. arboreum* should have favored the gall inducer, which did not come into contact with the toxic potential of the terpenes, and their anti-herbivore properties (Gershenzon, 1994). The strategy of disrupting the differentitation of terpenic idioblasts have been previously observed in the galling herbivores-*Lantana camara* systems (Moura, Isaias, & Soares, 2005; Moura, Isaias, & Soares, 2008).

4.2 Double Metabolites Accumulation in the Inner Compartment

As a host plant potentiality, lipid droplets were detected in the cells of the cortical parenchyma, originated from the NGL mesophyll, their intrinsic location. The lipids accumulated in the inner tissue layers may function as an energetic resource both for the galling Cecidomyiidae's nutrition and gall development. Even though attributed to Cynipidae (Bronner, 1992) and Lepidoptera galls (Vecchi, Menezes, Oliveira, Ferreira, & Isaias, 2013), lipidic droplets have been previously detected in some Cecidomyiidae galls of *Aspidosperma spruceanum* (Oliveira et al., 2010), *Copaifera langsdorffii* (Oliveira et al., 2011), and *Marcetia taxifolia* (Ferreira & Isaias, 2014). The accumulation of lipids in such galls has been related to the potential of the host plants for such accumulation (Oliveira et al., 2011; Ferreira & Isaias, 2014), as is true for *P. arboreum*.

The inner compartment of the galls on *P. arboreum*, i.e., the nutritive tissue, also accumulates proteins, similarly to the Cecidomyiidae galls on *Aspidosperma spruceanum* (Oliveira et al., 2010). Proteins are excellent nutritive resources for the galling herbivores, and may have accumulated as a cellular response to the increased oxidative and respiratory stresses established during the cecidogenetic process (Schönrogge, Harper, Lichtenstein, 2000). The increased level of proteins is followed by high levels of hexoses (Sturm & Tang, 1999), which are products of the activity of sucrose synthase and invertases (Roitsch & Gonzalez, 2004). The detection of invertases indicates the fast convertion of sucrose, and the activation of a mechanism of plant defense by increasing the synthesis of secondary metabolites (Wind et al., 2010; Sturm & Tang 1999). The double accumulation of nutritive compounds, such as lipids and proteins, in the nutritive tissue of a Cecidomyiidae gall is not the expected pattern, which should be carbohydrates storage (Bronner, 1992; Oliveira et al., 2010; Oliveira et al., 2011). This double stimuli is therefore a novelty for Cecidomyiidae galls in the Neotropics, and indicates a surplus for the galling herbivore nutrition.

4.3 Enzymatic Mediation of Carbohydrates Accumulation

The activity of sucrose synthase (SuSy) and invertases detected in the nutritive tissue of the galls on *P. arboreum* corroborated the metabolic similarity between Psylloidea and Cecidomyiidae galls. Both enzymes have been previously detected in galls induced by *Pseudophacopteron aspidospermii* (Malenovsky, Burckhardt, Queiroz, Isaias, & Oliveira, 2015) on *Aspidosperma australe* (Oliveira et al., 2010) and by a Cecidomyiidae on *A. spruceanum* (Oliveira & Isaias, 2010b). This enzymatic detection indicates a host plant metabolic requirement or potential rather than a dependence on the galling herbivore mode of feeding.

Moreover, current results demonstrate a common site for the activity of SuSy in the Cecidomyiidae galls on *P. arboreum*, and on *A. spruceanum* and *Copaifera langsdorffii*, which diverges in the *P. aspidospermii* galls on *A. australe*, where the activity of SuSy was restricted to the vascular bundles. Based on such comparison, we can conclude that the feeding sites of the galling herbivores does not determine the host plant cells metabolism, but may determine the sites of enzymes activity. The activity of SuSy is commonly responsible for the reversible cleavage of sucrose into fructose and UDP-glucose (Amor, Haigler, Johnson, Wainscott, & Delmer, 1995; Koch, 2004), but may be especially related to the synthesis of starch observed in the nutritive tissue of the galls on *P.*

arboreum. Furthermore, this enzyme provides substrate (UDP-glucose) for the formation of cell wall polysaccharides (Nolte & Koch, 1993; Salnikov, Grimson, Seagull, & Haigler, 2003), whose dynamics, together with the carbohydrates metabolism, have crucial roles in the development of the gall structure (Formiga et al., 2013, Oliveira et al., 2014, Carneiro et al., 2015), and consequently in the survival of the galling herbivore.

In terms of cell metabolism, the gall on *P. arboreum* functions as a new organ, in a strict intralaminar continuum with its host tissues, establishing a sink of photoassimilates, which culminate in the accumulation of metabolites involved in its own development, and in the feeding activity of its associated galling herbivore (Rehill & Schultz, 2003; Castro et al., 2012). The nutrients can be redirected and compartmentalized into the nutritive tissue by two pathways. The first one is the transport of sucrose, via phloem, from the non-galled portions of the host leaf towards the gall site, where it is promptly metabolized and converted into starch. The conversion of sugars into starch occurs in a SuSy dependent via, which produces UDP-glucose as demonstrated for tubers of potato (Baroja-Fernández et al., 2009). In the second pathway, the sucrose is irreversibly cleaved by the activity of invertases into glucose and fructose, which are used in cell respiration (Koch, 2004). The invertases are, indeed, the major sucrose-degrading enzyme in plants, as demonstrated in *Arabidopsis*, where their low levels affect plant growth (Barrat et al., 2009). In insect galls, the force of the sink towards the nutritive tissues seems to be maintained by the high cytological metabolism and the dynamics between the activity of SuSy and invertases (Oliveira et al., 2010), as can be inferred for *P. arboreum* galls.

4.4 ROS Accumulation and Reducing Sugars Suppression in Nutritive Cells

Due to the high activity of SuSy and invertases in the nutritive tissue, as well as the direct involvement of sugars in insect feeding, a gradient of carbohydrates should be expected (Bronner, 1992; Oliveira et al., 2014). However, the accumulation of reducing sugars is impaired in the nutritive tissue of the galls on *P. arboreum*, indicating that the sucrose is promptly metabolized to starch synthesis and monosaccharides used in cell respiration and gall growth. This supposed increase of cell respiration is corroborated by the accumulation of ROS in the nutritive tissue of the MG on *P. arboreum*. The relationship between the absence of reducing sugars and high accumulation of ROS in the inner tissues is theoretically proposed for Cecidomyiidae-induced galls. High levels of sugars are expected for Cecidomyiidae galls (Bronner, 1992), and should be crucial for disrupting the high oxidative stress detected by the presence of H_2O_2 in the inner compartment, as hypothesized by Isaias et al. (2015).

5. Conclusion

The galling Cecidomyiidae promotes the compartmentalization of specific secondary and primary metabolites of its host leaves, guaranteeing benefits to its own survivorship. The outer compartmentalization of the secondary metabolites maintained the pattern of the original host tissues of *P. arboreum* leaves, and may confer protection against natural enemies. The inner compartmentalization of primary metabolites mediated by enzymes reflects two important metabolic steps of gall development: (1) the sink of photoassimilates and the storage of starch mediated by SuSy activity, and (2) the accumulation of proteins and lipids in nutritive tissues, as well as a high respiratory metabolism mediated by invertases activity. The conversion of sucrose into monossacharides is a common metabolic step shared by the Cecidomyiidae and the Psylloidea galls. However such conversion occurs in distinct sites of enzymes activity, which is imposed by the different feeding habits of the two taxa of galling herbivores, and their target cells.

The intralaminar galls on *P. arboreum* have two functional inversions regarding the compartmentalization of metabolites. The outer compartment accumulates reducing sugars, expected to occur in the nutritive cells, and terpenoids, whose synthesis is disrupted in the inner gall tissues. We theoretically propose that a consequence of these new sites of accumulation is a deviation of functions, with reducing sugars taking part in cell respiration in the outer tissue layers, and mediating ROS metabolism all over gall cortex. The terpenoids, whose presence in the nutritive cells should intoxicate the galling larvae, and consequently impair gall development, may contribute to gall chemical defense against natural enemies in the outer cortical cells.

Acknowledgments

The authors thank Coordenação de Aperfeiçoamento de Pessoal de Nível Superior (CAPES), Conselho Nacional de Desenvolvimento Científico e Tecnológico (CNPq), and Fundação de Apoio à Pesquisa de Minas Gerais (FAPEMIG) for financial support.

References

Amor, Y., Haigler, C. H., Johnson, S., Wainscott, M., & Delmer, D. P. (1995). A Membrane-associated form of Sucrose Synthase and its Potential role in Synthesis of Cellulose and Callose in Plants. *Proceedings of the*

National Academy of Sciences, 92(20), 9353-9357. http://dx.doi.org/10.1073/pnas.92.20.9353

Baker, J. R. (1958). Note on the use of Bromophenol Blue for the Histochemical Recognition of Protein. *Quarterly Journal of Microscopical Science, 99,* 459-460.

Baroja-Fernández, E., Muñoz, F. J., Montero, M., Etxeberria, E., Sesma, M.T., ... Pozueta-Romero, J. (2009). Enhancing Sucrose Synthase activity in Transgenic Potato (*Solanum tuberosum* L) Tubers results in increased levels of Starch, ADPglucose and UDPglucose and total yield. *Plant Cell Physiology, 50*(9), 1651-1662. http://dx.doi.org/10.1093/pcp/pcp108

Barrat, D. H. P., Derbyshire, P., Findlay, K., Pike, M., Wellner, N., ... Smith, A. M. (2009). Normal Growth of *Arabidopsis* Requires Cytosolic Invertase but not Sucrose Synthase. *Proceedings of the National Academy of Sciences, 106*(31), 13124 -13129. http://dx.doi.org/10.1073/pnas.0900689106

Bedetti, C. S., Modolo, L.V., & Isaias, R. M. S (2014). The Role of Phenolics in the Control of Auxin in galls of *Piptadenia gonoacantha* (Mart) MacBr (Fabaceae: Mimosoideae). *Biochemical Systematics and Ecology, 55,* 53-59. http://dx.doi.org/10.1016/j.bse.2014.02.016

Blokhina, O., Virolainen, E., & Fagerstedt, K. V. (2003). Antioxidants, Oxidative damage and Oxygen Deprivation Stress: a Review. *Annals of Botany, 91,* 179-194. http://dx.doi.org/10.1093/aob/mcf118

Bronner, R. (1992). *The role of nutritive cells in the nutrition of cynipids and cecidomyiids.* In J. D Shorthouse & O. Rohfritsch, (Eds.), *Biology of insect-induced galls* (pp. 118-140). Oxford University Press, Oxford.

Brundett, M. C., Kendrick, B., & Peterson, C. A. (1991). Efficient Lipid Staining in Plant Material with Sudan Red 7B or Fluorol Yellow 088 in Polyethylene Glycol-glycerol. *Biotechnic & Histochemistry, 66*(3), 111-116. http://dx.doi.org/10.3109/10520299109110562

Carneiro, R. G. S., Castro, A. C., & Isaias, R. M. S. (2014). Unique Histochemical Gradients in Photosynthesis-deficient Plant Gall. *South African Journal of Botany, 92,* 94-104. http://dx.doi.org/10.1016/j.sajb.2014.02.011

Carneiro, R. G. S., Pacheco, P., & Isaias, R. M. S. (2015). Could the Extended Phenotype Extend to the Cellular and Subcellular Levels in Insect-induced Galls? *PLoS One,* http://dx.doi.org/101371/ journalpone012933.

Castro, A. C., Oliveira, D. C., Moreira, A. S. F. P., Lemos-Filho, J. P., & Isaias, R. M. S. (2012). Source-Sink Relationship and Photosynthesis in the Horn-shaped Gall and its Host Plant *Copaifera langsdorffii* Desf (Fabaceae). *South African Journal of Botany, 83,* 121-126. http://dx.doi.org/10.1016/j.sajb.2012.08.007

David, R. & Carde, J. (1964). Coloration différentielle des inclusions lipidiques et terpeniques des pseudophylles du Pin maritime au moyen du réactif Nadi. *Comptes rendus hebdomadaires des séances de l'Académie des sciences, 258,* 1338-1340.

Detoni, M. L., Vasconcelos, E. G., Rust, N. M., Isaias, R. M. S., & Soares, G. L. G. (2011). Seasonal Variation of Phenolic Content in Galled and Non-galled Tissues of *Calliandra brevipes* Benth (Fabaceae: Mimosoidae). *Acta Botanica Brasilica, 25*(3), 601-604. http://dx.doi.org/10.1590/S0102-33062011000300013

Doehlert, D. C. & Felker, F. C. (1987). Characterization and Distribution of Invertase Activity in Developing Maize (*Zea mays*) Kernels. *Physiology Plantarum, 70,* 51-57. http://dx.doi.org/10.1111/j.1399-3054.1987.tb08695.x

Ferreira, B. G. & Isaias, R. M. S. (2014). Floral-like Destiny Induced by a Galling Cecidomyiidae on the Axillary Buds of *Marcetia taxifolia* (Melastomataceae). *Flora, 209,* 391-400. http://dx.doi.org/10.1016/j.flora.2014.06.004

Feucht, W., Schmid, P. P. S., Christ, E. (1986). Distribution of Flavonols in Meristematic and Mature Tissues of *Prunus avium* Shoots. *Journal of Plant Physiology, 125,* 1-8. http://dx.doi.org/10.1016/S0176-1617(86)80237-1

Formiga, A. T., Oliveira, D. C., Ferreira, B. G., Magalhães, T. A., Castro, A. C., ... Isaias, R. M. S. (2013). The Role of Pectic composition of Cell Walls in the Determination of the New Shape-functional Design in Galls of *Baccharis reticularia* (Asteraceae). *Protoplasma, 250,* 899-908. http://dx.doi.org/10.1007/s00709-012-0473-8

Formiga, A. T., Soares, G. L. G., & Isaias, R. M. S. (2011). Responses of the Host Plant Tissues to Gall Induction in *Aspidosperma spruceanum* Müell Arg (Apocynaceae). *American Journal of plant Science, 2,* 823-834. http://dx.doi.org/10.4236/ajps.2011.26097

Gershenzon, J. (1994). Metabolic Costs of Terpenoid Accumulation in Higher Plants. *Journal of Chemical Ecology, 20*(6)1281-1328. http://dx.doi.org/10.1007/BF02059810.

Gomori, G. (1956). Histochemical Methods for Acid Phosphatase. *Journal of Histochemistry and Cytochemistry, 4*(5), 453-461. http://dx.doi.org/10.1177/4.5.453

Isaias, R. M. S, Oliveira, D. C. & Carneiro, R. G. S (2011). Role of *Euphalerus ostreoides* (Hemiptera: Psylloidea) in Manipulating Leaflet Ontogenesis of *Lonchocarpus muehlbergianus* (Fabaceae). *Botany, 89,* 581-592. http://dx.doi.org/10.1139/cjb-2013-0125

Isaias, R. M. S, Oliveira, D. C., Carneiro, R. G. S., & Kraus, J. E. (2014). Developmental anatomy of galls in the neotropics, arthropods stimuli versus host plant constraints. In G.W., Fernandes & J.C. Santos, (Eds.), *Neotropical Insect Galls* (pp. 15-34). Springer Netherlands.

Isaias, R. M. S., Oliveira, D. C., Moreira, A. S. F. P., Soares, G. L. G., & Carneiro, R. G. S. (2015). The Imbalance of Redox Homeostasis in Arthropod-induced Plant Galls: Mechanisms of Stress Generation and Dissipation. *Biochimica et Biophysica Acta, 1850,* 1509-1517. http://dx.doi.org/10.1016/j.bbagen.2015.03.007.

Jensen, W. A. (1962). *Botanical histochemistry.* WH Freeman, San Francisco.

Johansen, D. A. (1940). *Plant microtechnique.* McGraw-Hill, New York.

Koch, K. (2004). Sucrose Metabolism: Regulatory Mechanisms and Pivotal Roles in Sugar Sensing and Plant Development. *Current Opinion in Plant Biology, 7*(3), 235-246. http://dx.doi.org/10.1016/j.pbi.2004.03.014

Lev-Yadun, S. (2003). Stem Cells in Plants are Differentiated Too. *Current Topics in Plant Biology, 4,* 93-102.

Magalhães, T. A., Oliveira, D. C., Suzuki, A. Y. M., & Isaias, R. M. S. (2014). Patterns of Cell Elongation in the Determination of the Final Shape of *Baccharopelma dracunculifoliae* (Psyllidae) on *Baccharis dracunculifolia* DC (Asteraceae). *Protoplasma, 251,* 747-753. http://dx.doi.org/10.1007/s00709-013-0574-z

Malenovsky, I., Burckhardt, D., Queiroz, D. L., Isaias, R. M. S., & Oliveira, D. C. (2015). Descriptions of Two New Pseudophacopteron species (Hemiptera: Psylloidea: Phacopteronidae) inducing galls on Aspidosperma (Apocynaceae) in Brazil. *Acta Entomologica Musei Nationalis Pragae, 55*(2), 513-538.

Moura, M. Z. D., Isaias, R. M. S., & Soares, G. L. G. (2005). Ontogenesis of Internal Secretory Cells in Leaves of *Lantana camara* (Verbenaceae). *Botanical Journal of the Linnean Society, 148*(4), 427-431. http://dx.doi.org/10.1111/j.1095-8339.2005.00426.x

Moura, M. Z. D., Isaias, R. M. S., & Soares, G. L. G. (2008). Species-specific Changes in Tissue Morphogenesis Induced by two Arthropod Leaf Gallers in *Lantana camara* (Verbenaceae). *Australian Journal of Botany, 56*(2), 153-160. http://dx.doi.org/10.1071/BT07131

Nolte, D. K., & Koch, E. K. (1993). Companion-Cell Specific Localization of Sucrose Synthase in Zones of Phloem Loading and Unloading. *Plant Physiology, 101*(3), 899-905.

Nyman, T., & Julkunen-Titto, R. (2000). Manipulation of the Phenolic Chemistry of Willows by Gall-inducing Sawflies. *Proceedings of the national academy of Sciences, 97,* 13184-13187. http://dx.doi.org/10.1073/pnas.230294097

Oliveira, D. C., & Isaias, R. M. S (2010b). Cytological and Histochemical Gradients Induced by a Sucking Insect in Galls of *Aspidosperma australe* Arg Muell (Apocynaceae). *Plant Science, 178,* 350-358. http://dx.doi.org/10.1016/j.plantsci.2010.02.002

Oliveira, D. C., & Isaias, R. M. S. (2010a). Redifferentiation of Leaflet Tissues During Midrib Gall Development in *Copaifera langsdorffii* (Fabaceae). *South African Journal of Botany, 76,* 239-248. http://dx.doi.org/10.1016/j.sajb.2009.10.011

Oliveira, D. C., Carneiro, R. G. S., Magalhães, T. A., & Isaias, R. M. S. (2011). Cytological and Histochemical Gradients on Two *Copaifera langsdorffii* Desf (Fabaceae) Cecidomyiidae Gall Systems. *Protoplasma, 248*(4), 829-837. http://dx.doi.org/10.1007/s00709-010-0258-x

Oliveira, D. C., Christiano, J. C. S., Soares, G. L. G., & Isaias, R. M. S. (2006). Reações de Defesas Químicas e Estruturais de *Lonchocarpus muehlbergianus* Hassl (Fabaceae) à Ação do Galhador *Euphalerus ostreoides* Crawf (Hemiptera: Psyllidae). *Revista Brasileira de Botânica, 29,* 657-667. http://dx.doi.org/10.1590/S0100-84042006000400015

Oliveira, D. C., Magalhães, T. A., Carneiro, R. G. S., Alvim, M. N., & Isaias, R. M. S. (2010). Do Cecidomyiidae

Galls of *Aspidosperma spruceanum* (Apocynaceae) Fit the Pre-established Cytological and Histochemical Patterns? *Protoplasma, 242*, 81-93. http://dx.doi.org/10.1007/s00709-010-0128-6

Oliveira, D. C., Moreira, A. S. F. P., & Isaias, R. M. S. (2014). Functional gradients in insect gall tissues, studies on neotropical host plants. In G.W., Fernandes & J.C. Santos, (Eds.), *Neotropical Insect Galls* (pp. 35-49). Springer Netherlands. http://dx.doi.org/10.1007/978-94-017-8783-3_3

Price, P. W., Waring, G. L., & Fernandes, G. W. (1986). Hypotheses on the Adaptive Nature of Galls. *Proceedings of the Entomological Society of Washington, 88*, 361-363.

Price, P. W., Waring, G. L., & Fernandes, G. W. (1987). Adaptive Nature of Insect Galls. *Environmental Entomology, 16*, 15-24. http://dx.doi.org/10.1093/ee/16.1.15

Rehill. B. J., & Schultz, J. C. (2003). Enhanced Invertase Activities in the Galls of *Hormaphis hamamelidis*. *Journal of Chemical Ecology, 29*(12), 2703-2720. http://dx.doi.org/10.1023/B:JOEC.0000008014.12309.04

Roberts, M. F., & Wink, M. (1998). *Alkaloids: Biochemistry, Ecology, and Medicinal Applications*. Plenum Press New York. http://dx.doi.org/10.1007/978-1-4757-2905-4

Rhodes, M. J. (1994). Physiological roles for secondary metabolites in plants: some progress, many outstanding problems. *Plant Molecular Biology 24*(1), 1-20. http:// dx.doi.org/10.1007/BF00040570

Roitsch, T., & González, M. C. (2004). Function and Regulation of Plant Invertases: Sweet Sensations *Trends in Plant Science, 9*, 606-613. http://dx.doi.org/10.1016/j.tplants.2004.10.009

Rossetti, S., & Bonatti, P. M. (2001). In Situ Histochemical Monitoring of Ozone-and TMV- induced Reactive Oxygen Species in Tobacco Leaves. *Plant Physiology and Biochemistry, 39*, 433-442. http://dx.doi.org/10.1016/S0981-9428(01)01250-5

Róstas, M., Maag, D., Ikegami, M., & Inbar, M. (2013). Gall Volatiles Defend Aphids Against a Browsing Mammal. *BMC Evolutionay Biology, 13*, 193-204. http://dx.doi.org/10.1186/1471-2148-13-193

Salnikov, V. V., Grimson, M. J., Seagull, R. W., Haigler, C. H. (2003). Localization of Sucrose Synthase and Callose in Freeze-Substituted Secondary-Wall-Stage Cotton Fibers. *Protoplasma, 221*, 175-184. http://dx.doi.org/10.1007/s00709-002-0079-7

Sass, J. E. (1951) *Botanical microtechnique*. Iowa State College Press, Ames. http://dx.doi.org/10.5962/bhl.title.5706

Schönrogge, K., Harper, L. J., & Lichtenstein, C. P. (2000). The Protein Content of Tissues in Cynipid Galls (Hymenoptera: Cynipidae): Similarities Between Cynipid Galls and Seeds. *Plant, Cell & Environment, 23*(2), 215-222. http://dx.doi.org/ 10.1046/j.1365-3040.2000.00543.x

Sturm, A., & Tang, G. Q. (1999). The Sucrose-Cleaving Enzymes of Plants are Crucial for Development, Growth and Carbon Partitioning. *Trends in Plant Science, 4*(10), 401-407. http://dx.doi.org/ 10.1016/S1360-1385(99)01470-3

Vecchi, C., Menezes, N. L., Oliveira, D. C., Ferreira, B. G., & Isaias, R. M. S. (2013). The Redifferentiation of Nutritive Cells in Galls Induced by Lepidoptera on *Tibouchina pulchra* (Cham) Cogn Reveals Predefined Patterns of Plant Development. *Protoplasma, 250*, 1363-1368. http://dx.doi.org/10.1007/s00709-013-0519-6

Wagner, H., Bladt, S., & Zgainski, E. M. (1984). *Plant Drug Analysis*. Springer-Verlag, Berlin. http://dx.doi.org/10.1007/978-3-662-02398-3

Wind, J., Smeekens, S., Hanson, J. (2010) Sucrose: Metabolite and Signaling Molecule. *Phytochemistry, 71*, 1610-1614. http://dx.doi.org/10.1016/j.phytochem.2010.07.007

Wittich, P. E., Vreugdenhil, D. (1998). Localization of Sucrose Synthase Activity in Developmental Maize Kernels by in Situ Enzyme Histochemistry. *Journal of Experimental Botany, 49*(324), 1163-1171. http://dx.doi.org/10.1093/jxb/49.324.1163

Zrenner, R., Salanoubat, M., Willmitzer, L., & Sonnewald, U. (1995). Evidence for the Crucial Role of Sucrose Synthase for Sink Strength Using Transgenic Potato Plants *Solanum tuberosum* L. *Plant Journal, 7*(1), 97-107. http://dx.doi.org/10.1046/j.1365-313X.1995.07010097.x

13

Analysis of the Structure and Diversity of *Prosopis africana* (G. et Perr.) Taub. Tree Stands in the Southeastern Niger

Laouali Abdou[1], Boubé Morou[2], Tougiani Abasse[3] & Ali Mahamane[1,4]

[1] Université de Diffa, Faculté des Sciences Agronomiques, BP 78, Diffa, Niger

[2] Université Dan Dicko Dankoulodo de Maradi, Faculté des Sciences et techniques, Département de Biologie, BP 465, Maradi, Niger

[3] Institut National de Recherche Agronomique du Niger, BP 429, Niamey, Niger

[4] Université Abdou Moumouni, Faculté des Sciences et Techniques, Département de Biologie, Laboratoire Garba Mounkaila, BP 10662, Niamey, Niger

Correspondence: Laouali Abdou, Université de Diffa, Faculté des Sciences Agronomiques, BP 78, Diffa, Niger. E-mail: abdoulaouali2000@yahoo.fr

Abstract

All parts of *Prosopis africana* are used by rural people in Niger, and this exposes it to degradation and a regeneration problems. The objective of this study was to determine the structure and regeneration of *P. africana* stands in the southern regions of Maradi and Zinder, Niger. Data were collected in plots, following transects after stratified sampling. Trunk diameter of all woody species was recorded in 126 plots. The diversity was analyzed and diameter structure and regeneration rates were determined. *P. africana* was the predominant species in both Maradi and Zinder: frequency = 40.35% and 43.95% of all species, respectively in Maradi and Zinder; importance value index = 40.57% in Maradi and 48.60% in Zinder. The Shannon diversity index was 2.82 in Maradi and 2.40 in Zinder and the Sorensen similarity index between the two regions was 0.73. According to the diameter structure, the stands were degraded in Zinder but regenerating in Maradi. The density of *P. africana* per hectare for trees with trunk diameter ≥ 5 cm and < 5 cm, respectively was 16 and 51 in Maradi, and 30 and 12 in Zinder. The regeneration rate of *P. africana* was low compared with the general woody population. These results show the need for reforestation operations, using appropriate techniques, to avoid local extinction of the species.

Keywords: *Prosopis africana*, tree stand, structure, degradation, regeneration, Niger

1. Introduction

In the Sahel in general and in Niger in particular, trees provide many products and services to rural populations (Larwanou et al., 2010; Laouali et al., 2014), which exposes them to high anthropogenic pressure reducing their natural regeneration. As a result, many populations of woody species are in a regressive dynamics characterized by the increasing scarcity or lack of younger individuals (Endress et al., 2006; Bellefontaine et al., 2010; Sanogo et al., 2013). The exploitation and marketing of non-timber forest products in Africa to date are primarily designed to increase gatherers' individual incomes without concern for sustainable management. Sustainable management would require ensuring natural regeneration of the species, and promoting their domestication to reduce the pressure on natural populations (Ouédraogo et al., 2006). *Prosopis africana* is a particularly vulnerable species because all parts of the tree are used by rural communities (Faye et al., 2011). Its wood is dense (Sotelo Montes & Weber, 2009) and highly resistant, so it is used for making construction poles and planks, and mortars and pestles. The wood has a high calorific value (Sotelo Montes et al., 2011), so it is highly valued for charcoal by blacksmiths. The leaves, roots and especially the barkare used in traditional medicine. The leaves and pods are used for fodder and the seeds for food (Larwanou, 1994; Arbonnier, 2000; Agboola, 2004; Larwanou et al., 2012; Laouali et al., 2014). Unfortunately, this species is facing a regeneration problem (Ahoton et al., 2009, Niang-Diop et al., 2010; Laouali et al., 2015) and overexploitation to which are added climatic conditions increasingly difficult due to climate change. This will result in a regression of the species' population or its disappearance if sustainable management precautions are not taken.

In Niger, *P. africana* is represented by scattered individuals with some relic stands in the southern regions of

Dosso, Maradi and Zinder. Little information is available on their structure and regeneration. It is therefore necessary to have reliable data on the current state of these stands for better conservation and sustainable management of the species.

The main objective of this study is to characterize the *P. africana* tree stands in the southeastern Niger on based on the analysis of their structure and diversity.

2. Material and Methods

2.1 Study Area

The study was conducted in the southern regions of Maradi and Zinder (Figure 1). Relevant administrative departments are Gazaoua in Maradi region, Kantché and Magaria in Zinder region. The human population in the Aguié and Gazaoua departments (which formed the same department) expanded from 172,922 to 406,532 inhabitants or from 57.52 to 135.24 inhabitants / km^2 from 1988 to 2012. The human population in the Kantché and Magaria departments expanded from 518,452 to 976,924 inhabitants or from 82.95 to 156.3 inhabitants / km^2 from 1988 to 2012 (Institut National de la Statistique/Niger [INS], 2014). The socio-economic activities of these populations are dominated by agriculture, livestock, crafts, trade. The climate is sahelo-soudanian. In the south of Maradi, the average annual temperature is around 28 °C. The average annual rainfall from 1981 to 2010 at the Gazaoua station was 446.32 mm. The soils are mainly dunal and the flora is dominated by species in the Mimosaceae family (*Prosopis africana* (Guill. & Perr.) Taub., *Albizia chevalieri* Harms, *Faidherbia albida* (Del.) Chev....), the Caesalpiniaceae family (*Piliostigma reticulatum* (DC.) Hochst., *Cassia singueana* (Del.) Lock, *Bauhinia rufescens* Lam....), the Combretaceae family (*Combretum glutinosum* Perr., *Guiera senegalensis* J. F. Gmel, *Anogeissus leiocarpa* (DC.) Guill. & Perr. ...) and the Anacardiaceae family (*Sclerocarya birrea* (A. Rich.) Hochst., *Lannea microcarpa* Engl. & K. Krause …). In the south of Zinder, the average annual rainfall is around 525 mm. The average temperature is 22.5 °C. The soils are mainly sandy, loamy sand and clay loam. The flora is dominated by the Mimosaceae (*Faidherbia albida* (Del.) Chev., *Prosopis africana* (Guill. & Perr.) Taub.) and the Anacardiaceae (*Lanea microcarpa* Engl. & K. Krause) (United States Agency for International Development [USAID], 2006; Laouali et al., 2014). Each of these areas is shared between the central south sahelian and central north soudanian compartments, according to Saadou (1990) phytogeographic subdivisions on the basis of climatic conditions; the vegetation consists of Combretum thickets, steppes, lowland dry forests, gallery forests and savannas.

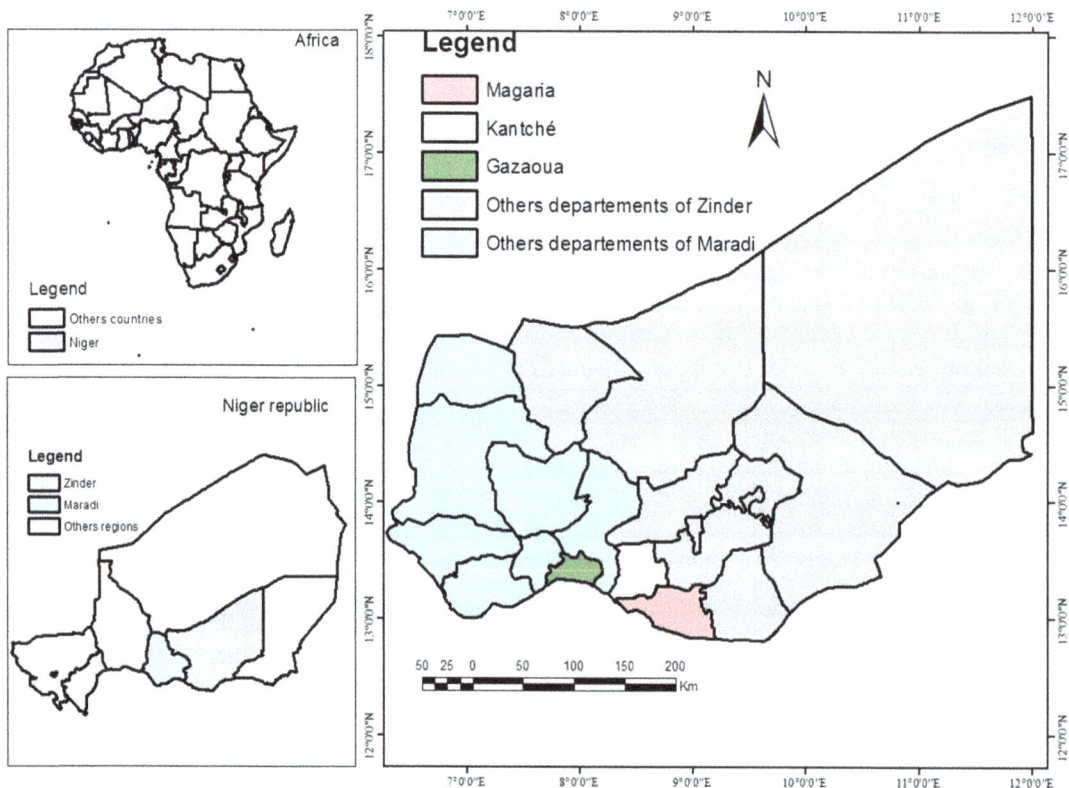

Figure 1. Location of the study area

2.2 Sampling

The sampling, guided by the presence of *P. africana* in the regions and administrative departments, was systematic across transects in the territories of villages. In the Gazaoua administrative department in the Maradi region, sampling was conducted in three villages (Elguéza, Guidan Adamou and Dan Damou). In the Zinder region, the surveys were conducted in eight villages (Bawada, Angoual kirya, Sabar, Gagéré, Kahin baka, Kadeye, Kokotaou and Tsagai) distributed in the Kantché and Magaria administrative departments. The administrative departments were selected after a field visit and a consultation of resource persons which allowed having more information about the presence of the species. Villages were selected following recommendations from the departmental offices of the Ministry of Environment.

The *P. africana* stands were identified prior to sampling. Different sampling schemes were used in the two regions because stand structure differed. In the Maradi region, the only one stand of *P. africana* studied was homogeneous and extended throughout the three village territories. In the Zinder region, there were many stands located in the village territories, and the stands were not contiguous. In the Maradi region, individuate plots (50 x 50 m) were arranged in seven linear and parallel sampling strategies (transects) with a length of 4 km each and separated by one km, individuated using a GPS. In each sampling strategy, there was a distance of 400 m between plots. A total of 62 plots were surveyed, and these covered the entire *P. africana* stand. Plots size is justified by the fact that the studied stands are parklands and the selected layout allowed reaching a sampling rate of 1.8%. In the Zinder region, plots (50 x 50 m) were placed along four sampling strategies (with a length of 1.5 km each) that extended radially in four directions (compass directions) from each village. A total of 64 plots distributed among the eight villages were surveyed.

2.3 Data Collection

Data collected during the inventory are including the number of individuals for each woody species, tree height, crown diameter and stem diameter at breast height. Data were recorded only for individuals with a trunk diameter ≥ 5 cm. Trees with trunk diameter <5 cm were considered young natural regeneration: these trees were counted.

2.4 Diversity and Species Importance Value Indices

To analyze the diversity between regions and inside regions, several indices were calculated.

The diversity index of Shannon and Weaver (1949) was calculated to study the Alpha diversity for assessing the weight of the species in land use in each region. This index varies depending on the number of species present. It is higher when there are more species, indicating greater diversity. It is calculated as bits per individual and its formula is:

$$H = -\sum_{i=1}^{S} pi \log_2 pi \tag{1}$$

S = total number of species and pi = relative frequency of species. Pielou evenness index was also calculated. Its formula is:

$$E = \frac{H}{Hmax} \quad \text{where } Hmax = \log_2 S \tag{2}$$

S is the total number of species. The Pielou evenness index varies between 0 and 1. It is 0 when there is a phenomenon of dominance and 1 when the distribution of individuals among species is homogenous.

The coefficient of Sorensen (1948) was calculated to assess beta diversity for comparing habitats of the two regions. This index expresses the degree of similarity between two sites and has the formula:

$$Is = \frac{2C}{2C+A+B} \tag{3}$$

A is the number of species found only in site 1; B is the number of species found only in site 2, and C is the number of species common to both sites. Beta diversity is the importance of species replacement or biotic change along environmental gradients. The interest of its study is to highlight the diversity across the region.

To appreciate the importance of tree species in general and that of *P. africana* in particular at the study regions, the importance value index (IVI) (Curtis & Macintosh, 1951) was calculated. This index is expressed by the formula:

$$IVI\ (\%) = relative\ density + relative\ dominance + relative\ frequency\ (for\ the\ species) \tag{4}$$

The relative dominance of one species (relative basal area) is the quotient of its basal area per total basal area of all species. The relative density of a species is the ratio between its absolute density and total absolute densities of all species multiplied by 100. The relative frequency of a species is the ratio between its specific frequency and the total specific frequencies of all the species multiplied by 100.

Basal area was calculated using the following formula:

$$G \ (m^2/ha) = \frac{\pi}{40000s} \sum_{i=1}^{n} di^2 \tag{5}$$

d is the trunk diameter (cm) of the tree i in the plot and s the area of the plot (ha). The trees were then divided into 18 diameter classes of 5 cm. Furthermore, to better analyze the data, the observed structure was modeled using the parameters of the theoretical Weibull distribution whose probability density function is (Rondeux, 1999):

$$f(x) = \frac{c}{b}\left(\frac{x-a}{b}\right)^{c-1} \ exp\left[-(\frac{x-a}{b})^c\right] \tag{6}$$

x is the trunk diameter of the tree; $f(x)$ is the probability density value at point x. a is the position parameter: it is 0 if all categories of trees are considered (from plantlets to the seed trees) during the inventory; it is not null if the trees have a diameter $\geq a$ ($a = 5$ in this study). b is the scale parameter, linked to the central value of the probability distribution of variable x = diameter. c is the shape parameter related to the diameter structure. A value of c < 1, distribution "inverted J" is characteristic of multispecies or uneven-aged stands, while c > 3.6 is characteristic of predominantly aged individuals stands. Moreover, 1 < c < 3.6 means stands with predominance of young individuals or small diameter. Minitab 16 software was used for this purpose. To test the adjustment of the structure observed in the Weibull distribution, a log-linear analysis was applied with the R 2.15.3 software. This structure determination concerns the total woody population on one hand and *P. africana* the other hand. To appreciate the regeneration, the density of adult trees and that of the young individuals were calculated for both the total woody population and *P. africana*, which were used to calculate the regeneration rates. The results were compared through a FISHER test performed with the Minitab 16 software. Thus, the regeneration rate of *P. africana* and that of the woody population were compared in each region. A comparison was also made between *P. africana* regeneration rates of the two regions and between the woody population regeneration rates of the two regions.

3. Results

3.1 Diversity Analysis

Values for Shannon diversity index (H), maximum diversity (H max) and Pielou evenness index (E) in the two regions are listed in Table 1.Diversity was relatively low compared to the maximum diversity at both regions. There was no significant difference in diversity and evenness between the Maradi and Zinder regions (P> 0.05). The value of the Sorensen similarity index is 0.73.

Table 1. Diversity and equitability indexes values

Region	H	Hmax	E
Maradi	2.82	4.70	0.60
Zinder	2.40	4.52	0.53

3.2 Dendrometric Characteristics and Regeneration

The inventories included 26 woody species in Maradi (Table 2) and 23 woody species in Zinder (Table 3). *P. africana* had the largest IVI at both regions: 121.72% (40.57% of the total IVI) in Maradi, and 145.79% (48.60% of the total IVI) in Zinder.

Table 2. Dendrometric parameters of species recorded in the Maradi region

Species	RF (%)	RDs (%)	RDm (%)	IVI (%)
Prosopis africana (Guill. & Perr.) Taub.	40.35	40.35	41.02	121.72
Piliostigma reticulatum (DC.) Hochst.	22.65	22.65	26.28	71.59
Faidherbia albida (Del.) Chev.	14.08	14.08	14.94	43.09
Albizia chevalieri Harms	1.88	1.88	3.66	7.41
Bauhinia rufescens Lam.	3.35	3.35	0.42	7.12
Dichrostachys cinerea (L.) Wight & Arn.	2.95	2.95	0.27	6.17
Diospyros mespiliformis Hochst. ex A. Rich.	1.34	1.34	2.62	5.31
Hyphaene thebaica (L.) Mart.	1.74	1.74	1.70	5.19
Lannea microcarpa Engl. & K. Krause	0.94	0.94	2.26	4.14
Balanites aegyptiaca (L.) Del.	1.61	1.61	0.71	3.93
Azadirachta indica A. Juss.	1.61	1.61	0.53	3.75
Tamarindus indica L.	0.40	0.40	2.47	3.28
Annona senegalensis Pers.	1.47	1.47	0.03	2.98
Pterocarpus erinaceus Poir.	0.40	0.40	2.04	2.85
Acacia nilotica (L.) Willd. ex Del. subsp. Nilotica	0.94	0.94	0.64	2.52
Guiera senegalensis J. F. Gmel	0.94	0.94	0.02	1.90
Combretum glutinosum Perr.	0.67	0.67	0.03	1.37
Cassia singueana (Del.) Lock	0.54	0.54	0.00	1.08
Strychnos innocua Del.	0.40	0.40	0.18	0.99
Maerua oblongifolia (Forssk.) A. Rich.	0.40	0.40	0.01	0.81
Moringa oleifera Lam.	0.40	0.40	0.00	0.81
Sclerocarya birrea (A. Rich.) Hochst.	0.27	0.27	0.07	0.60
Calotropis procera (Ait.) Ait. f.	0.27	0.27	0.00	0.54
Vitex doniana Sweet	0.13	0.13	0.07	0.34
Maerua angolensis DC.	0.13	0.13	0.00	0.27
Ziziphus mauritiana Lam.	0.13	0.13	0.00	0.27
Total: 26	**100**	**100**	**100**	**300**

Note. RF = Relative frequency; RDs = Relative density; RDm = Relative dominance; IVI = Importance value index.

Table 3. Dendrometric parameters of species recorded in the Zinder region

Species	RF (%)	RDs (%)	RDm (%)	IVI (%)
Prosopis africana (Guill. & Perr.) Taub.	43.95	43.95	57.89	145.79
Faidherbia albida (Del.) Chev.	32.58	32.58	22.69	87.86
Sclerocarya birrea (A. Rich.) Hochst.	4.24	4.24	4.88	13.37
Tamarindus indica L.	2.80	2.80	3.61	9.20
Combretum glutinosum Perr.	2.62	2.62	0.66	5.90
Annona senegalensis Pers.	2.62	2.62	0.12	5.35
Lannea microcarpa Engl. & K. Krause	1.71	1.71	1.58	5.01
Parkia biglobosa (Jacq.) R. Br. Ex G. Don	1.08	1.08	2.82	4.98
Albizia chevalieri Harms	1.71	1.71	1.51	4.94
Guiera senegalensis J. F. Gmel	2.17	2.17	0.01	4.34
Dichrostachys cinerea (L.) Wight & Arn.	1.17	1.17	0.32	2.67
Piliostigma reticulatum (DC.) Hochst.	0.81	0.81	0.59	2.21
Adansonia digitata L.	0.27	0.27	0.84	1.38
Daniellia oliveri (Rolfe) Hutch. & Dalz.	0.18	0.18	0.76	1.12
Terminalia avicennioides Guill. & Perr.	0.36	0.36	0.33	1.05
Balanites aegyptiaca (L.) Del.	0.27	0.27	0.35	0.89
Ficus platyphylla Del.	0.09	0.09	0.63	0.81
Ziziphus mauritiana Lam.	0.36	0.36	0.07	0.79
Bauhinia rufescens Lam.	0.27	0.27	0.02	0.56
Azadirachta indica A. Juss.	0.18	0.18	0.19	0.55
Cassia singueana (Del.) Lock	0.27	0.27	0.00	0.54
Hyphaene thebaica (L.) Mart.	0.18	0.18	0.10	0.46
Acacia nilotica (L.) Willd. Ex Del. Subsp. Nilotica	0.09	0.09	0.03	0.21
Total: 23	**100**	**100**	**100**	**300**

The density was higher in Zinder than Maradi for trees ≥ 5 cm in trunk diameter. Trees < 5 cm in trunk diameter were more common in Maradi (Table 4). This concerns woody population one hand and *P. africana* on the other hand.

Tableau 4. Woody population and *P. africana* density at both regions for trees ≥ 5 cm and < 5 cm in trunk diameter

Regions	Density (trees/ha)						
	Woody population				*P. africana*		
	≥ 5 cm	< 5 cm	Total		≥ 5 cm	< 5 cm	Total
Maradi	41	143	184		16	51	67
Zinder	65	77	142		30	12	42

Assuming that trees with trunk diameter ≤ 5 cm reflect relatively recent natural regeneration, the regeneration rate was significantly higher in Maradi than in Zinder (P <0.05), for both the woody population and for *P. africana*. For the regions separately, the regeneration rate of *P. africana* was significantly lower than that of woody population in Zinder (P <0.05), but these two rates were not significantly different in Maradi (P = 0.209) (Figure 2).

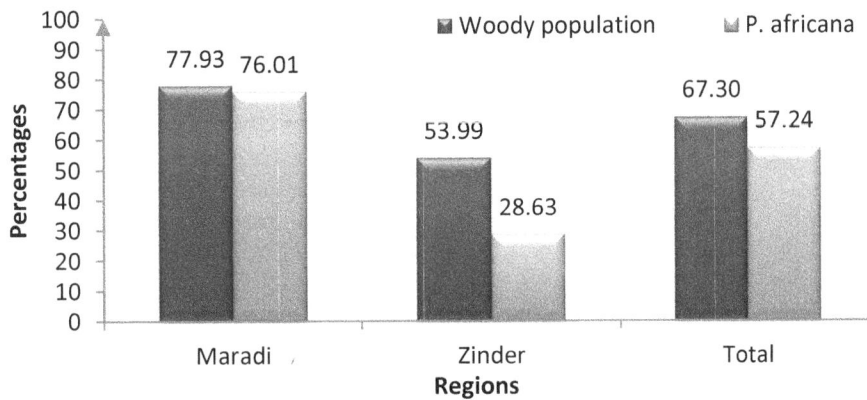

Figure 2. Regeneration rates of woody population and *P. africana* in Maradi, Zinder and both regions

3.3 Structure of Woody Population

The mean trunk diameter of *P. africana* was significantly higher in Zinder (38.13 ± 8.78 cm) than in Maradi (17.8 ± 13.6) (P <0.001).The distribution of trees in diameter classes in Maradi had a "reversed J" shape. It fit the theoretical Weibull distribution with shape parameter c = 0.93 for woody population and c = 0.82 for *P. africana* (Figure 3a and 3b). In Zinder, the distribution of trees in diameter classes had a bell-shape. It fit the theoretical Weibull distribution with shape parameter c = 2.96 for woody population and c = 3.88 for *P. africana* (Figure 3c and 3d). The results of the log-linear analysis indicate a good fit of the data to the Weibull distribution for both regions (P> 0.05).

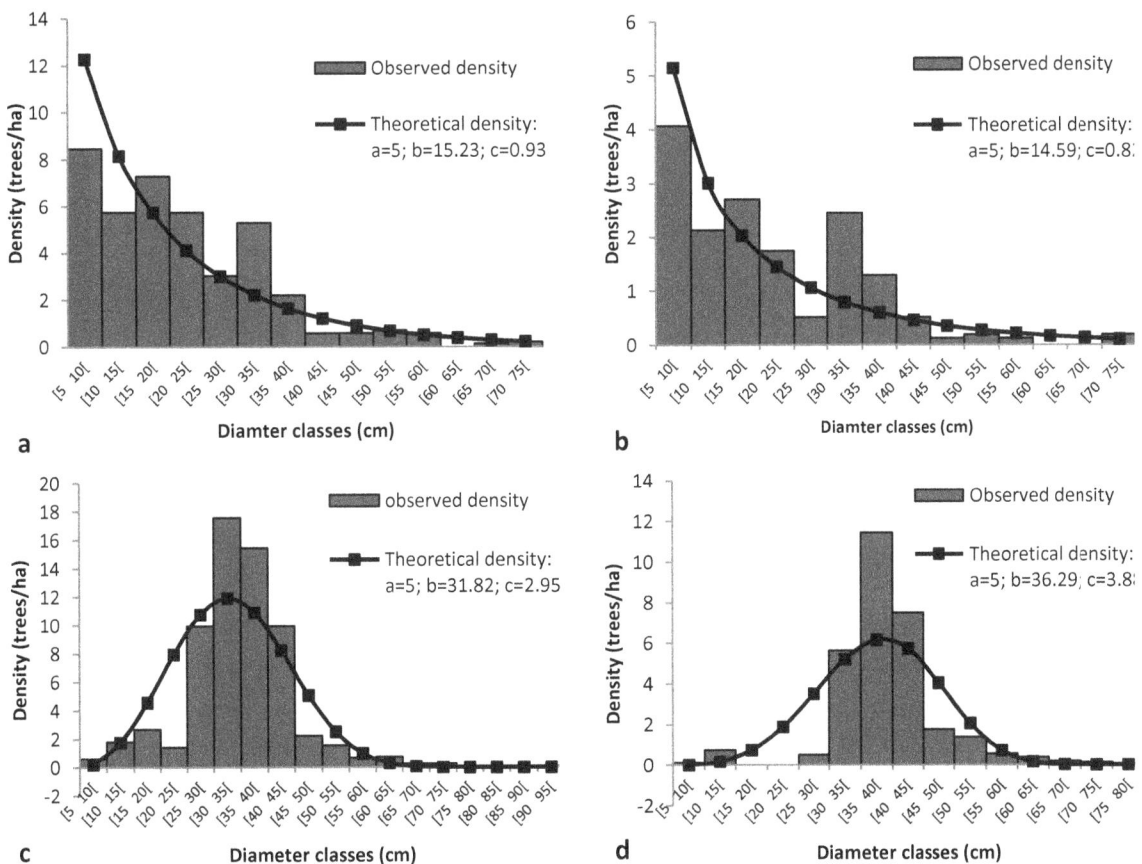

Figure 3. Structure of trunk diameter classes of (a) all woody species and (b) *P. africana* in the Maradi, and (c) all woody species and (d) *P. africana* (d) in Zinder

4. Discussion

Prosopis africana was the most common species in the two regions. This is linked to the importance given to the species by the local population. In fact, the population would have favored the conservation of this species in previous years, while the climatic conditions were more favorable and the demographic pressure lowest. Indeed, in practice, farmers agree to keep a tree in their fields if they can have a crop production gain or if the crop yield decrease can be compensated by the economic gains due to the presence of the tree (Pierre et al., 1992). This could also explain the relatively low species diversity. The relatively high similarity between the two regions can be explained by climatic conditions and socioeconomic activities, including agricultural practices, which are essentially the same.

The distribution of trees in diameter classes with a shape of "J reversed" observed in Maradi (Figures 3a and 3b) with the shape parameter $c = 0.82$ (< 1) for *P. africana* is characteristic of a fast regenerating stand. These results are similar to those found by others in vegetation study in Niger (Dan Guimbo et al., 2010; Abdourahamane et al., 2013; Boubacar et al., 2013).

The low proportion of *P. africana* trees in the large diameter classes in Maradi is due to a particularly important use of wood of this species. The wood of *P. africana* is strong and hard, so it is used as poles to support barns, houses, etc. and for mortars and pestles. For these uses, the wood of *P. africana* is preferred over the other species in this region (Laouali et al., 2014).

According to Ajonou et al. (2009), the bell shape distribution of trunk diameter reflects an unstable population characterized by an absence or very small proportion of individuals in one or more diameter classes. This is the situation in Zinder. The majority of *P. africana* trees were in the average diameter class (30 to 50 cm), and individuals in the smaller and larger diameter classes were poorly represented. This could be explained by systematic cutting of large diameter trees (as in Maradi) and the vulnerability of small diameter trees to browsing animals and farmers seeking fodder. The regeneration of a natural stand is more certain if it has a lot of trees in the smaller diameter classes (Morou, 2010), so regeneration of *P. africana* is more threatened in Zinder than in Maradi.

The difference in diameter found between the two regions could be related to rainfall which is slightly higher in Zinder, although overall, the climatic conditions are essentially similar. The low regeneration of of *P. africana* compared with the total woody population could be explained by its relatively poor seed germination (Ahoton et al., 2009, Niang-Diop et al., 2010; Laouali et al., 2015). The fact that regeneration was better in Maradi than Zinder could be explained by the promotion of assisted natural regeneration by several research projects. This started to be adopted by farmers in the southern part of Maradi in the late 1980s, while in the Zinder region, the adoption started later (USAID, 2006; Larwanou et al., 2010).

Not enough of previous studies on this topic in the same area was an obstacle in the discussion of the results but makes this study more interesting. Based on this study, we can say that in the still extant *P. africana* stands, the regeneration rate of the species is lower than that of other species, especially in the Zinder region, even though development projects have promoted awareness campaigns and the practice of assisted natural regeneration. This confirms that *P. africana* is threatened with local extinction. Given these results and the socio-economic importance of this species, restoration operations through reforestation campaigns should be considered in areas with suitable habitat for the species. Growth and survival of trees are better for *P. africana* provenances from drier locations in Niger (Weber et al., 2008), and this probably is due to a higher root/shoot ratio in these provenances (Weber et al., 2015). The climate is becoming hotter and drier and with variable rainfall in the West African Sahel (Buontempo, 2010), so it would be advisable to use seed from drier locations for the reforestation campaigns. In addition, monitoring forestry extension services should provide technical assistance to rural communities for all steps in the reforestation process (seed collection, germination, planting, maintenance, monitoring and evaluation).

References

Abdourhamane, H., Morou, B., Rabiou, H., & Mahamane, A. (2013). Caractéristiques floristiques, diversité et structure de la végétation ligneuse dans le Centre-Sud du Niger: cas du complexe des forêts classées de Dan Kada Dodo-Dan Gado. *International Journal of Biological and Chemical Sciences, 7*(3), 1048-1068. http://dx.doi.org/10.4314/ijbcs.v7i3.13

Agboola, D. A. (2004). *Prosopis africana* (Mimosaceae): stem, roots, and seeds in the economy of the savanna areas of Nigeria. *Economic Botany, 58(Supplement)*, S34-S42. Retrieved from http://www.jstor.org/stable/4256906

Ahoton, L. E., Adjakpa, J. B., M'po Ifonti M'po, & Akpo E. L. (2009). Effet des prétraitements des semences sur la germination de *Prosopis africana* (Guill., Perrot. Et Rich.) Taub., (Césalpiniacées). *Tropicultura, 27*(4), 233-238. Retrieved from http://www.tropicultura.org/text/v27n4/233.pdf

Ajonou, K., Bellefontaine, R., & Kokou, K. (2009). Les forêts claires du Parc national Oti-Kéran au Nord-Togo: structure, dynamique et impacts des modifications climatiques récentes. *Sécheresse, 20*(1), 1-10. http://dx.doi.org/10.1684/sec.2009.0217

Arbonnier, M. (2000). *Arbres, arbustes et lianes des zones sèches d'Afrique de l'Ouest.* CIRAD - MNHN - UICN, Montpellier (France).

Bellefontaine, R., Ferradous, A., Alifriqui, M., & Monteuuis, O. (2010). Multiplication végétative de l'arganier, *Argania spinosa*, au Maroc : le projet John Goelet. *Bois et Forêts des Tropiques, 304*(2), 47-59. Retrieved from http://bft.cirad.fr/cd/BFT_304_47-59.pdf

Boubacar, M. M., Inoussa, M. M., Ambouta, J. M. K., Mahamane, A., Jorgen, A. A., Harissou, Y., & Rabiou, H. (2013). Caractérisation de la végétation ligneuse et des organisations pelliculaires de surface des agroécosystèmes à différents stades de dégradation de la Commune rurale de Simiri (Niger). *Int. J. Biol. Chem. Sci., 7*(5), 1963-1975. http://dx.doi.org/10.4314/ijbcs.v7i5.15

Buontempo, C. (2010). *Sahelian climate: past, current, projections.* Met Office Hadley Centre, Devon. Retrieved from http://www.oecd.org/swac/publications/47092928.pdf

Curtis, J. T., & McIntosh, R. P. (1951). An upland forest continuum in the prairie-forest border region of Wisconsin. *Ecology, 32*(3), 476-496. http://dx.doi.org/10.2307/1931725

Dan Guimbo, I., Mahamane, A., & Ambouta, J. M. K. (2010). Peuplement des parcs à *Neocarya macrophylla* (Sabine) Prance et à *Vitellaria paradoxa* (Gaertn. C.F.) dans le sud-ouest nigérien : diversité, structure et régénération. *Int. J. Biol. Chem. Sci., 4*(5), 1706-1720. http://dx.doi.org/10.4314/ijbcs.v4i5.65568

Endress, B. A., Gorchov, D. L., & Berry, E. J. (2006). Sustainability of a non-timber forest product: effect of alternative leaf harvest practices over 6 years on yield and demography of the palm *Chamaedorea radicalis*. *Forest Ecology and Management, 234*, 181-191. http://dx.doi.org/10.1016/j.foreco.2006.07.020

Faye, M. D., Weber, J. C., Abasse, T. A., Boureima, M., Larwanou, M., Bationo, A. B., ... Sonogo Diaité, D. (2011). Farmers' preferences for tree functions and species in the West African Sahel. *For. Trees and Livelihoods, 20*, 113-136. http://dx.doi.org/10.1080/14728028.2011.9756702

INS. (2014). *Recensement général de la population et de l'habitat, 2012. Répertoire national des localités.* Institut National de la Statistique, Niger.

Laouali, A., Dan Guimbo, I., Larwanou, M., Inoussa, M. M., & Mahamane A. (2014). Utilisation de *Prosopis africana* (G. et Perr.) Taub. dans le sud du département d'Aguié au Niger: les différentes formes et leur importance. *Int. J. Biol. Chem. Sci., 8*(3), 1065-1074. http://dx.doi.org/10.4314/ijbcs.v8i3.20

Laouali, A., Dan Guimbo, I., Youchaou, A., Rabiou, H., & Mahamane A. (2015a). Etude de la germination de la graine et suivi de la croissance en pépinière de *Prosopis africana* (G. et Perr.) Taub., espèce menacée de disparition au Niger. *Annales de l'Université Abdou Moumouni de Niamey, Tome XVIII*-A, 1-12.

Larwanou, M. (1994). *Potentials of Prosopis africana (G. et Perr.) Taub leaf litter for soil nutrient enhancement and crop development.* M.Sc thesis. Department of Forest resources Management, University of Ibadan.

Larwanou, M., Moustapha, A. M., Rabé, M. L., & Dan Guimbo, I. (2012). Contribution de la Régénération Naturelle Assistée des ligneux dans l'approvisionnement en bois des ménages dans le département de Magaria (Niger). *Int. J. Biol. Chem. Sci., 6*(1), 24-36. http://dx.doi.org/10.4314/ijbcs.v6i1.3

Larwanou, M., Oumarou, I., Laura, S., Dan Guimbo, I., & Eyog-Matig, O. (2010). Pratiques sylvicoles et culturales dans les parcs agroforestiers suivant un gradient pluviométrique nord-sud dans la région de Maradi au Niger. *Tropicultura, 28*(2), 115-122. Retrieved from http://www.tropicultura.org/text/v28n2/115.pdf

Morou, B. (2010). *Impacts de l'occupation des sols sur l'habitat de la girafe au Niger et enjeux pour la sauvegarde du dernier troupeau de girafes de l'Afrique de l'Ouest.* Thèse de Doctorat Unique, Université Abdou Moumouni de Niamey.

Niang-Diop, F., Sambou, B., & Lykke, A. M. (2010). Contraintes de régénération naturelle de *Prosopis africana*: facteurs affectant la germination des graines. *Int. J. Biol. Chem. Sci., 4*(5), 1693-1705. http://dx.doi.org/10.4314/ijbcs.v4i5.65578

Ouédraogo, A., Thiombiano, A., Hahn-hadjali, K., & Guinko, S. (2006). Régénération sexuée de *Boswellia dalzielii* Hutch., un arbre médicinal de grande valeur au Burkina-Faso. *Bois et forêts des Tropiques, 289*(3), 41-48. Retrieved from http://bft.revuesonline.com/gratuit/BFT_PDF_289_41-48.pdf

Pierre, G., Waechter, P., & Yachinovsky, A. (1992). *Environnement et développement rural. Guide pour la gestion des ressources naturelles.* Editions Frisson-Roche, Agence de Coopération culturelle et technique, Ministère de la coopération et du Développement, Paris.

Rondeux, J. (1999). *La Mesure des arbres et des peuplements forestiers* (2nd ed.). Les presses agronomiques de Gembloux.

Saadou, M. (1990). *La végétation des milieux drainés nigériens à l'Est du fleuve Niger.* Thèse de Doctorat ès - Sciences Naturelles. Université de Niamey.

Sanogo, S., Sacandé, M., Van Damme, P., & N'Diaye, I. (2013). Caractérisation, germination et conservation des graines de *Carapa procera* DC. (Meliaceae), une espèce utile en santé humaine et animale. *Biotechnologie, Agronomie, Société et Environnement, 17*(2), 321-331. Retrieved from http://www.pressesagro.be/base/text/v17n2/321.pdf

Shannon, C. E., & Weaver, W. (1949). *The mathematical theory of communication.* Urbana, Univ. Illinois Press.

Sørensen, T. (1948). A method of establishing groups of equal amplitude in plant sociology based on similarity of species and its application to analyses of the vegetation on Danish commons. *Biol. Skr., 5,* 1-34.

Sotelo Montes, C., & Weber, J. C. (2009). Genetic variation in wood density and correlations with tree growth in *Prosopis africana* from Burkina Faso and Niger. *Ann. For. Sci., 66,* 713. http://dx.doi.org/10.1051/forest/2009060

Sotelo Montes, C., Silva, D. A., Garcia, R. A., Muñiz, G. I. B., & Weber, J. C. (2011). Calorific value of *Prosopis africana* and *Balanites aegyptiaca* wood: relationships with tree growth, wood density and rainfall gradients in the West African Sahel. *Biomass and Bioenergy, 35,* 346-353. http://dx.doi.org/10.1016/j.biombioe.2010.08.058

USAID. (2006). *Etude de la régénération Naturelle Assistée dans la Région de Zinder (Niger). Une première exploration d'un phénomène spectaculaire.* Retrieved from http://www.gtu.dk/Niger_Study_2006_re-greening.pdf

Weber, J. C., Larwanou, M., Abasse, T. A., & Kalinganire, A. (2008). Growth and survival of *Prosopis africana* provenances tested in Niger and related to rainfall gradients in the West African Sahel. *For. Ecol. Manage., 256,* 585-592. http://dx.doi.org/10.1016/j.foreco.2008.05.004

Weber, J. C., Sotelo Montes, C., Kalinganire, A., Abasse, T., & Larwanou, M. (2015) Genetic variation and clines in growth and survival of *Prosopis africana* from Burkina Faso and Niger: comparing results and conclusions from a nursery test and a long-term field test in Niger. *Euphytica, 205,* 809-821. http://dx.doi.org/10.1007/s10681-015-1413-4

Effects of Planting Method on Enhanced Stand Establishment and Subsequent Performance of Forage Native Warm-Season Grasses

Vitalis W. Temu[1], Maru K. Kering[1] & Laban K. Rutto[1]

[1] Agricultural Research Station, Virginia State University, Petersburg, Virginia, USA

Correspondence: Vitalis W. Temu, Agricultural Research Station, Virginia State University, 238 M.T. Carter Bldg, P.O. Box 9061 Petersburg, Virginia 23806, USA. E-mail: vtemu@vsu.edu

Abstract

To develop strategies for successful establishment of forage native warm-season grasses (NWSGs) in southeastern USA, early agronomic performance of transplanted and seeded stands of big bluestem (BB, *Andropogon gerardii* Vitman), eastern gamagrass (GG, *Tripsacum dactyloides* L.), indiangrass [IG, *Sorghastrum nutans* (L.). Nash], and switchgrass (SG, *Panicum virgatum* L.) were compared. In early June, about 6-week old high tunnel grown seedlings were transplanted to assigned clean seedbeds. Plant spacing (cm) was 30-within and 45-between rows. Seeded plots received ≥ 11 kg seeds ha^{-1} planted at ≤ 2 cm deep in 45-cm wide rows, a month after transplanting to match rainfall availability. As needed, tall-growing broadleaf weeds were removed physically by cutting with a hand hoe. Plants were allowed uninterrupted first year growth with no fertilizer applied. Early in the following spring, dead standing biomass was mowed down to allow the emerging tillers more access to sunlight. During the second and third year after planting, plots were machine harvested twice between July and September for yield assessment. Percentage ground covered by plant material and species basal diameters were also recorded. Data were analyzed as a randomized complete block design for effects of planting method, species, and stand age. Except for GG, transplanting resulted in greater (>3,000 kg DM ha^{-1}) forage yield and more so during the second harvest year. Total second year yields were similar for BB and GG and averaged 9,600 and 6,300 kg DM ha^{-1} for transplanted and seeded, respectively. Indiangrass and SG yields (kg DM ha^{-1}) were 11,500 and 8,300 and 13,000 and 10,000 for transplanted and seeded plots, respectively. The NWSG ground cover was greater in the transplanted than seeded plots, while the reverse was true for weed cover. Data indicate that, in less than two years, transplanting under comparable growing conditions can produce harvest-ready uniform NWSG stands in weed infested areas. For practical recommendations, however, data on comparable responses of transplanted vs seeded stands to actual grazing at a similar timing is necessary.

Keywords: Native warm-season grass, establishment, transplant, forage yield, basal diameter, cover, defoliation

1. Introduction

1.1 Summer Forage Shortages

In Virginia, as in most other southeastern states of America, inability of introduced warm-season forage grasses (IWSGs) to withstand the July-August dry spell causes summer forage shortages and can potentially impact the productivity of ruminant livestock. Besides their inability to withstand the summer droughty conditions, the productivity of these introduced warm-season forages such as bermudagrass (*Cynodon dactylon* (L), bahiagrass (*Paspalum notatum* Flueggé), and *P. dilatatum* Poir.) is dependent on rainfall distribution and application of external fertilizers. This is partly so because their root systems are usually too shallow to access the receding moisture front at deep soil profiles during dry periods. Livestock producers are, therefore, forced to sustain their animals on costly hay and concentrate feeds, which negatively impact their profit margins. Unfortunately, this reliance on purchased hay somehow eliminates the producers' control over the quality and composition of feeds in use, thus making it even harder for those interested in organic production. This makes forage systems composed of plant species adapted to a diversity of growing conditions more likely to support year-round forage-based ruminant production.

1.2 Advantages of Native Warm-Season Forage Grasses

Unlike IWSG, native warm-season grasses (NWSGs) are relatively more drought tolerant, less dependent on fertilizers, and usually remain vegetative longer into the summer dry-spell (Kiss et al., 2007; Mulkey et al., 2008). So, incorporating NWSGs into the pasture systems will most likely improve summer forage production and availability in the region. For over a decade now, the mid-south and southeastern regions of the US have seen a growing interest in NWSGs mainly due to their superior summer forage potential (Angima et al., 2009; Temu et al., 2014a), wildlife habitat quality features (Temu et al., 2014b), and low-input demand. Of the NWSGs, big bluestem (BB, *Andropogon gerardii* Vitman), eastern gamagrass (GG, *Tripsacum dactyloides* L.), indiangrass [IG, *Sorghastrum nutans* (L.). Nash], and switchgrass (SG, *Panicum virgatum* L.) are the most favored (Vogel, 2000; Jones et al., 2007). These NWSGs grow notably better than the IWSGs owing to their superior morphological and physiological adaptations. Once well established, most NWSGs are able to exploit large soil volumes for moisture and nutrients owing to their deep and extensive root systems (Huang, 2000). They are also able to sustain high photosynthetic activity during hot and dry summer months due to their C_4 photosynthetic pathway. This ability to sustain growth under harsh growing conditions may explain the observed high annual yields of 2637 kg ha^{-1} for unfertilized BB, IG, and SG (Mulkey et al., 2008), and 8472 kg ha^{-1} from mixed stands of BB, IG, and LB (Temu et al., 2014a). In another study, SG yield of 8000 kg ha^{-1} is reported with 55% of it being obtained in July and August (Jung et al., 1985).

Being tall-growing, NWSGs make even better pastures for goats, which prefer feeding from tall-, over short-growing plants. On these tall grasses, goats are also less likely to pick gastrointestinal nematode (GIN) eggs off the ground. Like in other hot humid environments, GINs have a tremendous economic impact on small ruminant production in the southeastern US. Usually, GINs reduce animal performance and losses in production of up to 50% have been reported in sheep (Sykes, 1994). Therefore, the potential contribution of NWSG pastures towards reduced GIN loads in grazing small ruminants, cannot be over emphasized. In addition, NWSG stands, because of their bunch-like structure, have voids close to the ground that make room for legumes, which improve the forage nutritive value. In pastures, legumes also improve soil fertility through biological nitrogen fixation and increase the crude protein concentration of the forage biomass (Muir et al., 2011).

Besides forage production, NWSGs because of their growth characteristics; tall growing, open canopy structure, wider canopy cover, and relatively smaller basal area (White et al., 2005; Harper & Moorman, 2006) offer good wildlife habitats. Their compatibility with non-grass species also supports the formation of diverse plant communities that allows for a broader wildlife food base. For grassland birds and small mammals, voids among tall grasses facilitate movement through the stand and with upward of 20% bare ground which is good for visibility and access to food (Lusk et al., 2006; Jones et al., 2007; Temu et al., 2015). Most NWSG pastures have multiple ecological benefits. For example, in a typical NWSG forage stand, northern bobwhites retain vigilance to predators, visibility to foods on the ground, easy mobility, and concealment while foraging under the canopy.

1.3 Slow Adoption of Native Warm-Season Forage Grasses

Despite their beneficial attributes, incorporation of NWSGs into pasture systems is still very slow, in the Southeast, mainly because they are costly and difficult to establish. They usually exhibit poor germination, low seedling vigor that easily succumb to competitive annual weeds, and could be at least two years before you have a reasonable stand. Unfortunately, low seedling vigor during seeded establishment of NWSGs is not avoidable because their initial resource allocation favors root system development at the expense of vegetative growth. This differential resource allocation also accounts for poor initial NWSG ground coverage in seeded stands, which without timely weed control and gap-fillings, gives way to poor initial stands and undesirable plant species encroachment into the pasture. However, once well stablished, NWSGs do preferentially channel resources towards above-ground growth, which is a good forage attribute.

Strategies for achieving faster establishment of thick NWSG stands without costly weed control and/or gape-filling measures will likely improve their standing as alternative summer forage resources. Unlike their seeded counterparts, transplanted NWSGs have been found to attain maturity and produce seeds within the first year, (Goedhart & Warne, 2006). However, there is paucity of information about their initial ground coverage and readiness for forage harvesting. This study, therefore, was designed to evaluate the use of seedling transplants as a strategy for achieving faster establishment of the selected four NWSGs with respect to their early establishment ground cover and forage yield during the second year of production.

2. Materials and Methods

2.1 Location and Field Preparations

The study was conducted at Virginia State University's research farm (Randolph Farm) located in Chesterfield county, Virginia at 37° 13" 43' N; 77° 26" 22' W, and 45 m above sea level. The soils at the farm are Bourne series fine sandy loam (mixed, semiactive, thermic Typic Fragiudults). The area has a 20 year June, July, and

August average precipitation of 92, 113, and 121 mm with day temperatures of 30.2, 32.1, and 31.2 ℃, respectively (Satellite N.O.A.A., 2013). Prior to starting the experiment, the field had been under rotational corn-soybean row-cropping for several years. In mid-May, the field was first sprayed with a broad spectrum herbicide, Glyphosate {N-(phosphonomethyl) glycine}, to kill the cover crop and later disk plowed and harrowed repeatedly to produce a fine seedbed.

2.2 Seedling Preparation and Planting

For about 5-6 weeks starting late April, degradable paper strip cups (5×5 cm top, 2×2 cm bottom and 5 cm deep), arranged in perforated flats on polyethylene lined table tops were seeded with BB, GG, IG, and SG and kept moist. Seedlings were allowed to grow in a high tunnel. In early June, 64, 6×7 m plots were marked and randomly assigned to BB, GG, IG, and SG in a randomized complete block design (RCBD). Two days after a heavy rainfall in mid-June, seedlings of each species were transplanted into pre-assigned plots, spaced 30 cm within and 45 cm between rows, using a tractor operated seedling planter. The planting was such that seedlings were upright with their entire root mass completely covered with at least an inch thick layer of soil. Missing and improperly placed seedlings were manually replaced and corrected immediately after planting. Following the return of summer rains in mid-July, about six weeks after transplanting, a control plot for each species and in each block (replication) was seed-drilled at about 2 cm depth for comparison. Ernst Conservation Seeds, Inc. high seed rate recommendations for establishing a forage stand were followed.

2.3 Field Management

Because there was no rainfall for over a month after transplanting and fields were not irrigated, some poorly covered seedlings died. Following the return of summer rains in mid-July, a little over a month after transplanting, dead seedlings were immediately replaced. Later in the season, broadleaf weeds were manually removed by cutting as needed, using a hand hoe and throughout the study to minimize their competitiveness against the native grass seedlings. For the remainder of the season, plants grew uninterrupted and no fertilizer was applied. In late March of 2013, the standing dead biomass accumulated since being planted was mowed down so that the new growth could have ample access to solar radiation.

2.4 Forage Yield Measurements

During the second year after planting, 1.5 m wide strips were harvested twice from July to September, to assess (kg DM ha^{-1}) during the first year (Year1) of production. A CIBUS F Plot Forage Harvester (Winterstaiger Ag, Dimmelstrasse, Austria) equipped with a Harvestmaster weighing system (Juniper Systems, Inc, USA) with a 0.01 kg accuracy was used to determine fresh plot weights. Cutting height was set at 18 cm. From each plot harvest, a representative forage sample was collected and transported in plastic bags to a field laboratory, where it was weighed before and after drying to a constant weight at 65 ℃ in a forced-air oven to determine percentage moisture content. The sample moisture content values were used to convert the respective fresh plot weights into dry matter, with which per hectare yield estimates (kg DM ha^{-1}) were calculated. Forage yields during the second year (Year2) of production were obtained similar to that of year1. However, respective harvest dates differed by about 1-2 weeks due to weather.

2.5 Cover Estimates

In vegetation survey, ground cover is defined as the percentage of material other than bare ground covering the land surface (Anderson, 1986). In the current study, visual estimates of the proportion of land surface covered by the vertical projection of above ground plant parts were recorded, a day or two before harvest. These assessments were made towards the end of May for the two years. During 2013, the first year in production, proportions of ground area covered by the native grasses, weeds, and litter, within a 1-m square quadrat, were recorded. For faster area estimation, the quadrat-sides were painted in alternate 10-cm color bands such that a 10×10-cm cell represented a 1% cover. All dead plant parts found recumbent on the ground surface, including those still attached to the mother plant, were considered litter. Generally, the stands had no bare ground patches and, therefore, there were no records for the same. In each plot, three randomly placed quadrats were sampled. During the second year in production (2014), six basal diameter readings of the native grass stubbles, the cross-sectional area of plants near the ground, as indicators of respective basal cover, were recorded along a line transect across the rows. Two diameter readings were recorded within each of three 1-m line segments at least a meter apart and excluding the outermost rows.

2.6 Design and Data Analysis

The data were subjected to analysis of variance (ANOVA) as a RCBD with planting method, years in production, and species as fixed effects. Means were compared by the Fisher's Least Significant Difference test at α = .05.

During the data analyses, the first and second cuts were combined into respective year total yields. The plot yield data were organized by number of years in production and analyzed for effects of planting method and species. The latter compared species yields within the first or second year in production to their same-established-method counterparts. Yields were also separately compared between planting methods within years in production. To assess treatment effects on stand establishment, yield ratios obtained by expressing the first-year yields as proportions of their respective second-year values were also compared. The cover and basal diameter data, however, were pooled across years in production for the analyses.

3. Results and Discussion

3.1 Effects of Planting Method

During the 2012 summer months, the study area experienced severe droughty conditions that impacted survival of the seedling transplants. In fact monthly rainfall totals were only about 45 and 66 mm in June and August, respectively, and although the July total was significantly greater (205 mm), it was still poorly distributed (Satellite N.O.A.A., 2013). Results of forage yield by the NWSG stands during the first and second year in production are summarized in Table 1. Throughout the section and unless otherwise indicated, mean differences have been declared not significant when $P > .05$. There were significant planting method × year and/or species interactions in the yield rankings, so results are presented separately by species and year. During the first year in production, both first and second harvest yields for each species were consistently greater ($P < .01$) from the transplanted than their matching seeded plots. At the second harvest, for example, the percent yield differences ranged from slightly below 40 for SG to nearly 60 for BB and IG. During the second year, first harvest trends in the corresponding yield ranks and the percentage differences due to planting method were more or less sustained, but not in the second harvest, except for BB. Second-harvest yields for the other three species were not statistically different and only averaged 29 units for GG, < 20 for the rest, and even as low as 10 for IG. However, total yields remained significantly greater from the transplanted ($P < .02$) than the seeded plots.

These yield differences were partly attributable to the growing advantage the transplants had over the seeded plants. A greater seedling survival in the transplanted plots, which translated into greater initial plant population density, also contributed to their resulting better yields, somewhat agreeing with an earlier report on GG (Springer et al., 2003). Similar results regarding greater forage biomass from transplanted grasses compared to seeded ones have also been reported (Houser, 1983; Borman, et al., 1991). While the seeded NWSGs had their root development phase amidst severe weed completion, the transplants established their root systems ahead of weeds emergence. In terms of number of years after planting, first and second year yields are comparable for BB and SG. Although the values were below the reported averages of 7.7 and 9.2 kg DM ha^{-1}, respectively, for the two species (Propheter et al., 2010). The noted yield differences may also be due to the fact that fertilizers were not applied in this study unlike that reported by Propheter et al. (2010) which received 168 to 180 kg N ha^{-1} as urea. However, even without fertilizers, total yields in the current study from transplants were still notably greater than those reported earlier. These results suggest that the performance of newly established NWSG stands may be limited more by weed challenges rather than soil fertility and that the growing advantage associated with transplanting would more or less guarantee success.

3.2 Effects of Species on Forage Yield

For each harvest and years in production, there were significant species yield differences. While a planting method × harvest order interaction was not detected, significant harvest order × year interactions were. Therefore, results of species yield ranking for each harvest event and or their combinations are described separately. Throughout the study, mean SG yield values were greater than those of other species, although not always significantly so. For example, mean second harvest yield for SG was only statistically greater than that of BB during the first and second year, and greater than GG but not different from that of IG during the second year (Table 1). In each year, however, mean total forage yield for SG, was significantly greater than that of any other species ($P < .02$). For the transplanted SG, respective total yields, during the first and second year, were 10,452 and 13,474 kg DM ha^{-1}. Although these yields were on the lower end of reported five-year averages of 10.4 to 19.1 Mg ha^{-1} in the Southeast (Fike et al., 2006), that may be attributable to differences in number of harvests per year and the use of fertilizers. Yields in year 1 were similar for transplants and seeded IG, BB, and GG plots, averaging 3955 kg DM ha^{-1} (Seeded) and 7870 kg DM ha^{-1} (transplants). However, during the second year, IG had higher biomass, but with relatively narrower yield differences between seeded and transplanted SG plots than the other species.

To some extent, the observed relatively narrow SG yield differences due to planting methods suggest that aggressive tillering ability for the performer variety enabled the seeded stands to quickly minimize the potential effects of their initial stand density on biomass production. This also indicates that, among the four NWSGs, SG establishment was less susceptible to weed competition, and therefore, under comparable growing conditions, the productivity of fall-seeded new stands of similar varieties may benefit more from careful early defoliation management that promotes tillering than it might with intensive weed control alone. Defoliation usually favors the growth of auxiliary tillers through improved light environment (Assuero & Tognetti, 2010) and the removal of the reproductive tillers'' apical dominance (Richards, et al., 1988; Briske and Richards, 1995; Tomlinson and O'connor, 2004). However, when deciding on appropriate defoliation strategies, species differences in response to defoliation as in the case of SG in the current study should be taken into consideration.

3.3 Effects of Years in Production on Forage Yield

While first harvest yields for each species, within a planting method, showed no significant differences due to number of years in production, the corresponding second harvests and total yields were consistently greater during the second than the first year. However, while it was not unusual for yields of the second year stands to surpass their respective first-year performances (Cornelius, 1946), total yields during these harvest years were notably greater from transplanted- than seeded plots indicating that transplanting favored faster stand establishment. These faster stand establishments would also explain the notable yield increase of the second year stands relative to their first year performance. While the second-year seeded plot-yields were as much as 118% higher than their first-year values, respective yield increase for the transplants was only 45%. This notable closeness of transplanted first year stand performance to their second year values was also consistent with the enormously greater percentages of native grass ground cover observed (Table 2). However, the fact that second year yields, for each planting, exceeded their respective first year records indicates the importance of careful initial defoliation management for newly established NWSG stands to attain higher forage potentials in future.

Table 1. Effects of planting method on forage yields (kg DM ha^{-1}) from pure stands of four native warm-season grasses[†] during the first (Year1) and second (Year2) year of production recorded in 2013 and 2014, respectively

Species/Method	Yield (harvest order and total) by year of production						
	First cut		Second cut		Total		Yield increase[‡]
	Year1	Year2	Year1	Year2	Year1	Year2	
	----------------------------------kg DM ha^{-1}----------------------------------						%
Big Bluestem							
Seeded	3587^{b}_{Ba}[¶]	3694^{b}_{BCa}	432^{b}_{Bb}	3274^{b}_{Ba}	4019^{b}_{Bb}	6968^{b}_{Ca}	73
Transplanted	6962^{a}_{Ba}	5842^{a}_{BCa}	999^{a}_{Bb}	4058^{a}_{Ba}	7961^{a}_{Bb}	9900^{a}_{Ca}	24
P>α[§]	<.01	.02	<.01	0.04	<.01	.01	
Eastern gamagrass							
Seeded	3227^{b}_{Ba}	2680^{b}_{Ca}	790^{b}_{Ab}	3035^{a}_{Ba}	4017^{b}_{Ba}	5716^{b}_{Ca}	42
Transplanted	6198^{a}_{Ba}	4994^{a}_{Cb}	1522^{a}_{Ab}	4260^{a}_{Ba}	7720^{a}_{Ba}	9254^{a}_{Ca}	20
P>α	<.01	<.01	<.01	.12	<.01	.01	
Indiangrass							
Seeded	3205^{b}_{Ba}	3441^{b}_{Ba}	625^{b}_{Ab}	4910^{a}_{Aa}	3830^{b}_{Bb}	8350^{b}_{Ba}	118
Transplanted	6410^{a}_{Ba}	6068^{a}_{Ba}	1518^{a}_{Ab}	5402^{a}_{Aa}	7928^{a}_{Bb}	11470^{a}_{Ba}	45
P>α	<.01	<.01	<.01	.31	.01	.01	
Switchgrass							
Seeded	5408^{b}_{Aa}	5355^{b}_{Aa}	975^{b}_{Ab}	4969^{a}_{Aa}	6383^{b}_{Ab}	10324^{b}_{Aa}	61
Transplanted	8875^{a}_{Aa}	7624^{a}_{Aa}	1576^{a}_{Ab}	5850^{a}_{Aa}	10452^{a}_{Ab}	13474^{a}_{Aa}	29
P>α	<.01	.02	<.01	.08	.01	.01	

[†]Big bluestem (*Andropogon gerardii*), eastern gamagras (*Tripsacum dactyloides*), indiangrass (*Sorghastrum nutans*), and switchgrass (*Panicum virgatum*). [‡]The difference between Year1 and Year2 total yields as a percentage of the total in Year2. [¶]Means followed by the same letter, superscript for paired planting methods, or same subscript (uppercase for species, within a column and lowercase for paired year1&2, within a row), are not significantly different at α = .05. [§]Probability of mean difference between planting methods for the respective species.

There was also a significant difference in how the first and second harvest yields compared, within the first and second harvest year. During the first year, the yields were greater at the first- than the second harvest by a range of 75 to 88% while values for their counterparts during the second year were only ≤ 30%. In fact, for seeded GG and IG, yields were even greater at the second- than first harvest. These yield differences are attributable to changes in the proportions of weed biomass in the harvested material and crown expansion. While the first harvests in year1 included significant proportions of annual weed biomass, their proportions in the second harvest were negligible. The native grasses regrowth had a growing advantage over weeds and could exploit larger soil volumes for moisture and nutrients to quickly outcompete the weeds. Additionally, growth of perenials in the second year were from bigger crowns and started well before most weed seeds could even germinate and that minimized their proportions of the latter in the harvested biomass.

3.4 Effects of Planting Methods on Ground Cover

Treatment differences in stand establishment were also reflected in visual estimates of the proportions of ground covered by live or fallen dead vegetation parts. Table 2 summarizes results on means of ground cover values for the NWSGs, weeds, and litter, during the first year in production. Significant planting method × species interactions were detected and so results of mean comparisons, for the native grasses are discussed separately. Percentage ground cover by the NWSGs was greater in transplanted plots than their seeded counterparts (P < .01). While NWSGs cover for all transplants ranged from 58 to 93%, it is only seeded SG that had values within this range (Table 2). Other species show values of 42% (BB and IG), and 6 (GG). Again, except for SG, values for the proportions of ground covered by weeds were greater in seeded plots, ranging 51 - 94%. Weed cover values in transplanted plots were only about 30% for IG and < 10 for BB and GG. Among the seeded plots, values for ground cover by litter were ≤ 7% except for GG which had no litter, but the highest cover of weeds. Except for SG where litter cover was comparable for both planting methods (5%), the cover by litter was greater in all transplanted plots than their seeded counterparts in other species and values averaged 13%, significantly greater than in SG.

Table 2. Effects of species and planting method[†] on mean early-summer percentage ground cover by live vegetation and litter (visual estimates) for four native warm-season grasses[‡] during the first year in production, 2013

Species/Method	Ground cover[‡‡]		
	Native grass	Weeds	litter
Big Bluestem	------------------------------%------------------------------		
Seeded	$40^b{}_{AB}$[§]	$57^a{}_{BC}$	$3^b{}_{BC}$
Transplanted	$81^a{}_{AB}$	$7^b{}_B$	$12^a{}_{AB}$
P >Fα[#]	<.01	<.01	.05
Gamagrass			
Seeded	$6^b{}_C$	$94^a{}_A$	$0^b{}_C$
Transplanted	$77^a{}_B$	$8^b{}_B$	$15^a{}_A$
P >Fα	<.01	<.01	.01
Indiangrass			
Seeded	$43^b{}_B$	$51^a{}_B$	$6^b{}_{AB}$
Transplanted	$58^a{}_C$	$29^b{}_A$	$13^a{}_A$
P >Fα	<.01	<.01	0.18
Switchgrass			
Seeded	$66^b{}_A$	$27^a{}_C$	$7^a{}_A$
Transplanted	$93^a{}_A$	$2^b{}_B$	$5^b{}_B$
P >Fα	<.01	<.01	.04

[†]Six week-old seedlings, raised in a high tunnel, were transplanted by machine in June onto rain-soaked, disked fine seedbeds with seeds of the same species being drilled at 1-2 cm depth in matching plots later in the summer. [‡] Big bluestem (*Andropogon gerardii*), eastern gamagras (*Tripsacum dactyloides*), indiangrass (*Sorghastrum nutans*), and switchgrass (*Panicum virgatum*). [‡‡]Proportions of ground covered by a vertical projection of plant parts of the native grass, all other live plants (weeds), or dead fallen plant material. [§]Within a column means followed by the same letter, superscript for species or subscript for planting method, do not differ significantly at α =.05. [#]Probability of mean difference between planting methods by species.

The observed superiority of the transplanted plots over the seeded ones in native grass cover was indicative of a greater population of fast growing tillers that outcompeted the weeds. Thus the anticipated growth advantage that transplants would have over weeds was actually achieved and that resulted in their dominance during the recovery, similar to earlier findings (Borman et al., 1991). The results also indicate that the transplanted stands were better prepared for recovery spring-growth with greater numbers of dormant tillers and energy reserves in their crowns. This demonstrated better performance of the transplanted stands compared to their seeded counterparts is also partly attributable to their age differences since the transplants were at least 6 weeks old at planting and it took over a month for the control plots to be seeded. There was no attempt to eliminate this age difference since fall planting is actually a common practice to avoid severe weed competition. Similar reports on better performance of transplanted native grass stands than their seeded counterparts also exist (Brown & Johnston, 1976). Having greater proportions of litter in the transplanted plots was actually consistent with their exhibited robust growth that resulted with relatively more forage yields (Table 1) and senescent leaves as indicated by litter cover (Table 2).

3.5 Effects of Species and Planting Method on Basal Diameters

When subsequent stand recovery growth from the preceding harvest regimes were compared based on early-spring basal diameters, differences due to planting method and species were detected (Table 3). Basal diameters in the transplanted plots exceeded their seeded counterparts by up to 8 cm for GG plots and about 4 cm for the other species. Among the transplanted NWSG stands, mean basal diameter was greater for GG (19 cm) than any other species ($P < .01$), the least being 13 cm for BB, but with no difference between IG and SG. In the seeded plots, values were still the least for BB (9 cm), but not statistically different from GG (11) and all other differences were only numeric ($P > .05$). Specific differences in basal diameter could result from variable competitive abilities of the native grass species against the already established weeds similar to earlier observations (Schmidt et al., 2008). While the current data still indicate that transplanting may be a better establishment method to direct seeding for the NWSGs, results further showed that upon successful establishment, GG will produce a thicker stand relatively sooner while BB may suffer weed competition longer.

Table 3. Effects of species and planting method[†] on early-spring basal diameters and early-summer ground cover of native warm-season grasses[‡] in pure stands during the second year in production, 2014

Method	Big Bluestem	Gamagrass	Indiangrass	Switchgrass	$P > F\alpha$[‡‡]
	Basal diameter[§]				
	------------------------------- cm ----------------------------				
Seeded	9bB[¶]	11bAB	12bA	13bA	.01
Transplanted	13aC	19aA	17aB	16aB	<.01
$P > F\alpha$	<.01	0.01	.02	<.01	
	Ground cover[#]				
	-----------------------------% ----------------------------				
Seeded	49bB	23bC	46bB	64bA	<.01
Transplanted	93aA	92aA	77aB	96aA	<.01
$P > F\alpha$	<.01	<.01	<.01	<.01	

[†]Six week-old seedlings raised in a high tunnel, were transplanted by machine, in June, onto rain-soaked, pre-tilled and leveled fine seedbeds and seeds of the same species drilled at 1-2 cm depth in earmarked matching plots later in the summer. [‡]Big bluestem (*Andropogon gerardii*), eastern gamagras (*Tripsacum dactyloides*), indiangrass (*Sorghastrum nutans*), and switchgrass (*Panicum virgatum*). [‡‡]Probability of difference between means of planting methods, within species. [¶]Means of paired planting methods followed by the same letter lowercase letter, within column, or those of species followed by the same uppercase letter, within row, are not significantly different at α = .05. [§]Average of six diameters of the sprouting native warm-season grass crowns along a perpendicular transect across rows. [#]Late-May visual estimates of the proportion of land surface covered by the vertical projection above ground plant parts of the native warm-season grasses.

3.6 Effects of Species and Planting Method on Ground Cover

During 2014, the second year in production, visual estimates of early-summer percentage ground cover also showed the effects of planting method and species. In ground cover, the transplanted plots outnumbered their seeded counterparts by 30-60 units (Table 3). The fact that the transplanted stands sustained superiority over their seeded counterparts suggests that the growth advantage they had at planting was sustained into the second year. This difference in performance was partly more attributable to inability of the seeded stands to suppress the already established annual weed populations. Thus the annual weeds retained their competitive advantage over resources, which may have impacted the rate at which the seeded bunches expanded. Faster establishment of transplanted grasses over their seeded counterparts have also been reported (Hauser, 1983). In the current study, however, mean differences in the transplanted plots were only significant between IG (77) and all others, which nearly averaged 94 ($P < .01$). Among the seeded plots, percentage ground cover was greater for SG (64) than any other species and was the least for GG (23). Comparable percent cover was exhibited by BB and IG whose values averaged at 47% (Table 3).

4. Conclusions

The observed narrower differences between the first and second year forage yields from transplanted plots demonstrate the reliability of seedling transplants in achieving enhanced establishment of NWSG stands in less than two years. The results also show that, under similar growing conditions, the forage productivity for transplanted SG stands may not remain superior over their seeded counterparts beyond three growing cycles. The demonstrated greater weed suppression in the transplanted NWSG stands compared to their seeded counterparts has implications on costs for weed control and forage quality attributes associated with the proportions of undesirable species in pastures. For the studied NWSGs, transplanting has shown ability to alter species rankings in relative growth response of their newly established stands to defoliation. There is a need to establish whether the demonstrated ability of transplanted NWSG stands to suppress weeds could be sustained if fertilizers were applied. There is also a need to establish how transplanting may affect species performance during establishment of mixed NWSG stands.

Acknowledgments

The authors are grateful to the USDA Evans Allen program for funding the study, the management of the Agricultural Research Station in the College of Agriculture at Virginia State University for housing the project as well as providing logistical and material support to the research team. The authors are also indebted to Michael P. Brandt, David Johnson, Steven LeMaster, Kevin Kidd, and Christos Galanopoulos for their help with field operations and data collection at different stages of the research. This article is a publication No. 326 of the Agricultural Research Station, Virginia State University.

References

Anderson, E. W. (1986). A guide for estimating cover. *Rangelands*, 236-238.

Angima, S. D., Kallenbach, R. L., & Riggs, W. W. (2009). Optimizing hay yield under lower nitrogen rates for selected warm-season forages. *J Integr Biosci, 7*, 1-6.

Borman, M. M., Krueger, W. C., & Johnson, D. E. (1991). Effects of Established Perennial Grasses on Yields of Associated Annual Weeds. *J. Range Manage, 44*(4), 318-322. http://doi.org/10.2307/4002390

Briske, D. D., & Richards, J. H. (1995). Plant responses to defoliation: a physiological, morphological and demographic evaluation. *Wildland plants: physiological ecology and developmental morphology. Society for Range Management, Denver, CO*, (pp. 635-710).

Brown, R. W., & Johnston, R. S. (1976). *Revegetation of an alpine mine disturbance: Beartooth Plateau, Montana* (Vol. 206). Intermountain Forest & Range Experiment Station.

Cornelius, D. R. (1946). Establishment of Some True Prairie Species Following Reseeding. *Ecology, 27*(1), 1-12. http://doi.org/10.2307/1931012

Ernst Conservation Seeds, Inc. (n.d.). Native Warm-season Grasses for High Quality Biomass Forage, Including Livestock Bedding & Mushroom Compost. Retrieved from http://www.ernstseed.com/files/documents/native_warm.pdf

Fike, J. H., Parrish, D. J., Wolf, D. D., Balasko, J. A., Green, J. T., Rasnake, M., & Reynolds, J. H. (2006). Long-term yield potential for switchgrass for biofuel systems. *Biomass Bioenergy, 30*, 198-206. http://doi.org/10.1016/j.biombioe.2005.10.006

Goedhart, J., & Warners, D. (2006). Assessment of prairie pot transplants as a restoration tool. Piece Cedar Creek Institute Biology Dept, Calvin College Grand Rapids, Michigan. Retrieved from http://www.cedarcreekinstitute.org/sites/default/files/Calvin%20 %20Jennifer%20Goedhart%20and%20David%20Warners.pdf

Harper, C. A., & Moorman, C. E. (2006, October). Qualifying native warm-season grasses and early succession habitat. In *11th Triennial National Wildlife & Fisheries Extension Specialists Conference (2006)* (p. 10).

Hauser, V. L. (1983). Grass establishment by bandoleer transplants, and germinated seeds. American Society of Agricultural and Biological Engineers. *ASAE, 26*(1), 0074-0078.

Huang, B. (2000). Role of root morphological and physiological characteristics in drought resistance of plants (pp. 39-64). In R. Wilkinson (ed.) *Plant-environment interactions*. Marcel Dekker, Inc. New York, NY. http://dx.doi.org/10.1201/9780824746568.ch2

Jones, J., Coggin, D. S., Cummins, J. L., & Hill, J. (2007). *Restoring and Managing Native Prairies*. A Handbook for Mississippi Landowners. Wildlife Mississippi. Starkville, MS.

Jung, G. A., Griffin, J. L., Kocher, R. E., Shaffer, J. A., & Gross, C. F. (1985). Performance of switchgrass and bluestem cultivars mixed with cool-season species. *Agron. J., 77*, 846-850. http://dx.doi.org/10.2134/agronj1985.00021962007700060005x

Kiss, Z., Fieldsend, A. F., & Wolf, D. D. (2007). Yield of switchgrass (*Panicum virgatum* L.) as influenced by cutting management. *Acta Agron. Hungar., 55*, 227-233. http://dx.doi.org/10.1556/AAgr.55.2007.2.10

Lusk, J. J., Smith, S. G., Fuhlendorf, S. D., & Guthery, F. S. (2006). Factors Influencing Northern Bobwhite Nest-Site Selection and Fate. *Journal of Wildlife Management, 70*, 564-571. http://dx.doi.org/10.2193/0022-541X(2006)70[564:FINBNS]2.0.CO;2

Muir, J. P., Pitman, W. D., & Foster, J. L. (2011). Sustatinable, low-input, warm-season, grass-legume grassland mixtures: mission (nearly) impossible?. *J. Grass and Forage Sci., 66*, 301-315. http://dx.doi.org/10.1111/j.1365-2494.2011.00806.x

Mulkey, V. R., Owens, V. N., & Lee, D. K. (2008). Management of warm-season grass mixtures for biomass production in South Dakota USA. *Bioresource Technology, 99*, 609-617. http://dx.doi.org/10.1016/j.biortech.2006.12.035

Page, H. N., & Bork, E. W. (2005). Effect of planting season, bunchgrass species, and neighbor control on the success of transplants for grassland restoration. *Restoration Ecology, 13*(4), 651-658. http://dx.doi.org/10.1111/j.1526-100X.2005.00083.x

Propheter, J. L., Staggenborg, S. A., Wu, X., & Wang, D. (2010). Performance of annual and perennial biofuel crops: yield during the first two years. *Agron. J, 102*(2), 806-814. http://dx.doi.org/10.2134/agronj2009.0301

Richards, J. H., Mueller, R. J., & Mott, J. J. (1988). Tillering in tussock grasses in relation to defoliation and apical bud removal. *Annals of Botany, 62*(2), 173-179.

Schmidt, C. D., Hickman, K. R., Channell, R., Harmoney, K., & Stark, W. (2008). Competitive abilities of native grasses and non-native (Bothriochloa spp.) grasses. *Plant Ecology, 197*(1), 69-80. http://dx.doi.org/10.1007/s11258-007-9361-2

Satellite, N. O. A. A. (2013). Information Service. National Climatic Data Center. *US Dept of Commerce.*

Sykes, A. R. (1994). Parasitism and production in farm animals. *Anim. Prod., 59*, 155-172. http://dx.doi.org/10.1017/S0003356100007649

Temu, V. W., Baldwin, B. S., Reddy, K. R., & Riffell, S. K. (2015). Harvesting Effects on Species Composition and Distribution of Cover Attributes in Mixed Native Warm-Season Grass Stands. *Environments, 2*(2), 167-185. http://dx.doi.org/10.3390/environments2020167

Temu, V. W., Baldwin, B. S., Reddy, K. R., Riffell, S. K., & Burger, L. W. (2014b). Wildlife habitat quality (sward structure and ground cover) response of mixed native warm-season grasses to harvesting. *Environments, 1*(1), 75-91. http://dx.doi.org/10.3390/environments1010075

Temu, V. W., Rude, B. J., & Baldwin, B. S. (2014a). Yield response of native warm-season forage grasses to harvest intervals and durations in mixed stands. *Agron., 4*, 90-107. http://dx.doi.org/10.3390/agronomy4010090

Tomlinson, K. W., & O'connor, T. G. (2004). Control of tiller recruitment in bunchgrasses: uniting physiology and ecology. *Functional Ecology, 18*(4), 489-496.

Vogel, K. P. (2000). Improving warm-season forage grasses using selection, breeding, and biotechnology. *Native warm-season grasses: research trends and issues*, (nativewarmseaso), 83-106. http://dx.doi.org/10.1111/j.0269-8463.2004.00873.x

White, B., Graham, P., & Pierce, R. A. II. (2005). Missouri Bobwhite Quail. Habitat Appraisal Guide: Assessing your farm's potential for bobwhites featuring a revised habitat appraisal tool (pp. 13-14). University of Missouri Extension MP902.

15

Analysis of the Performance of two Rangeland Protocols, Monitoring and Assessment

Soumana Idrissa[1,2]

[1] Faculté des Sciences Agronomiques, Université de Diffa, BP 78, Diffa, Niger

[2] Institut National de la Recherche Agronomique du Niger (INRAN), BP 429, Niamey, Niger

Correspondence: Soumana Idrissa, Université de Diffa & Institut National de la Recherche Agronomique du Niger. BP 78, Diffa, Niger. E-mail: smaiga15@yahoo.fr

Abstract

This study compared and contrasted data from the stick and modified Braun-Blanquet monitoring protocols in three areas with different land use histories: an unrestored barren area, a young and old restored areas. The study areas are part of extensive degraded of birch woodland and willow shrubland that have partly been re-vegetated. Vegetation and site characteristics were assessed in the three areas using the two protocols and soil sampling to characterize the ecological status of a land that has been re-vegetated. The analysis of the two protocols data indicates similar tendency which is the improvement of the ecological condition of the restored areas compared to the unrestored area. The soil carbon and nitrogen contents increased when the pH decreased with the restoration age. The improvement is better at the old restored area which has received more fertilization compared to the young restoration. Stick method estimated greater cover of vascular plants, litters, mosses and rocks, and lower amount of bare ground than modified Braun-Blanquet. The two protocols provided similar estimates cover of lichens and sedges. Stick method also provided three supplementary indicators which were not included in modified Braun-Blanquet: plants base, basal and canopy gaps. Another observation that could be proved by further studies, stick seemed to be more precise and economical than modified Braun-Blanquet. The indicators provided by the two protocols were related to the three attributes of ecosystems and the rangelands health indicators. This study is a preliminary that cannot be able to recommend one method, but it advocates stick method to assess and monitor vegetation dominated by herbaceous layer as grassland and modified Braun-Blanquet for the one dominated by woody layer.

Keywords: rangeland ecology, restoration, protocol, biodiversity, resilience, Iceland

1. Introduction

Monitoring biodiversity and detecting changes on natural resources are often been quantified by collecting data on vegetation composition and structure. Assessing plant composition and structure is necessary, but not enough to predict long-term resilience (Herrick et al., 2006b). Since ecosystem resilience depends on the functioning of ecological processes, it is crucial to base the assessment on them. However, directly assessing ecological processes is difficult due to their complexity and the interactions among them (Pellant et al., 2000). For that reason, simple indicators that relate ecological processes have been developed for monitoring rangeland condition (Ludwig et al., 2004). The typical example of indicators is the " Indicators of Rangeland Health' (IRH) developed by the United States land management agencies within a protocol titled 'Interpreting Rangeland Indicators Health" to assess rangelands condition (see Pellant et al., 2000, 2005; Pyke et al., 2002; Herrick et al., 2006a; Herrick et al., 2006b; Herrick et al., 2012). In this protocol, 17 indicators of plants cover and diversity, soil, water component, etc., are used to assess three ecosystem attributes on which all lands use depend: Site and soil stability, hydrologic function, and biotic integrity (Toevs et al., 2011). Soil and site stability refers to the capacity of a site to limit redistribution and loss of soil resources (e.g. nutrients and organic matter) by wind and water. Hydrologic function is the capacity of a site to capture, store, and safely release water from rainfall, run-off, and snowmelt. Biotic integrity is defined as the capacity of a site within an ecosystem to support natural processes within a normal, or expected, range of variability. Collectively, these three attributes define rangeland health, i.e., how ecological processes (water cycling, energy flow, and nutrient cycling) are functioning within a normal range of variation to support specific plant and animal communities. In practice, quantifiable biological and physical components of ecosystems that are correlated to those attributes are assessed as indicators of ecological processes and site integrity. Biological components of ecosystem include plants cover and composition, functional groups cover and composition, biological crusts, etc. The physical components of ecosystems consist of percentages of bare ground, rocks, etc. Several protocols including modified Braun-Blanquet and line-point intercept methods or recently "stick method", give measures that are used as indicators of some ecosystem attributes (Pellant et al., 2000; Ludwig et al., 2004; Tongway & Hindley 2004; Herrick et al., 2005; Pellant et al., 2005; Riginos & Herrick 2010). The modified Braun-Blanquet protocol for sampling vegetation is adapted from the Zurich -Montpellier school of phytosociology, one of the classic methods of studying vegetation (Braun-Blanquet, 1932). Braun-Blanquet protocol, even if it has been challenged to be subjective (Egler, 1954), the approach is still widely used and is argued to represent scientifically sound, versatile and efficient assessment method in botany (Werger, 1974). It is developed to identify and describe plants communities, used to monitor effects of changes on plant species within these communities, and to assess restoration or reclamation success of disturbed plant communities (Bonham et al., 2004). The "stick method" is a modification of the line-point intercept method, developed in context with a monitoring tool for rangeland assessment (see Riginos & Herrick, 2010). This protocol is suggested for rangelands assessment, it seems to be precise, easy to learn and to apply, and provide easily attributes that relate to productivity, infiltration or runoff and soil loss. Different studies have been carried out to describe several assessment protocols and show their strengths and weaknesses (Stohlgren et al., 1998; Prosser et al., 2003; Anderson & Fehmi, 2005; Carlsson et al., 2005; Godínez-Alvarez et al., 2009), additionally this study proposed to determine the differences between the protocols described above and how the indicators they provided can be linked to the ecological status of a given land.

This study intends to compare the two protocols, the stick method and modified Braun-Blanquet, for assessing the ecological status of a land that has been revegetated. Specifically, the purposes of the study is to: (1) compare and contrast the two monitoring protocols in three areas with different land use histories: an unrestored barren area, young revegetated area and old revegetated area, located within the same ecological site, (2) assess the succession trend in the three areas; (3) relate the indicators provided by the two protocols to the three key attributes of ecosystem and the Rangeland Health Indicators (RHI) for interpretation; (4) and evaluate the relevance of these simple indicators for sustainable land management.

2. Material and Methods

2.1 Study Area

The study was conducted in southern Iceland (Figure 1), 20 m above sea level, at three areas, with similar environmental characteristics and different land use histories: (1) unrestored area and (2) young restoration area (three years old) located at Varmadalur, and (3) old restoration area (seven years old) located at Selalækur (Figure 2). The climate of southern Iceland close to the study area is oceanic-boreal with a mean temperature from 1990 to 2004 of -0.97°C in January and 11.3°C in July, and a mean annual precipitation of 970.38 mm (Icelandic Meteorological Office, unpublished data from Hella weather station). The soils of Iceland, mostly Andosols (WRB; Vitric Andosol) or Andisols (Soil Taxonomy; Vitricryand), formed on volcanic deposit lava were exposed to wind and water erosion (Arnalds et al., 2001, 2013). The study sites are part of extensive degraded areas that have partly been revegetated. The cumulative effect of natural disturbances such as the cooling weather, the active volcanos, increased aeolian deposits; and human activities like deforestation, overgrazing added to the susceptibility of the soil to erosion, amplified the degradation (Arnalds, 2000). The nearby Hekla volcano is very active. It has erupted 20 times in historical times, producing both tephra and andesitic lava flows and has occasionally strewn tephra over the study area (Elmarsdottir et al., 2003). The history of the study area probably resembles that of large areas in Iceland where wood gathering and heavy grazing has destroyed native birch woodlands and willow shrublands. The soil surface of the study area is typical gravelly sand classified as lag gravel (Arnalds et al., 2001). The lag gravel soil, seem to result, from the degradation of the birch woodlands and willow shrublands, which were the original vegetation of Iceland at the time of settlement (Gunnlaugsdottir, 1985). Restoration actions started at southern Iceland about 100 years ago with the objective to increase vegetation cover on eroded areas and improve the pasture for grazing animals by fencing and aerial sowing grass. Treating degraded land with fertilizers, were initiated about 60 years ago with a rate of about 100-150 kg N ha^{-1}year^{-1} (see Elmarsdottir et al., 2003; Gretarsdottir et al., 2004). In this study, treatment were done by application of about 200 kg/ha of inorganic NP (25: 6). The three years old restored area had received three applications of fertilizer and the seven years old restored area four applications.

Figure 1. Location of the study in south Iceland; 1 = unrestored area, 2 = young restored area (3 years) and 3 = old restored area (7 years)

Figure 2. Cover of plants and bare ground at the three study areas: A = unrestored area, B = young restored area (3 years) and C = old restored area (7 years). (Photos: I. Soumana, 9-12 July 2013)

2.2 Sampling Design

Vegetation and site characteristics were assessed at four randomly selected points in each area. From each pre-determined point, four transects of 25 m were established in the direction of the four cardinal points for vegetation and sites characteristics surveying using the "stick method". At each pre-determined point, a 10 m × 10 m plot was established, in the north-east quadrant. Five 0.25 m² quadrats were randomly selected within the plot, for vegetation and sites characteristics surveying using modified Braun-Blanquet (Figure 3).

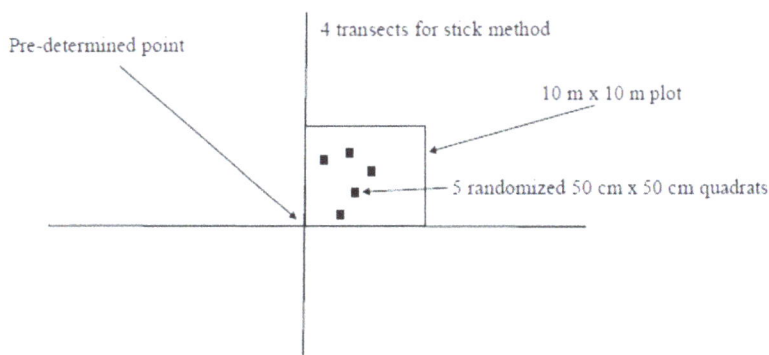

Figure 3. Placement of the transects for "stick method" and the 10 m × 10 m plot for modified Braun-Blanquet relative to each of the predetermined points

2.3 Sampling Vegetation and Site Characteristics with "Stick Method"

Along the four transects of each pre-determined point, a stick of one meter, was laid systematically on the ground at every five meters for recording vegetation and environmental variables. Foliar cover of plants functional groups along the 1-m stick by dropping a metal rod of one mm diameter vertically towards the soil at every 20 cm and all shrub, grass, forb, sedge, moss and lichen, contacted by the rod were recorded, for a total of 25 points/transect and 100 points/predetermined point. At the soil surface, contacts of the rod with plants base, litter, bare ground and rock; and base and canopy gaps through the stick were recorded. The total height of the vegetation which covers the stick, was also estimated visually Plant bases are places where plants are rooted to the ground, it reduced soil erosion by slowing water flowing. When plants are not close together, there are gaps either between plants base or between plants canopy that allow water and wind to pick up enough speed to carry away the soil (see Riginos and Herrick, 2010).

2.4 Sampling Vegetation and Site Characteristics with Modified Braun-Blanquet

Braun-Blanquet five levels of abundance have subsequently been modified into six, eight or ten levels by splitting one, two or three scales in order, to improve the accuracy of the estimated data (Daubenmire, 1959; van der Maarel, 1979). In this study, plants functional groups were estimated in the quadrats of 0.25 m² using the following eight cover classes: 1 = <1%; 2 = 1–5%, 3 = 6–10%; 4 = 11–15%; 5 = 16–25%; 6 = 26–50%; 7 = 51–75%; and 8 = 76–100%. The cover of total plants and other vegetation, percentages of bare ground, rock, litter and the height of the tallest branch were also recorded. The two protocols provide measurement of similar indicators, but Modified Braun-Blanquet does not include measures of plants base, basal and canopy gaps offered by the stick method.

2.5 Soils Sampling

Soils surface layer were sampled in the centre of each 0.25 m² quadrat, with an auger, to the depth of five cm, and then the five samples from each quadrat were mixed to make a composite sample. Soils were dried at 30°C and passed through a 2 mm sieve to prepare them for analysis. Furthermore, the soil samples were checked for moisture content at the time of analysis for adjusting results. Total carbon (g/kg) and nitrogen (g/kg) contents were determined by dry combustion using Vario Max C/N-Macro Elemental Analyser. Soil pH was measured with electrodes in a 1:5 soil-water suspension (Blakemore et al., 1972).

2.6 Data Analysis

Statistical analysis was done on the mean cover of total vascular plants, functional groups, litter, rocks, bare ground, plants base and basal gaps recorded in the three treatments. Before analysis, the cover scores from modified Braun-Blanquet were transformed to percentages by using the central value of each cover class and averaged over all the five quadrats of each 10 m × 10 m plot (cf Aradottir, 2012). The amount of shrub, grass, forb, sedge moss and lichen, base, litter, bare ground, rock, base and canopy gaps recorded on transects by the stick, were also averaged for each pre-determined point. Thus, there were four data points for each protocol in each area (treatment), for a total of 12 points. The pooled data from the two protocols was used to test for effects of assessment protocol, restoration age (treatment) and their interaction by analysis of variance (ANOVA, Generalized Linear Model) where restoration age was nested within the sample areas. The relationships between measurements by the two protocols were also analysed using the correlation of Pearson (r). For the using of Analysis of Variance (ANOVA) and Pearson Correlation, the normalities of the pooled data were tested by using the test of Kolmogorov-Smirnov. When the normality and equal variances were not met, the data were $\ln(x + 1)$, lnx, square-root or ASINH transformed. Transformation by $\ln(x + 1)$ was used for amount of rock and bare ground, lnx for litter cover, square-root for sedge cover and ASINH for moss, lichen, grass, forb and shrub covers. One way ANOVA was used to test the differences of soil pH, total nitrogen and carbon contents, and C/N ratio among treatments (restoration ages). The ANOVAs, normality and correlation tests were done with Minitab v.14. (Dytham, 2011). Principal Component Analysis (PCA), a multivariate test which weights the variables to maximize the variance of the response variable (Dytham, 2011), was used to visualize the differences between the two protocols in ordination space. PCA was also done separately on the two data sets, to observe how well they reflect difference in functional groups cover and composition, and changes in site characteristics.In the PCAs, cover of grass, forb, sedge, moss, shrub and lichen were used as variables of abundance. The PCAs were done using PC-ORD v.5.0 (McCune & Grace, 2002).

3. Results

3.1 Variation of Functional Groups Abundance and Site Characteristics with Increased Restoration Age and Between Protocols

GLM analysis done on the pooled data revealed significant effects of protocol types, restoration ages and their interaction for cover of total plant, rock, bare ground, moss, litter, grass, forb and shrub (Figure 4). Modified Braun-Blanquet protocol significantly gave lower cover of total plant, rock, moss, litter, grass, forb and shrub for all the treatments ($p < 0.001$) compared to the stick method. On the other hand, stick method gave significantly lower cover of bare ground for all the three treatments ($p < 0.001$) compared to the modified Braun-Blanquet. In contrast, there were no significant effects ($p > 0.05$) of protocols, restoration ages, and their interaction for cover of lichen and sedge excepted for lichen, which showed only significant effect for restoration ages ($p = 0.006$). In fact, for all the treatments, stick method seemed to capture more vegetation, plants functional groups, rock, and litter; and modified Braun-Blanquet appeared to detect high cover of bare ground. Compared to the unrestored area, the two protocols revealed significantly higher cover of total plants, moss, litter, grass, forb, lichen and shrub; and significantly lower cover of bare ground and rocks at the restored areas with increased age of the restoration treatment. Only the cover of sedge was not different between the three areas and the two protocols. In fact, despite the variation of cover estimates of plants and site characteristics, the two protocols showed similar tendencies.

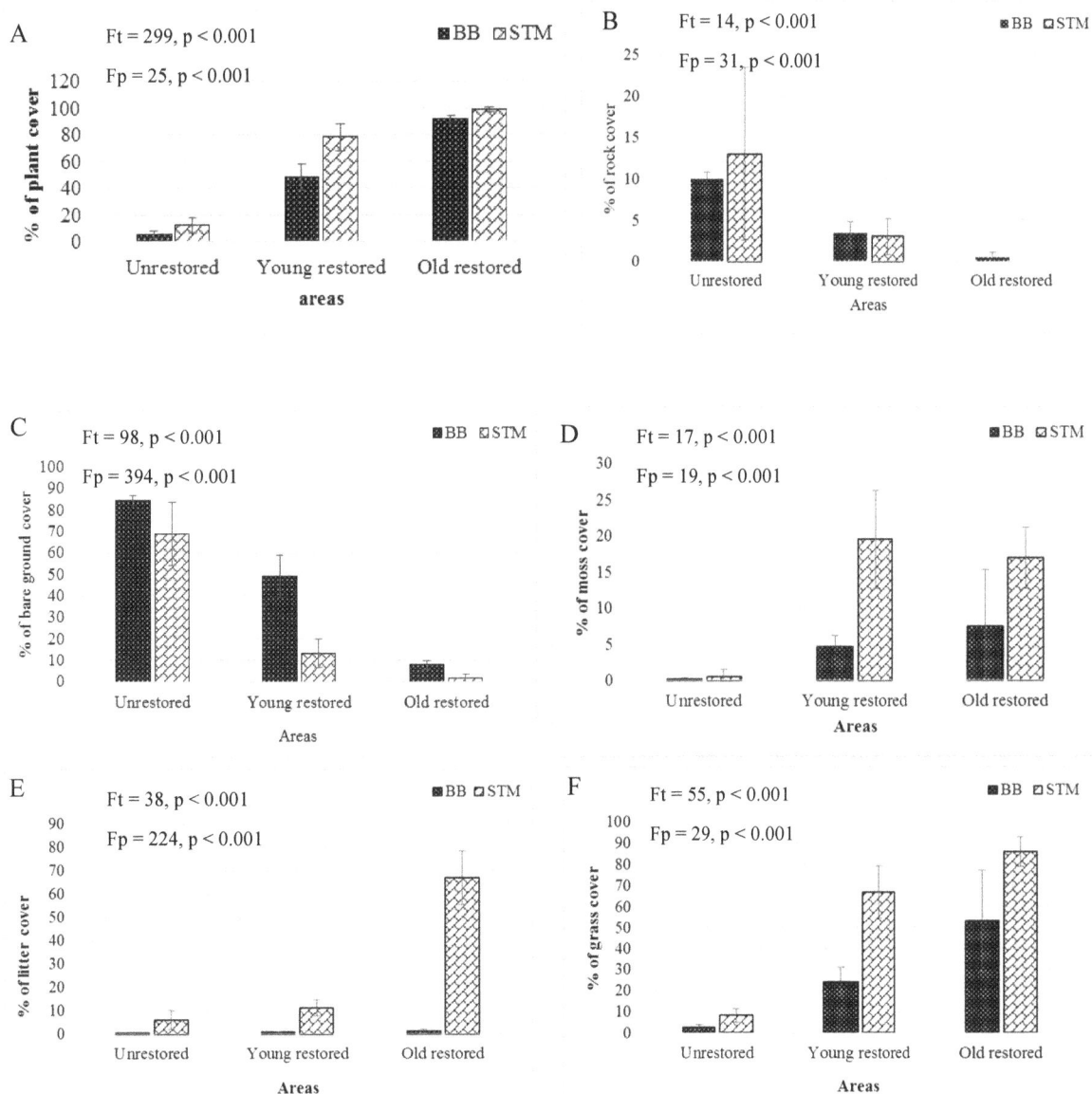

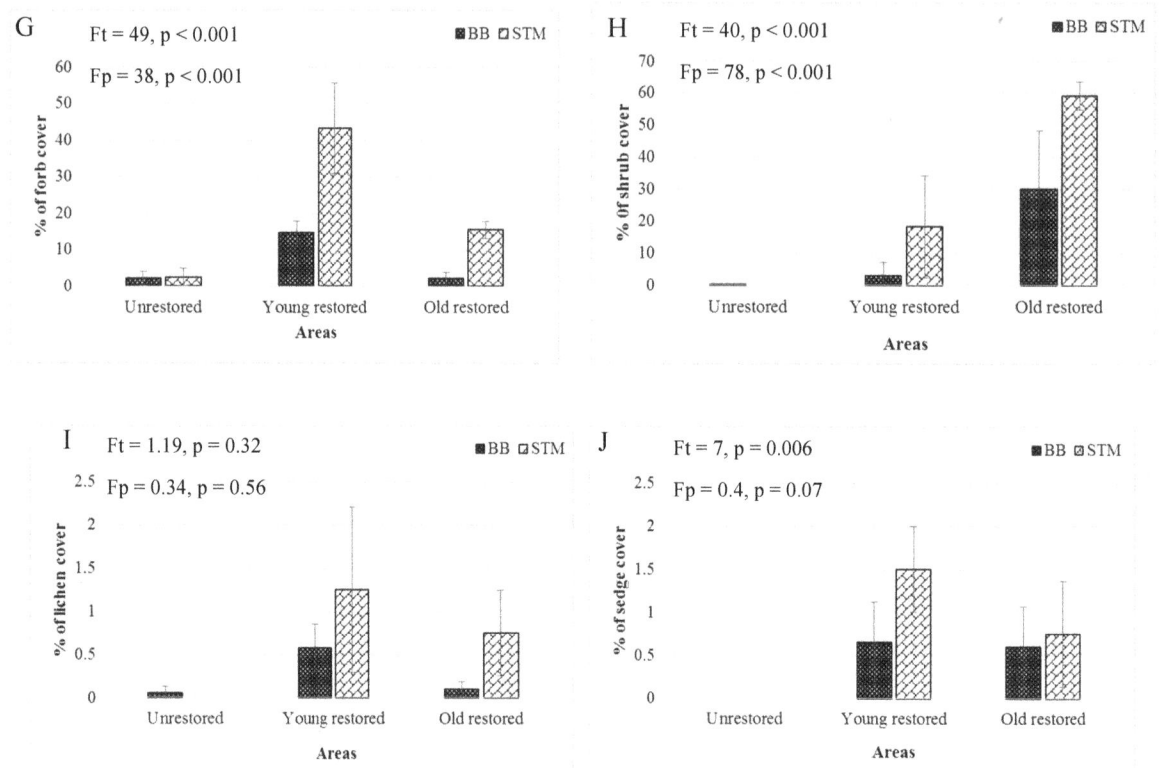

Figure 4. Estimated cover (mean and standard error) of (A) total plants, (B) rocks, (C) bare ground, (D) moss, (E) litter, (F) grass, (G) forb, (H) Shrub, (I) sedge and (J) lichen in different study areas for stick method (STM) and modified Braun-Blanquet protocol (BB)l. Results from nested ANOVA on protocol types (Fp), restoration (treatments)(Ft) and their interaction (Ft×Fp) for each cover (A-J). When P < 0.05 the effect is significant

3.3 Effects of Restoration Activities on the Land

The two first axis of the PCA ordination of pooled data from stick method and modified Braun-Blanquet assessment explained cumulatively 86.81% of the variance (Figure 5). In the graph, only plots recorded in the unrestored area were located in the same place, the other plots were scattered in the ordination space. Thus, data from the unrestored and restored areas showed respectively high homogeneity and variability of vegetation cover with samples, between treatments and protocols.

Axis 1 of the two PCAs of the stick method and Modified Braun-Blanquet when analysed separately (Figure 6) explained respectively 61 and 36 % of the total variances. The two graphs revealed similar tendency, plots from restored areas versus the unrestored area were separated along axis 1, and reflected distinctly a recovery gradient. Analysis of the data from the stick method PCA (Figure 6A), showed strong positive correlations between axis 1 and total plant cover, height, plants base, litter, carbon, nitrogen and C/N ratio. Strong negative correlations were observed between the same axis and rocks, bare ground, basal gaps and pH. Similar correlations were observed between factorial axis and environmental variables in the modified Braun-Blanquet PCA (Figure 6B). Axis 1 of the both ordinations was interpreted as a gradient of recovering plants cover, height and base, carbon and nitrogen contents, C/N ratio and litter; and reducing pH, rocks, and bare ground.

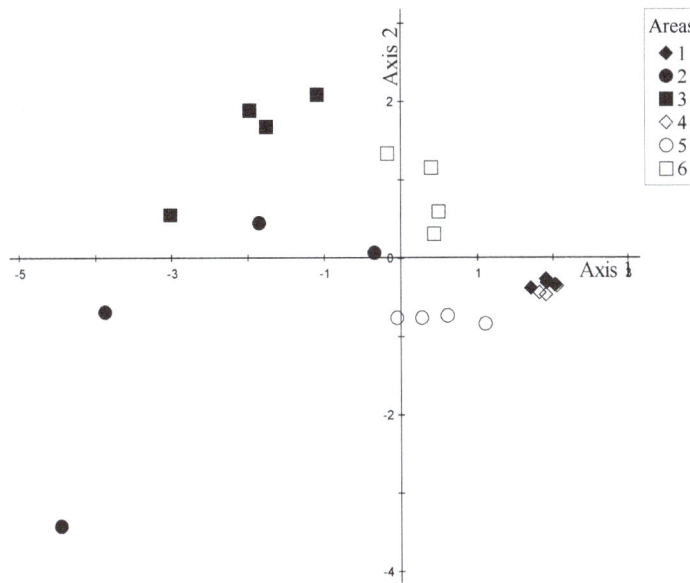

Figure 5. Principal components ordination from pooled data, stick method (filled plots) and Braun-Blanquet (empty plots), diamonds = unrestored plots, circles = yound restored plots and boxes = old restored plots; Eigenvalue and variance of axis 1 are respectively 0.79 and 61%, and eigenvalue and variance of axis 2 are 0.12 and 21%

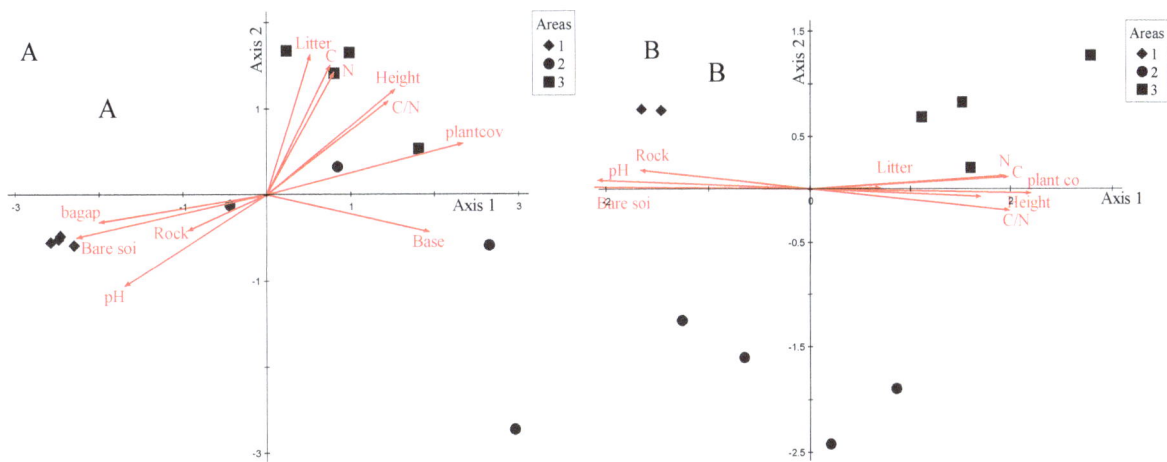

Figure 6. Principal components ordinations of stick method and modified Braun-Blanquet data: graph A = PCA with stick method; graph B = PCA with modified Braun-Blanquet; diamonds = unrestotred plots, circles = yound restored plots and boxes = old restored plots; Graph A: Eigenvalue and variance of axis 1 are 0.79 and 63%, and eigenvalue and variance of axis 2 are 0.19 and 31%; Graph B: Eigenvalue and variance of axis 1 are 0.71 and 36% and eigenvalue and variance of axis 2 are 0.24 and 29%. Plantcov = % of plants cover, height = height of plants (cm), % of plants base, C = soil surface carbon content, N = soil surface nitrogen contents, C/N ratio, litter = % of litter cover, rocks = % of rocks cover, Bare soi = % of bare ground cover, bagap= % of basal gaps

3.4 Comparison of the two Protocols

Strong correlation was observed between the stick method and modified Braun-Blanquet protocols for cover of total vegetation (r = 0.95), rocks (r = 0.86), bare ground (r = 0.91), moss (r = 0.87), grass (r = 0.93) and forb (r = 0.73). On the other hand, there was no relationship between the protocols for cover estimates of shrub, sedge, lichen and litter. Comparison between the stick method and modified Braun-Blanquet protocols (Figure 7) showed only similar cover estimates of sedge and lichen. Stick method gives higher cover values of total plant, rocks, grass, moss, litter, forb and shrub than modified Braun-Blanquet. On the other hand, modified Braun-Blanquet tended to estimate higher bare ground cover than stick method.

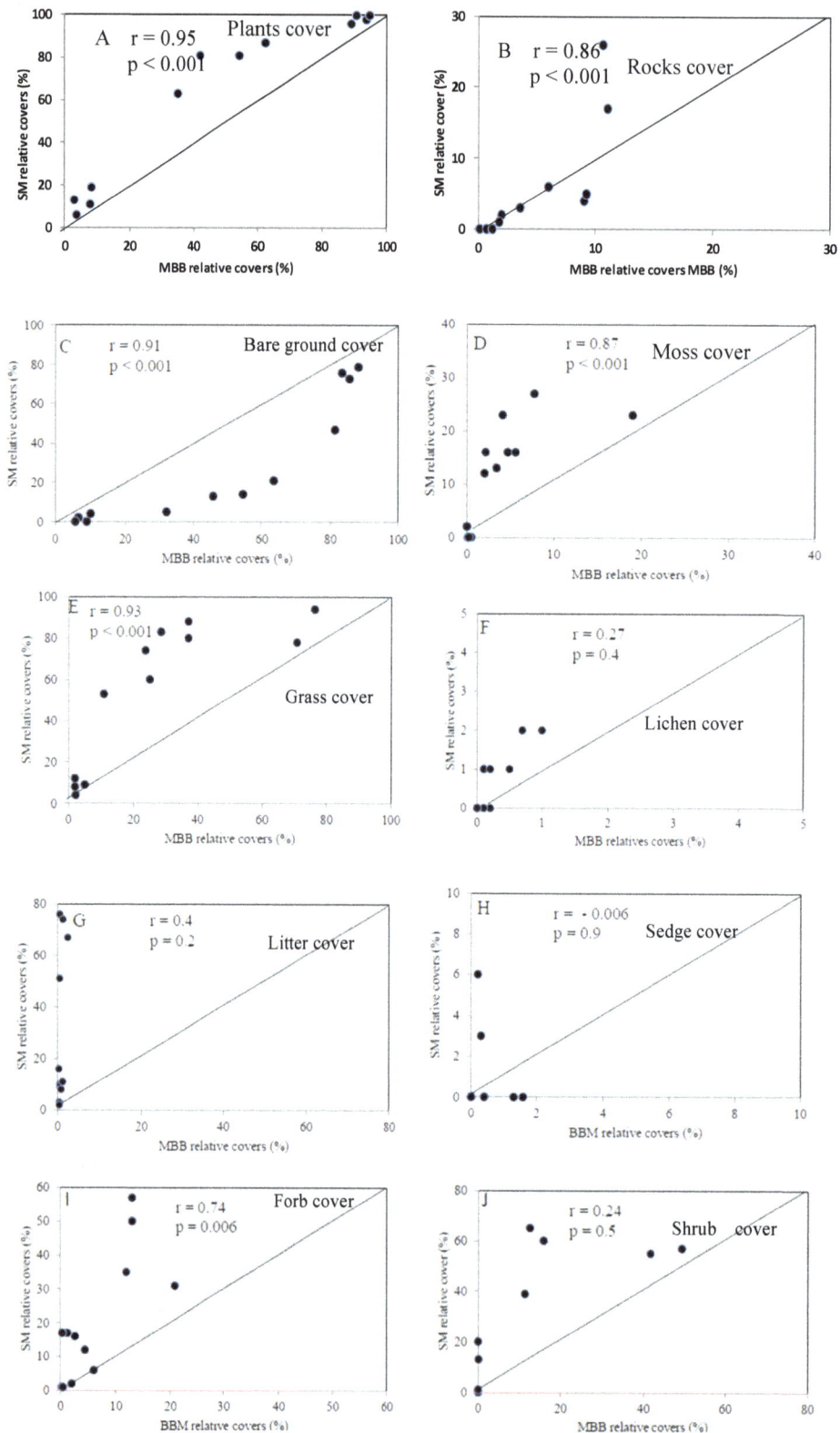

Figure 7. correlation of Person (r) between average cover per point measured by the stick method (SM) and the average cover per point for the modified Braun-Blanquet (MBB) protocol for (A) Total plant cover, (B) Rock, (C) bare ground, (D) moss, (E) grass, (F)lichen, (G) litter, (H) sedge, (I) forb and (J) shrub; when p < 0.05, the correlation is significant

3.5 Effects of Restoration Activities on Soil Surface Properties

Compared to the unrestored area, soil carbon (C) and nitrogen (N) contents and the C/N ratio increased significantly with restoration ages while pH decreased significantly with restoration ages ($p < 0.001$) (Figure 8).

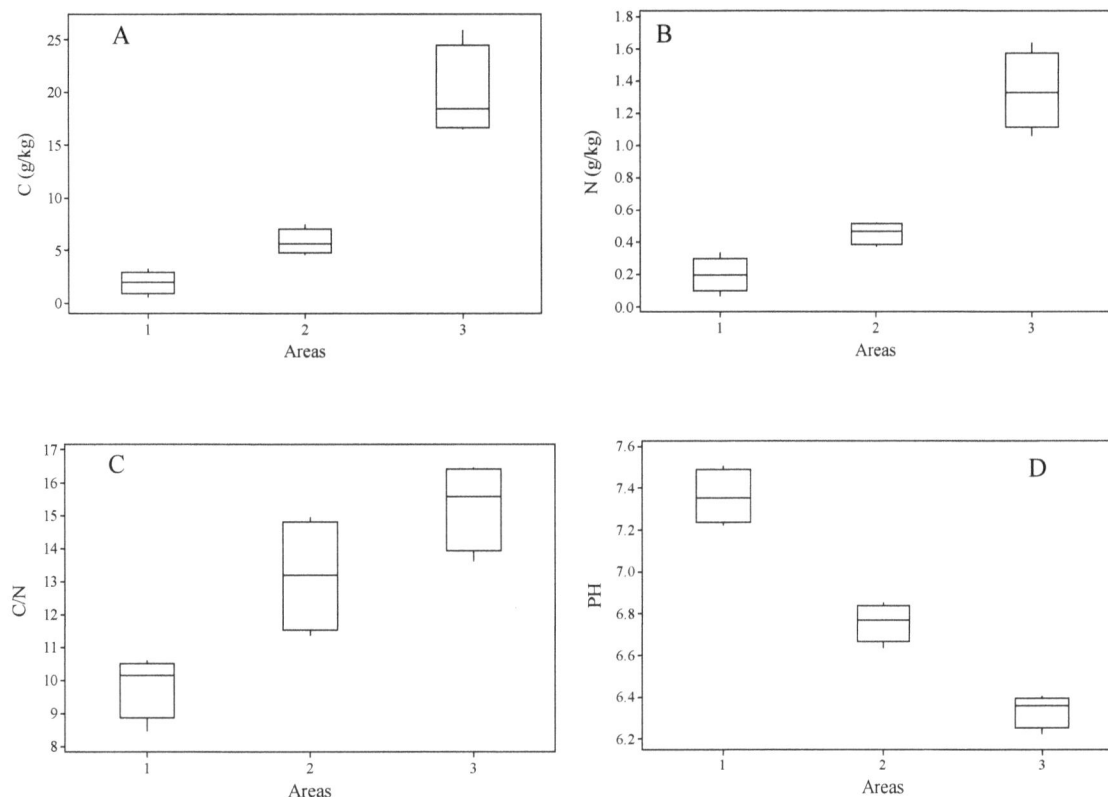

Figure 8. Variation of soil properties shown in box plots between the three areas; 1 = unrestored, 2 = young restored, 3 = old restored; Carbon (A), Nitrogen (B), C/N ratio (C) and pH (D); dash in box = median; the interquartile range = minimum and maximum

4. Discussion

4.1 Successional Trend and Interpretation of the Quantitative Indicators

Compared to the unrestored area, the restored areas condition was changed by the restoration action, which led to increase cover of vascular plants, lichen and moss, plants base, litter, soil carbon and nitrogen contents, and C/N ratio; and decrease cover of bare ground, rocks, basal gaps and soil pH. When the carbon and the soil contents were increasing with the restoration age in the restored areas, the pH was decreasing. The quantitative indicators showed the recovery of the fertilized areas. The differences between the three areas could be attributed to the age of restoration and the number of fertilizer applications. The abiotic and biotic conditions of the unrestored area can also constrain seedling survival and plants growth (Elmarsdottir et al., 2003). Fertilization may remove the constrained by improving the soil fertility at the restored areas. This may enhance the stability and the hydrological functions of the soil. Stabilized and high infiltrated soil could facilitate the turnover and the establishment of the plant species that increase plants productivity through the availability of safe microsite and the capture of wind-blown seeds (Gretarsdottir et al., 2004). Plants biomass production may increase foliar cover of vascular plants and base, and litter production that reduce the surface of bare ground and the amount of rocks. The enhancement of foliar cover of vascular plants can create a microclimate that can allow the establishment and the expansion of the understorey layer such as lichen and moss.

Table 1. Assessment protocols, quantitative indicators, keys attributes of ecosystems, and Rangeland Health Indicators (RHI) (Pellant et al. 2005; Riginos & Herrick 2010); HF = hydrological functional, SS = soil stability, BI = biotic integrity

Assessment protocols	Stick method			Modified Braun-Blanquet			
Quantitative Indicators	HF	SS	BI	HF	SS	BI	Rangeland Health indicators
% Total vascular plants cover	√		√	√		√	Bare ground, Annual production, gullies, plant mortality, number of function groups, plant communities, water flow
% bare ground	√	√			√	√	Rills, water flow, pedestals, gullies, wind-scoured areas, blowouts or deposition areas, litter movement, bare ground, soil resistance to erosion, soil loss and compaction layer, litter movement
% plant Base	√	√					Soil resistance to erosion, soil loss, invasive plant, compaction layer, litter amount, annual production, invasive plant, reproductive capability of perennial plants.
% Litter	√	√	√	√	√	√	Soil resistance to erosion, soil loss, compaction layer, plant mortality, litter amount, annual production, plant mortality, litter amount, productivity, invasive plants, reproductive capability of perennial plants.
% Rock	√	√		√	√		Water flow, pedestals, bare ground, wind-scoured areas, blowouts or deposition areas, soil resistance to erosion, soil loss and degradation
% Basal gaps	√	√	√				Rills, water flow, pedestals, gullies, wind-scoured areas, blowouts or deposition areas, litter movement, bare ground, soil resistance to erosion, soil loss and compaction layer, litter movement
% Plant compositions	√		√	√		√	Annual production, plant mortality, function groups, plant communities, invasive plants, reproductive capability
% Functional groups	√		√	√		√	Soil resistance to erosion, soil loss, compaction layer, plant mortality, litter amount, annual production, plant mortality, litter amount, productivity, invasive plants, reproductive capability of perennial plants
% Lichen	√	√	√	√	√	√	Biological soil crusts distribution and degree of development
% Moss	√	√	√	√	√	√	Biological soil crusts distribution and degree of development
% Grass cover	√		√	√	√		Above ground production, water flow, soil surface loss, soil resistance to erosion, compaction layer, litter movement

According to Elmarsdottir et al. (2003), application of fertilizer without additional seeding on degraded lands can enhance favourable microsite availability and the turnover of native species, and expand plants cover. Site treatments such as seeding, planting turfs, fertilizing, organic mulching or soils physical treatment were known to accelerate succession by improving biotic and abiotic conditions of degraded lands (Aronson et al., 2006; Prach & Hobbs 2008; Řehounková & Prach, 2008; Aradottir, 2012).

Sustainability of the restored area depends on the recovering of the biotic integrity, hydrological functions and soil stability (Herrick et al., 2012). These attributes are the foundation of resilience; i.e. the capacity of the site to recover after perturbation (Holling, 1973). The simple quantitative indicators can be measured as surrogates to the attributes of resilience and the rangeland indicators health (RHI) (Pellant et al., 2005; Riginos & Herrick, 2010; Kachergis et al., 2011) (Table 1). Biotic integrity as surrogate to energy capture and nutrient cycling can be simply measured by cover of plants, lichen and moss, soil carbon and nitrogen contents, etc.; hydrological functions can be simply measured by cover of bare ground, rocks, basal gaps etc.; soil stability can be estimated by plants cover, litter distribution, soil carbon and nitrogen contents, etc. Therefore, restored area with high cover of vascular plants, lichen and moss and high contents of nitrogen and carbon has a great biotic integrity. These also infer low bare ground, basal gaps and rocks which address improved hydrological functions and soil stability. Some of the indicators can act for more than one attribute, e.g. degraded area with a high cover of bare ground allows water flow and soil loss, which reflects low foliar cover and infer reduced soil stability and biotic integrity. Accordingly, the quantitative indicators can be assessed to address nearby ecosystem functions. This information may probably include biodiversity, plants mortality, soil condition, nutrient and energy fluxes, etc. that are likely to address future changes. Herrick et al. (2012) suggested to use those simple indicators, which reflect both earlier and future changes to monitor short and long term effects of management. These informations could be extrapolated to a large area, by using remote sensing and Geographic Information System (GIS) tools. Such simple indicators were needed for assessment and monitoring land management because they can act for more than one attributes of ecosystems and cover a large landscape (Ludwig et al., 2004). Temporal measurements of these indicators can also be stocked in a data base, integrated in conceptual models such as state and transition model (S & T) to guide lands management by identifying thresholds and trends, and adjusting strategies (Karl & Herrick, 2010).

4.2 Comparison of the two Monitoring Protocols

Similar trends were observed in the recovery of the restored areas when data from monitoring protocols were analysed separately. Both protocols showed greater plants foliar cover and base, functional groups abundance, soil carbon and nitrogen contents and C/N ratio and lower bare ground, rocks, plants basal gaps and soil pH at the restored areas. Consequently, the indicators revealed the gradual improvement of the ecological condition of the restored areas, which is better at the old restoration area than at the young restored area. Analysis of the pooled data showed variations of assessment data between the three treatments, samples and the two methods despite the surveying of the same areas. These variations could be attributed to the difference of the data provided by the two protocols. Compared to modified Braun-Blanquet, the stick method gave lower cover of bare ground while modified Braun-Blanquet tended to give lower cover of total plants, litter, moss shrub, forb, grass, lichen, and litter (Figure 4). This variation could be also attributed to the difference of the sampling locations at the same pre-determined point, but variation between protocols and plots could be more important than between surveying locations (Anderson & Fehmi, 2005). Moreover plants were recognized to have spatial patterns than being distributed uniformly. In fact, changes in surveying location could allow change in vegetation data in the same plants community (Carlsson et al., 2005). Observer behaviour in placing the sample and the rod, following the transect, and ocular estimation level (Tonteri, 1990) could also affect the data. Experience of the observer in vegetation sampling has been shown to improve the accuracy of the data (Kercher et al., 2003; Carlsson et al., 2005; Milberg et al., 2008).

As the two protocols provided similar tendencies, they led to a comparable interpretation of the data. Hence, the differences observed should be considered as bias that could be linked to the differences between the two protocols and the spatial variability of the vegetation. These factors could interact and influence the precision of the data. Studies to compare the accuracy of ocular estimation of cover such as modified Braun-Blanquet, Daubenmire and modified Whittaker plots, etc to other methods of surveying vegetation (Kercher et al., 2003; Leis et al., 2003; Anderson & Fehmi, 2005; Godínez-Alvarez et al., 2009; Laliberté et al., 2010) have shown that ocular estimation can lower estimate plants cover, but it seems to have great potential to detect maximum species number of assessing areas compared to the line-point intercept. Consequently, ocular estimates methods seem to be good to monitor and assess biodiversity (Stohlgren et al., 1998; Godínez-Alvarez et al., 2009). Ocular estimates methods also seem to be consistent for assessing shrub cover than line-point intercept protocol (Brun & Box, 1963; Floyd & Anderson, 1987). Godínez-Alvarez et al. (2009) supported to use ocular estimates method to assess vegetation dominates by shrubs. The stick method is a modification of line-point intercept. Compared to ocular estimations methods, line-point intercept seems to be more precise to measure foliar cover (Godínez-Alvarez et al., 2009; Kercher et al., 2003). Accordingly, the stick method may also be more accurate to estimate cover of vascular plants, lichen, moss, rocks and litter than modified Braun-Blanquet. It provided three supplementary indicators which were not available by modified Braun-Blanquet: basal and canopy gaps, and plants base that are related to wind and water erosion, and infer hydrological functions, biotic integrity and soil stability. It is also a science-based monitoring protocol that can be simply used by local communities without assistance (Riginos & Herrick, 2010), which makes it very useful in land assessment as in the "Farmers Heal the Land" project. Other observations to be proved by further studies, stick method seem to be more economic in time than modified Braun-Blanquet.

5. Conclusion and Recommendation

The main purpose of land management assessment and monitoring is to provide indicators that can reliably assess the condition of the land. The investigation showed that the two assessing protocols provide the same tendency. The estimate indicators can be related to the three attributes of ecosystems that include soil stability, hydrological functions and biotic integrity, and the indicators of rangeland health. The information from the two protocols could be extrapolated to a large area using remote sensing and GIS tools. The results can also be integrated in conceptual models such as S & T models to identify management trends, thresholds for the control of the management. This study revealed the robustness of the two protocols to assess and monitor land management. Compared to the modified Braun-Blanquet, the stick method seems to estimate greater cover of vascular plants, moss, litter and rocks. The two protocols provide similar estimates cover of lichen and sedge. In spite of that, the stick method may better assess lands condition and monitor revegetated lands because it also provides supplementary indicators like plants base, basal and canopy gaps that are not provided by modified Braun-Blanquet protocol.

Acknowledgements

This paper is a part of my final report of the United Nation University-Land Restoration Training programme (UNU-LRT), I would like to acknowledge the UNU-LRT programme, the Soil Conservation Service of Iceland and the "Institut National de la Recherche Agronomique du Niger" for having allowed me to attend this training.

References

Anderson, A. B., & Fehmi, J. S. (2005). Comparison of Two Survey Methods for Estimating vegetative Cover. *Transactions of the Illinois State Academy of Science, 97*, 165-178.

Aradottir, A. L., Robertson, A., & Moore, E. (1997). Circular statistical analysis of birch colonization and the directional growth response of birch and black cottonwood in south Iceland. *Agricultural and Forest Meteorology, 84*, 179-186. http://dx.doi.org/10.1016/s0168-1923(96)02385-4

Aradottir, A. L. (2005). Restoration of birch and willow woodland on eroded areas. Proceedings of the AFFORNORD conference on Effects of afforestation on ecosystems, landscape and rural development. Reykholt, Iceland, June 18-22, 2005. Nordic Council, Copenhagen.

Aradottir, A. L. (2012). Turf transplants for restoration of alpine vegetation: does size matter? *Journal of Applied Ecology, 49*, 439-446. http://dx.doi.org/10.1111/j.1365-2664.2012.02123.x

Aradottir, A. L., & Oskarsdottir, G. (2013). The use of native turf transplants for roadside revegetation in a subarctic area. *Icelandic Agricultural Sciences, 26*, 59-67.

Aradottir, A. L., & Hagen, D. (2013). Ecological Restoration: Approaches and Impacts on Vegetation, Soils and Society. *Advances in Agronomy, 120*, 173-222. http://dx.doi.org/10.1016/b978-0-12-407686-0.00003-8.

Arnalds, O. (2000). Desertification: an appeal for a broader perspective. In Rangeland desertification. In O. Arnalds & S. Archer (Eds.), Kluwer Academic Dordrecht/Boston/London (pp. 5-15).

Arnalds, O., Thorarinsdottir, E. F., Metusalemsson, S., Jonsson, A., Gretarsson, E., & Arnason, A. (2001). *Soil erosion in Iceland*. Soil Conservation Service and Agriculture Research Institute of Iceland.

Arnalds, O., Orradottir, B., & Aradottir, A. L. (2013). Carbon accumulation in Icelandic desert Andosols during early stages of restoration. *Geoderma, 193*, 172-179. http://dx.doi.org/10.1016/j.geoderma.2012.10.018.

Aronson, J., Clewell, A. F., Blignaut, J. N., & Milton, S. J. (2006). Ecological restoration: A new frontier for nature conservation and economics. *Journal for Nature Conservation, 14*, 135-139. http://dx.doi.org/10.1016/j.jnc.2006.05.005

Aronson, J.C Floret, E Floc'h, C Ovalle and R Pontanier (1993). Restoration and Rehabilitation of Degraded Ecosystems in Arid and Semi-Arid Lands. I. A View from the South. *Restoration Ecology, 1*, 8-17. http://dx.doi.org/10.1111/j.1526-100X.1993.tb00004.x

Blakemore, I. C, Searle, P. L., & Daly, B. K. (1972). *Methods of Chemical Analysis of Soils. New Zealand Soil Bureau Report 10A*. Government Printer, Wellington, New Zealand. Retrieved from http://digitallibrary.landcareresearch.co.nz/cdm/ref/collection/p20022coll2/id/139

Bestelmeyer, B. T, Tugel, A. J., Peacock Jr, G. L., Robinett, D. G., Shaver, P. L., Brown, J. R., ... Havstad, K. M. (2009). State-and-transition models for heterogeneous landscapes: a strategy for development and application. *Rangeland Ecology & Management, 62*, 1-15.

Bonham, C. D., Mergen, D. E., & Montoya, S. (2004). Plant cover estimation: a contiguous Daubenmire frame. *Rangeland Ecology & Management, 26*, 17-22. http://dx.doi.org/10.2111/RANGELANDS-D-10-00077.1

Brun, J. M., & Box, T. W. (1963). A comparison of line intercepts and random point frames for sampling desert shrub vegetation. *Journal of Range Management, 16*, 21-25.

Cairns Jr, J., & Heckman, J. R. (1996). Restoration ecology: the state of an emerging field. *Annual Review of Energy and the Environment, 21*, 167-189. http://dx.doi.org/10.1146/annurev.energy.21.1.167

Callaway, R. M., & Walker, L. R. (1997). Competition and facilitation: a synthetic approach to interactions in plant communities. *Ecology, 78*, 1958-1965. http://dx.doi.org/10.1890/0012-9658(1997)078

Carlsson, A. L., Bergfur, M. J., & Milberg, P. (2005). Comparison of data from two vegetation monitoring methods in semi-natural grasslands. *Environmental Monitoring and Assessment, 100*, 235-248.

Clements, F. E. (1936). Nature and structure of the climax. *Journal of Ecology, 24*, 252-284.

Connell, J. H., & Slatyer, R. O. (1977). Mechanisms of succession in natural communities and their role in community stability and organization. *American Naturalist*, 1119-1144. Retrieved from https://www.jstor.org/stable/2460259

Daubenmire, R. F. (1959). A canopy -cover method of vegetational analysis. *Northwest Science, 33*, 43-46.

Day, R., & Quinn, G. (1989). Comparisons of treatments after an analysis of variance in ecology. *Ecological Monographs, 59*, 433-463. http://dx.doi.org/10.2307/1943075/full

Del Moral, R., Saura, J. M., & Emenegger, J. N. (2010). Primary succession trajectories on a barren plain, Mount St. Helens, Washington. *Journal of Vegetation Science, 21*, 857-867. http://dx.doi.org/10.1111/j.1654-1103.2010.01189.x

Dudley, N., Morrison, J., Aronson, J., & Mansourian, S. (2005). Why Do We Need to Consider Restoration in a Landscape Context? In S. Mansourian, D. Vallauri & N. Dudley (eds.), F*orest Restoration in Landscapes: Beyond Planting Trees* (pp. 51-58). Springer, New York (USA).

Dytham, C. (2011). *Choosing and using statistics: a biologist's guide.* Wiley-Blackwell Malden, MA, USA.

Egler, F. E. (1954). Philosophical and practical considerations of the Braun-Blanquet system of phytosociology. *Castanea, 19*, 45-60. http://dx.doi.org/10.2307/20165286/pdf

Elmarsdottir, A., Aradottir, A. L., & Trlica, M. (2003). Microsite availability and establishment of native species on degraded and reclaimed sites. *Journal of Applied Ecology, 40*, 815-823. http://dx.doi.org/10.1046/j.1365-2664.2003.00848.x

Emery, S. M., & Rudgers, J. A. (2010). Ecological assessment of dune restorations in the great lakes region. *Restoration Ecology, 18*, 184-194. http://dx.doi.org/10.1111/j.1526-100X.2009.00609.x

Floyd, D. A., & Anderson, J. E. (1987). A comparison of three methods for estimating plant cover. *The Journal of Ecology*, 221-228. http://www.jstor.org/stable/2260547

Fukami, T., & Nakajima, M. (2011). Community assembly: alternative stable states or alternative transient states?. *Ecology letters, 14*, 973-984. http://dx.doi.org/10.1111/j.1461-0248.2011.01663.x

Galatowitsch, S. M. (2012). *Ecological restoration.* Sinauer Associates, Sunderland, Massachusetts USA. http://dx.doi.org/10.1002/jwmg.747/abstract

Gleason, H. A. (1917). The structure and development of the plant association. *Bulletin of the Torrey Botanical Club, 44*, 463-481.

Godínez-Alvarez, H., Herrick, J., Mattocks, M., Toledo, D., & Van Zee, J. (2009). Comparison of three vegetation monitoring methods: their relative utility for ecological assessment and monitoring. *Ecological Indicators, 9*, 1001-1008.

Gretarsdottir, J., Aradottir, A. L, Vandvik, V., Heegaard, E., & Birks, H. (2004). Long-Term Effects of Reclamation Treatments on Plant Succession in Iceland. *Restoration Ecology, 12*, 268-278. http://dx.doi.org/10.1111/j.1061-2971.2004.00371.x.

Gunnlaugsdottir, E. (1985). Composition and dynamical status of heathland communities in Iceland in relation to recovery measure. *Acta Phytogeographica Suecica, 75*, Uppsala, Sweden. http://dx.doi.org/10.1111/j.1756-1051.1987.tb00930.x/abstract

Havstad, K., & Herrick, J. (2003). Long-term ecological monitoring. *Arid Land Research and Management, 17*, 389-400. http://dx.doi.org/10.1080/713936102

Herrick, J, Bestelmeyer, B., Archer, S, Tugel, A., & Brown, J. (2006). An integrated framework for science-based arid land management. *Journal of Arid Environments, 65*, 319-335.

Herrick, J. E., Urama, K. C., Karl, J. W., Boos, J., Johnson, M. V. V., Shepherd, K. D., ... Guerra, J. L. (2013). The global Land-Potential Knowledge System (LandPKS): Supporting evidence-based, site-specific land use and management through cloud computing, mobile applications, and crowdsourcing. *Journal of Soil and Water Conservation, 68*, 5A-12A. http://dx.doi.org/10.2489/jswc.68.1.5A

Herrick, J. E., Duniway, M. C., Pyke, D. A., Bestelmeyer, B. T., Wills, S. A., Brown, J. R., ... Havstad, K. M. (2012). A holistic strategy for adaptive land management. *Journal of Soil and Water Conservation, 67*, 105-113. http://dx.doi.org/10.2489/jswc.67.4.105A

Herrick, J. E., Justin, W., Van Zee, K., Havstad, M., Burkett, L. M., & Whitford, W. G. (2005). Monitoring manual for grassland, shrubland and savanna ecosystems. Vol. II: Design, supplementary methods and interpretation. Jornada Experimental Range, Las Cruces, NM: Distributed by University of Arizona Press., Arizona press. Retrieved from http://jornada.nmsu.edu/monit-assess/manuals

Herrick, J. E., Schuman, G. E., & Rango, A. (2006). Monitoring ecological processes for restoration projects. *Journal for Nature Conservation, 14*(3), 161-171.

Hobbs, R. J., & Cramer, V. A. (2008). Restoration ecology: interventionist approaches for restoring and maintaining ecosystem function in the face of rapid environmental change. *Annual Review of Environment and Resources, 33*, 39-61. http://dx.doi.org/10.1146/annurev.environ.33.020107.113631

Hobbs, R. J., & Norton, D. A. (2004). Ecological filters, thresholds, and gradients in resistance to ecosystem reassembly. In Assembly rules and restoration ecology: Bringing the gap between theory and pratice. In V. M. Temperton, R. J. Hobbs, T. Nuttle & S. Halle (Eds), Island Press, Washington, DC (USA), 72-95. Retrieved from http://research-repository.uwa.edu.au

Holling, C. S. (1973). Resilience and stability of ecological systems. *Annual Review of Ecology and Systematics, 4*, 1-23. http://dx.doi.org/10.1146/annurev.es.04.110173.000245

Howell, E. A., Harrington, J. A., & Glass, S. B. (2012). *Introduction to restoration ecology*. Washington DC, USA: Island Press.

Kachergis, E., Rocca, M. E., & Fernandez-Gimenez, M. E. (2011). Indicators of ecosystem function identify alternate states in the sagebrush steppe. *Ecological applications, 21*, 2781-2792. http://warnercnr.colostate.edu

Karl, J. W., & Herrick, J. E. (2010). Monitoring and assessment based on ecological sites. Rangelands *Ecology and Management, 32*, 60-64. http://dx.doi.org/10.2111/RANGELANDS-D-10-00082.1

Karlsdóttir, L., & Aradóttir, Á. L. (2006). Propagation of Dryas octopetala L. and Alchemilla alpina L. by direct seedling and planting of stem cuttings. *Icelandic Agricultural Sciences, 19*, 25-32.

Kercher, S. M., Frieswyk, C. B., & Zedler, J. B. (2003). Effects of sampling teams and estimation methods on the assessment of plant cover. *Journal of Vegetation Science, 14*(6), 899-906.

Lal, R. (2004). Soil carbon sequestration impacts on global climate change and food security. *Science, 304*(5677), 1623-1627.

Lal, R. (2004). Soil carbon sequestration to mitigate climate change. *Geoderma, 123*(1), 1-22.

Laliberté, E., Norton, D. A., Tylianakis, J. M., & Scott, D. (2010). Comparison of two sampling methods for quantifying changes in vegetation composition under rangeland development. *Rangeland ecology & management, 63*(5), 537-545.

Leis, S. A. (2015, February). Comparison of vegetation sampling procedures in a disturbed mixed-grass prairie. In *Proceedings of the Oklahoma Academy of Science* (Vol. 83, pp. 7-15).

Ludwig, J. A., Tongway, D. J., Bastin, G. N., & James, C. D. (2004). Monitoring ecological indicators of rangeland functional integrity and their relation to biodiversity at local to regional scales. *Austral ecology, 29*(1), 108-120.

McCune, B. P., & Grace, J. B. (2002). Analysis of ecological communities. MJM software design, Gleneden beach, Oregon, USA.

MEA. (2005). Millenium Ecosystem Assessment. Ecosystems and Human Well-being: Biodiversity Synthesis. World Resources Institute, Washington DC, USA.

Milberg, P., Bergstedt, J., Fridman, J., Odell, G., & Westerberg, L. (2008). Observer bias and random variation in vegetation monitoring data. *Journal of Vegetation Science, 19*(5), 633-644.

Miller, M. E. (2008). Broad-scale assessment of rangeland health, Grand Staircase-Escalante National Monument, USA. *Rangeland Ecology & Management, 61*, 259-262. www.bioone.org/doi/pdf/10.2111/07-107.1

Odum, E. P. (1969). The strategy of ecosystem developement. *Science, 64*, 262-270.

Palmer, M., Allan, J. D., Meyer, J., & Bernhardt, E. S. (2007). River restoration in the twenty‐first century: data and experiential knowledge to inform future efforts. *Restoration Ecology, 15*(3), 472-481.

Parker, V. T., & Pickett, S. T. (1997). Restoration as an ecosystem process: implications of the modern ecological paradigm. In K. M. Urbanska, N. R. Webb & P. J. Edwards (eds.), *Restoration Ecology and Sustainable Development* (pp. 17-32). Cambridge University Press, Cambridge.

Pellant, M., Shaver, P., Pyke, D. A., & Herrick, J. E. (2000). Interpreting indicators of rangeland health, version 3. Technical Reference 1734-6. U.S. Department of the Interior, Bureau of Land Management, National Science and Technology Center, Denver, CO. BLM/WO/ST-00/001+1734/REV05. Retrieved from http://jornada.nmsu.edu/monit-assess/manuals

Pellant, M., Shaver, P., Pyke, D. A., & Herrick, J. E. (2005). Interpreting indicators of rangeland health, version 4. Technical Reference 1734-6. U.S., Department of the Interior, Bureau of Land Management, National Science and Technology Center, Denver, CO. BLM/WO/ST-00/001+1734/REV05. Retrieved from http://jornada.nmsu.edu/monit-assess/manuals

Petursdottir, T., Aradottir, A. L., & Benediktsson, K. (2013). An Evaluation of the Short - Term Progress of Restoration Combining Ecological Assessment and Public Perception. *Restoration Ecology, 21*(1), 75-85. http://dx.doi.org/10.1111/j.1526-100X.2011.00855.x.

Pickett, S. T. A., Collins, S. L., & Armesto, J. J. (1987). Models, mechanisms and pathways of succession. *The Botanical Review, 53*(3), 335-371.

Prach, K, Marrs, R., Pyšsek, P., & Van Diggelen, R. (2007). Manipulation of succession. In L. R. Walker, J. Walker & R. Hobbs (Eds.), *Linking ecological restoration and ecological succession* (pp. 121-149). Springer, New York (USA).

Prach, K., & Hobbs, R. J. (2008). Spontaneous succession versus technical reclamation in the restoration of disturbed sites. *Restoration Ecology, 16*, 363-366.

Prosser, C. W., Skinner, K. M., & Sedivec, K. K. (2003). Comparison of 2 techniques for monitoring vegetation on military lands. *Journal of Range Management*, 446-454.

Raevel, V., Violle, C., & Munoz, F. (2012). Mechanisms of ecological succession: insights from plant functional strategies. *Oikos, 121*(11), 1761-1770.

Řehounková, K., & Prach, K. (2008). Spontaneous vegetation succession in gravel–sand pits: a potential for restoration. *Restoration Ecology, 16*(2), 305-312.

Riginos, C., & Herrick, J. E. (2010). Monitoring Rangeland Health: A Guide for Pastoralist Communities and Other Land Managers in Eastern Africa, Version II, Nairobi, Kenya: ELMT-USAID/East Africa.

Ruiz-Jaen, M. C., & Mitchell Aide, T. (2005). Restoration success: how is it being measured?. *Restoration Ecology, 13*(3), 569-577. http://dx.doi.org/10.1111/j.1526-100X.2005.00072.x

SER. (2004). Society for Ecological Restoration International Science & Policy Working Group. The SER International Primer on Ecological Restoration. Tucson, Arizona, USA. Retrieved from http://www.ser.org/resources/resources-detail-view/ser-international-primer-onecological-restoration

Sheley, R. L., James, J. J., Vasquez, E. A., & Svejcar, T. J. (2011). Using rangeland health assessment to inform successional management. *Invasive Plant Science and Management, 4*(3), 356-366.

Silver, W. L., Ostertag, R., & Lugo, A. E. (2000). The potential for carbon sequestration through reforestation of abandoned tropical agricultural and pasture lands. *Restoration ecology, 8*(4), 394-407.

Steen, D. A., Conner, L. M., Smith, L. L., Provencher, L., Hiers, J. K., Pokswinski, S., ... Guyer, C. (2013). Bird assemblage response to restoration of fire - suppressed longleaf pine sandhills. *Ecological Applications, 23*(1), 134-147.

Steen, D. A., Smith, L. L., Conner, L. M., Litt, A. R., Provencher, L., Hiers, J. K., ... Guyer, C. (2013). Reptile assemblage response to restoration of fire-suppressed longleaf pine sandhills. *Ecological Applications, 23*(1), 148-158.

Stohlgren, T. J., Bull, K. A., & Otsuki, Y. (1998). Comparison of rangeland vegetation sampling techniques in the Central Grasslands. *Journal of Range Management*, 164-172.

Suding, K. N., Gross, K. L., & Houseman, G. R. (2004). Alternative states and positive feedbacks in restoration ecology. *Trends in Ecology & Evolution, 19*(1), 46-53.

Toevs, G. R., Karl, J. W., Taylor, J. J., Spurrier, C. S., Karl, M. S., Bobo, M. R., & Herrick, J. E. (2011). Consistent indicators and methods and a scalable sample design to meet assessment, inventory, and monitoring information needs across scales. *Rangelands, 33*(4), 14-20.

Tongway, D. J., & Hindley, N. L. (2004). *Landscape function analysis manual: procedures for monitoring and assessing landscapes with special reference to minesites and rangelands.* CSIRO Sustainable Ecosystems Canberra, ACT. Retrieved from http://www.cse.csiro.au

Tonteri, T. (1990). Inter-observer variation in forest vegetation cover assessments. *Silva Fennica, 24*, 189-196.

UNEP (2002). United Nations Environment Programme. Report of the sixth meeting of the conference of the parties to the Convention on Biological Diversity (UNEP/CBD/COP/6/20). Decision VI/26, UNEP. Retrieved from http://www.biodiv.org/decisions/?mZcop-06/

Van Der Maarel, E. (1979). Transformation of cover-abundance values in phytosociology and its effects on community similarity. *Vegetation, 39*, 97-114.

Walker, L. R., Bellingham, P. J., & Peltzer, D. A. (2006). Plant characteristics are poor predictors of microsite colonization during the first two years of primary succession. *Journal of Vegetation Science, 17*(3), 397-406.

Walker, L. R., & del Moral, R. (2003). *Primary succession and ecosystem rehabilitation.* Cambridge University Press, New York, USA. http://dx.doi.org/10.1017/CBO9780511615078

Walker, L. R., & del Moral, R. (2008). Transition dynamics in succession: implications for rates, trajectories and restoration. In K. Suding & R. J. Hobbs (eds), *New Models for Ecosystem Dynamics and Restoration* (pp. 33-50). Island Press, Washington DC, USA.

Walker, L. R., & del Moral, R. (2009). Lessons from primary succession for restoration of severely damaged habitats. *Applied Vegetation Science, 12*, 55-67.

Walker, L. R., Walker, J., & del Moral, R. (2007). Forging a new alliance between succession and restoration. In L. R. Walker, J. Walker & R. Hobbs (eds.), *Linking ecological restoration and ecological succession* (pp. 1-18). Springer, New York, USA. Retrieved from http://www1.inecol.edu.mx/repara/download

Werger, M. (1974). The place of the Zürich-Montpellier method in vegetation science. *Folia Geobotanica et Phytotaxonomica, 9*, 99-109. http://www.jstor.org/stable/4179782

Whisenant, S. (1999). *Repairing damaged wildlands: a process-orientated, landscape-scale approach.* Cambridge University Press, New York, USA. http://dx.doi.org/10.1017/CBO9780511612565

Yates, C. J., & Hobbs, R. J. (1997). Woodland restoration in the Western Australian wheatbelt: a conceptual framework using a state and transition model. *Restoration Ecology, 5*, 28-35. http://dx.doi.org/10.1046/j.1526-100X.1997.09703.x

16

Diversity and Dynamics of Populations of Mites in nectarine Trees (*Prunus persica* var. *nucipersica*) (Rosaceae)

Fernando Berton Baldo[1], Adalton Raga[1], Jeferson Luiz de Carvalho Mineiro[2] & Jairo Lopez de Castro[3]

[1] Laboratory of Economic Entomology, Biological Institute Experimental Center, Brazil

[2] Laboratory of Acarology, Biological Institute Experimental Center, Brazil

[3] Regional APTA, Regional Center of Agribusiness Technological Development of the Southwest of the State of São Paulo, Brazil

Correspondence: Fernando Berton Baldo, Laboratory of Economic Entomology, Biological Institute Experimental Center, Rodovia Heitor Penteado Km 3, CEP 13092-543, Campinas-SP, Brazil.
E-mail: fernandobaldo@gmail.com

Abstract

The international literature does not provide much information about the incidence of species of mites in nectarine cultivars. The purpose of the present study was to determine diversity and dynamics of populations of mites and their interactions in different nectarine cultivars in the southwestern region of the State of São Paulo, Brazil. These mites were split into 15 families, 22 genders and 28 species. *Aculus fockeui* (Nalepa & Trouessant) (Eriophyidae) was the most abundant species, with 90.2 % of the mites collected. The populations of *A. fockeui* displayed specific periods with greater number of individuals. Phytoseiidae showed the highest richness of species. *Ricoseius loxocheles* (De Leon) and *Euseius ho* (De Leon) were the most abundant predators. *Euseius ho* population showed a positive correlation with rainfall.

Keywords: Acari, *Aculus fockeui*, *Prunus* sp., Rosaceae, stone fruit

1. Introduction

In cultivations of peaches and nectarines, species of mites may cause economic losses to production (Raseira & Centellas-Quezada, 2003). In *Prunus* spp., the eriophyids *Aculops berochensis* (Keifer & Delley), *Aculus fockeui* (Nalepa & Trouessant), *Diptacus gigantorhyncus* (Nalepa) and *Eriophyes insidiosus* (Keifer & Wilson) are mentioned as being of economic importance (Castagnoli & Oldfield, 1996).

Aculus fockeui is considered the most critical species in cultivation of peach trees because it causes yellowish stains on leaves, small deformations or curling. In severe infestations, this eriophyid mite reduces fruit production and quality (Flechtmann, 1979; Jeppson, Keifer & Baker, 1975; Keifer, Baker, Kono, Delfinado, & Styer, 1982).

In the far west region of the State of São Paulo, Tetranychidae mites were seen in several peach cultivars (Montes, Raga, Boliani, Mineiro, & Santos, 2011, 2012). In nectarine and peach trees, *Tetranychus urticae* (Koch) attacks the underside of leaves, and causes damage that lead to leaf shedding (Gallo et al., 2002). Large populations of *A. fockeui*, *Mononychellus planki* (McGregor)*, T. urticae* and phytoseiidae *Euseius citrifolius* Denmark & Muma and *Euseius concordis* (Chant) were found in leaves of several peach cultivars in the west of the State of São Paulo (Montes et al., 2012).

Unlike the number of records in the literature about mites in peach trees, there is little information about mite fauna in nectarine trees. As a consequence, the purpose of the present research was to determine the mite fauna in nectarine trees, to assess the relationships among the key species of phytophagous and predatory mites, and to analyze the influence of rainfall and temperature on the populations of such mites.

2. Material and Methods

2.1 Study Area

The research was conducted at the Germplasm Bank of Temperate and Subtropical Climate Fruit Trees of the Agronomic Institute, located at the municipality of Capão Bonito, State of São Paulo, Brazil

(S24°2'24.65",W48°23'3.54"O, 730 m).

The nectarine trees were part of a mixed orchard set up in 1993, comprising peach, plum and nectarine tops on root-stocks of the Okinawa cultivar, covering a total area of 2,415 m², with 34 different genetic materials with two plants per cultivar. Plants showed approximately three meters high, spacing between plants was 5.0 x 3.0 m (668 plants per ha) (Barbosa, Ojima, Campo-DallOrto, & Martins, 1993). During the study, no pesticide was used for pest management. The orchard was performed only by pruning and mechanical cleaning for weed control.

Three cultivars of economically representative nectarine trees for fruit-growing in São Paulo were used in the present research: Colombina (Fla. 1937-S), Josefina (IAC N 1579-1) and Rubro-Sol (Rigitano, Ojima, & Campo-Dall'orto, 1975; Ojima et al., 1986).

2.2 Sampling

Mites were sampled biweekly from January 2004 to March 2006. The sampling included random collection of 30 leaves of the middle third of the plants from each cultivar. The collected leaves of each cultivar were placed in individual paper bags. The leaves of each sample were transferred to 500-ml glass bottles with 70 % alcohol for preservation. Next the mites were extracted at the laboratory. Hoyer's medium was used to set most mites on the slides, except for eriophyid mites mounted in modified Berlese medium (Jeppson et al., 1975).

The mites were identified with the help of a phase-contrast optical microscope with 100x augmentation. "Voucher species" were deposited in the "Geraldo Calcagnolo" Reference Collection of Mites of Agricultural Importance (ICMBio 35919-1) at the Acarololgy Laboratory of the Experimental Center of the Biological Institute (CEIB), in Campinas, SP, Brazil.

The data on temperature and precipitation were obtained from the Meteorological Station of APTA (São Paulo Agency of Agribusiness Technology) Capão Bonito Regional Office, installed approximately 250 m far from the study area.

2.3 Statistical Analyses

Pearson linear correlation analyses were conducted with 5 % significance level to assess the influence of precipitation and temperature on the population dynamics of the species of phytophagous and predatory mites, and also possible interactions among these groups, using the BioEstat 5.0 statistical program (M. Ayres, J. M. Ayres, D. L. Ayres, & Santos, 2007).

The mite abundance of each cultivar was analyzed by Tukey test (p>0.05) using the ASSISTAT 7.7 program (Silva & Azevedo, 2006). To do so, the data of the four most representative species were used during the study, two phytophagous species and two predatory species.

The PAST program was used to perform H' diversity analysis and cluster analysis (Hammer, Harper, & Ryan, 2001).

The relative frequency was noted in percentage, obtained by the ratio of the number of samples with the occurrence of the respective species and the total number of samples. The abundance corresponds to the total number of individuals of each recovered species from the nectarine leaves during the whole period of the experiment.

3. Results

A total of 19,297 mites was found, split in 14 families, 22 genera and 28 species. In general, the number of mites recovered from Colombina (4,801) and Josefina (4,326) cultivars was similar, accounting for 25 % and 22 % of the total mites, respectively. In the case of Rubro-Sol cultivar (10,170), the number of specimens was larger than all the other cultivars together, and accounted for 53 % of all mites recovered (Table 1).

Colombina and Josefina cultivars displayed the same number of species (19), whereas for Rubro-Sol cultivar, 22 species were found (Table 1). Colombina and Josefina cultivars were similar, with similar diversity indexes (H') (0.55 and 0.54 respectively) and different from Rubro-Sol cultivar (0.31). With the t-test, it was possible to verify that the diversity index of Rubro-Sol cultivar was significantly different from the other cultivars: t=10,607 and p=4,266 between Colombina, t=-12,896 and p=9,683 for Josefina cultivar, whereas the t-test between Colombina and Josefina cultivars indicated: t=0.614 and p= 0.538, which confirm the similarity between the two cultivars.

Table 1. Frequency, abundance, diversity (H') and species richness recovered from cultivars of nectarines Colombina (C), Josefina (J) and Rubro-sol (RS). Capão Bonito, SP, Brazil

Family	Species	Frequency (%)			Abundance			Total
		C	J	RS	C	J	RS	
Acaridae	*Tyrophagus putrescentiae* (Schrank)	6.25	8.33	14.58	4	6	12	22
Cunaxidae	*Armascirus* sp.	2.08	-	-	1	-	-	1
Echimyopodidae	*Blomia* sp.	4.17	-	6.25	15	-	4	19
Eriophyidae	*Aculus fockeui* (Nalepa & Trouessant)	54.17	45.83	45.83	4,029	3,797	9,583	17,409
Iolinidae	*Parapronematus acaciae* Baker	2.08	6.25	4.17	2	4	2	8
Melicharidae	*Proctolaelaps* sp.	-	6.25	2.08	-	3	1	4
Oripodidae	*Oripoda* sp.	2.08	-	-	1	-	-	1
Phytoseiidae	*Amblyseius chiapensis* De Leon	4.17	4.17	4.17	2	3	2	7
	Amblyseius compositus Denmark & Muma	4.17	8.33	8.33	6	6	5	17
	Amblyseius herbicolus (Chant)	14.58	31.25	14.58	9	37	10	56
	Euseius alatus De Leon	12.50	10.42	14.58	10	9	21	40
	Euseius concordis (Chant)	2.08	-	-	1	-	-	1
	Euseius ho (De Leon)	27.08	22.92	31.25	25	29	43	97
	Iphiseiodes matatlanticae Mineiro, Castro & Moraes	4.17	2.08	2.08	3	1	1	5
	Iphiseiodes saopaulus (Denmark & Muma)	14.58	16.67	18.75	7	9	16	32
	Neoseiulus anonymus (Chant & Baker)	-	2.08	-	-	1	-	1
	Neoseiulus idaeus Denmark & Muma	-	-	2.08	-	-	1	1
	Phytoseiulus macropilis (Banks)	4.17	8.33	2.08	7	4	4	15
	Ricoseius loxocheles (De Leon)	6.25	10.42	12.50	4	73	101	178
	Typhlodromus transvaalensis (Nesbitt)	4.17	2.08	-	2	1	-	3
Stigmaeidae	*Agistemus* sp.	-	-	2.08	-	-	1	1
Tarsonemidae	*Fungitarsonemus* sp.	-	-	2.08	-	-	1	1
	Tarsonemus confusus Ewing	-	4.17	6.25	-	2	4	6
Tenuipalpidae	*Brevipalpus* sp.	-	2.08	2.08	-	1	1	2
Tetranychidae	*Tetranychus urticae* Koch	68.75	41.67	45.83	672	338	325	1,335
Tydeidae	*Lorryia* sp.	-	-	2.08	-	-	2	2
	Pretydeus sp.	2.08	-	-	1	-	-	1
Winterschmidtiidae	*Oulenzia* sp.	-	2.08	2.08	-	2	30	32

	Colombina	Josefina	Rubro-sol
Species richness	19	19	22
Diversity H'	0.55	0.54	0.31

The scheme shown in the dendrogram of Figure 1 corroborates the results seen in the diversity index H' and the t-test, with formation of two clusters. The first cluster is formed by species of mites found only in Rubro-Sol cultivar. The second cluster is formed by species of mites of the two other cultivars, which were similar in the foliar mite fauna.

Similarity

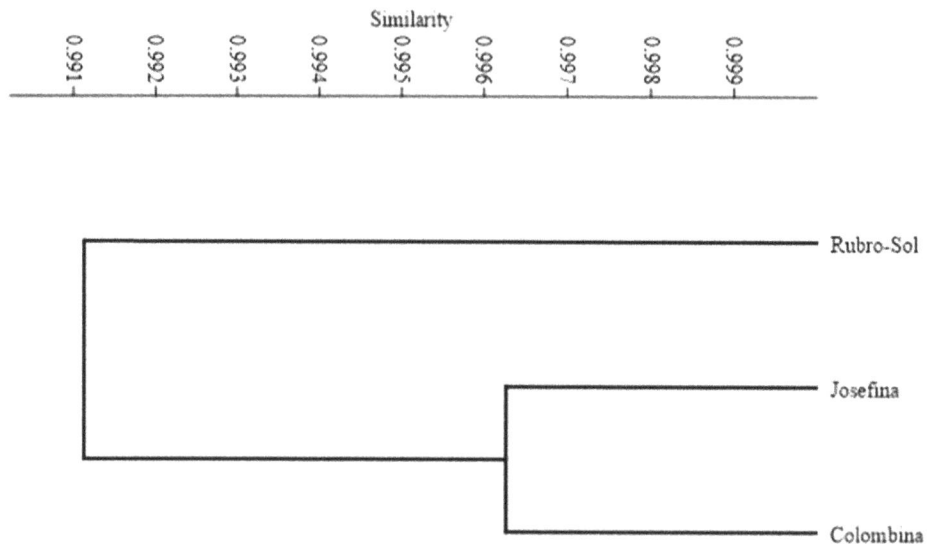

Figure 1. Dendrogram based on the similarity of mite diversity recovered from nectarine cultivars ʹColombinaʹ, ʹJosefinaʹ and ʹRubro-solʹ. Capão Bonito, SP, Brazil

3.1. Mite diversity

Due to the large diversity of mites obtained from the three cultivars under study, it was possible to group them in conformance with their respective eating habits, if known. In the present study, mites with phytophagous and predatory eating habits are those of the greatest interest because of the known potential for damage of phytophagous mites, and the potential biological control agent represented by predatory mites to control populations of phytophagous mites (Watanabe, Moraes, Gastaldo, & Nicolella, 1994; Gerson, Smiley, & Uchoa, 2003; Gerson, 2014).

Among the mite community, the group of predatory mites showed the highest richness (16), where more than 50 % of the species were common among the three cultivars (Table 1). However, the abundance of these predatory species proved to be distinct among the nectarine cultivars.

In Colombina and Josefina cultivars, 11 species (92 %) were Phytoseiidae mites. Only one individual of *Armascirus* sp. (Cunaxidae) was caught in Colombina, and another specimen of *Proctolaelaps* sp. (Melicharidae) in Josefina. The Rubro-Sol cultivar had three families of predatory mites (Table 1): Phytoseiidae (10); Melicharidae (*Proctolaelaps* sp.) and Stigmaeidae (*Agistemus* sp.) (1).

The Phytoseiidae family displayed the greatest number of species among the predatory mites (13), with approximately 2 % (453) of the total mites recovered in the study. *Ricoseius loxocheles* (De Leon) was the most abundant species (178), representing 39 % of the phytoseiidae of the present study, and it was the most abundant among the Phytoseiidae obtained from Josefina (42%) and Rubro-Sol (49%) cultivars (Table 1). *Euseius ho* was the second most numerous species (97), representing 21% of the Phytoseiidae mite found in the study. In Colombina cultivar, *E. ho* represented 33 % of the phytoseiid mite, whereas in Josefina and Rubro-Sol, *E. ho* represented 17 % and 21 %, respectively (Table 1).

Ricoseius loxocheles was the most abundant among the phytoseiidae under study; however, during the sampling, the occurrence was only 6 % in Colombina, 10 % in Josefina and 12 % in Rubro-Sol. *Euseius ho* was the most frequent phytoseiid mite, displaying 27 %, 23 % and 31 % in Colombina, Josefina and Rubro-Sol respectively. Other phytoseiid mite such as *Amblyseius herbicolus* (Chant), *Iphiseiodes saopaulus* (Denmark & Muma) and *Euseius alatus* De Leon, were also the predators most commonly recovered from nectarine trees, but these predators were found in a smaller number.

The group of strictly phytophagous mites displayed only three species related to three families (Table 1). The occurrence of these mites among the cultivars under study was distinct. In Colombina, only the following species were detected: *A. fockeui* and *T. urticae*, whereas Josefina and Rubro-Sol also displayed the tenuipalpid mite *Brevipalpus* sp. However, the occurrence of *Brevipalpus* sp. was only of one individual in each cultivar. Only *A. fockeui* and *T. urticae* were abundant among phytophagous mites (Table 1).

Aculus fockeui eriophyid was the most abundant mite. It accounted for 90.2 % of the entire mite fauna recorded in this study (17,409). Although *A. fockeui* was the most abundant mite in Rubro-Sol (9,583), the populations in Colombina and Josefina were also very high in comparison to the rest of the species of mites registered, with recovery of 4,029 and 3,797 individuals, respectively.

The second most abundant phytophagous mite was *T. urticae* (1,335), which accounted for 6.9 % of the entire fauna of mites recovered in the study. This mite was observed in the Colombina (672), Josefina (338) and Rubro-Sol (325) cultivars. *Tetranychus urticae* can be considered the most frequent mite in the study, found in 69 % of the samples of Colombina, 2 % of Josefina and 46 % of Rubro-Sol (Table 1).

In spite of being very abundant, the frequency of *A. fockeui* has a peculiar seasonal distribution and it is more frequent in Colombina cultivar, found in 54 % of the dates sampled. In the other cultivars, *A. fockeui* displayed at least one individual in 46 % the collection dates.

The other species of mites registered in the work (9) had distinct eating habits. Out of these species of mites, at least one individual was found in one of the cultivars (Table 1). *Tyrophagus putrescentiae* (Schramk) (Acaridae) was the most frequent species in Rubro-Sol cultivar (15 %) and *Oulenzia* sp. (Winterschmidtiidae) was more abundant, with a total of 32 individuals: 30 in Rubro-Sol, two in Josefina and no records in Colombina.

3.2. Population dynamics

The population fluctuation of the two most abundant species of predatory mites found in this study displayed a distinct standard of occurrence during the time length under study. *Euseius ho* was the Phytoseiidae most frequently observed (27 %) in the present study. In Colombina cultivar, it was the most abundant and frequent, since it was detected in 27 % of the samples. *Euseius ho* was more frequent in January 2004, 2005 and 2006, which coincides with the times when rainfall and temperature were higher (Figures 2 and 3). Josefina was the only cultivar in which *E. ho* was not the most frequent predator, found in 23 % of the samples. In this cultivar, *A. herbicolus* was the most frequent predator. Still in Josefina, *E. ho* was more frequent between January and June 2004 and January 2006 (Figure 3). Rubro-Sol was the cultivar in which *E. ho* was the most frequent among predators (31 %). This predator was seen more often in the first and last months of 2004, and from November 2005 to January 2006 (Figure 3).

There was a significant positive correlation ($p=0.0223$) between rainfall and the population of *E. ho* in Colombina cultivar, which suggests that this weather variable is a key factor in the population dynamics of this phytoseiid. This event can be observed in a comparison between Figures 2 and 3. In the other cultivars, no correlations were detected between *E. ho* and the weather variables analyzed.

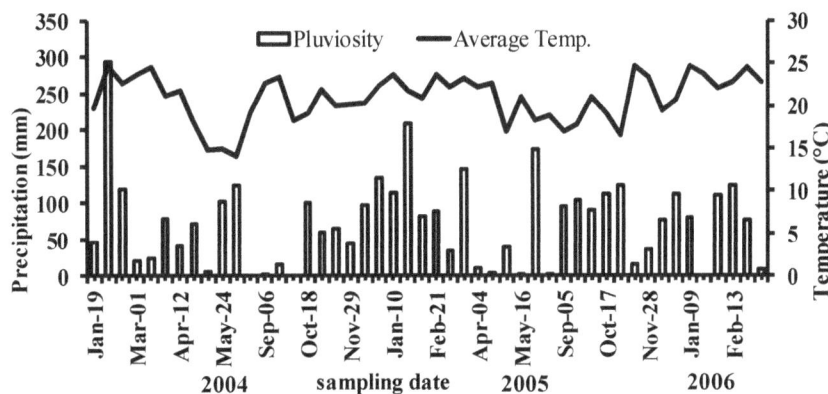

Figure 2. Cumulative rainfall and average temperature during the intervals of collections. Capão Bonito, SP, Brazil

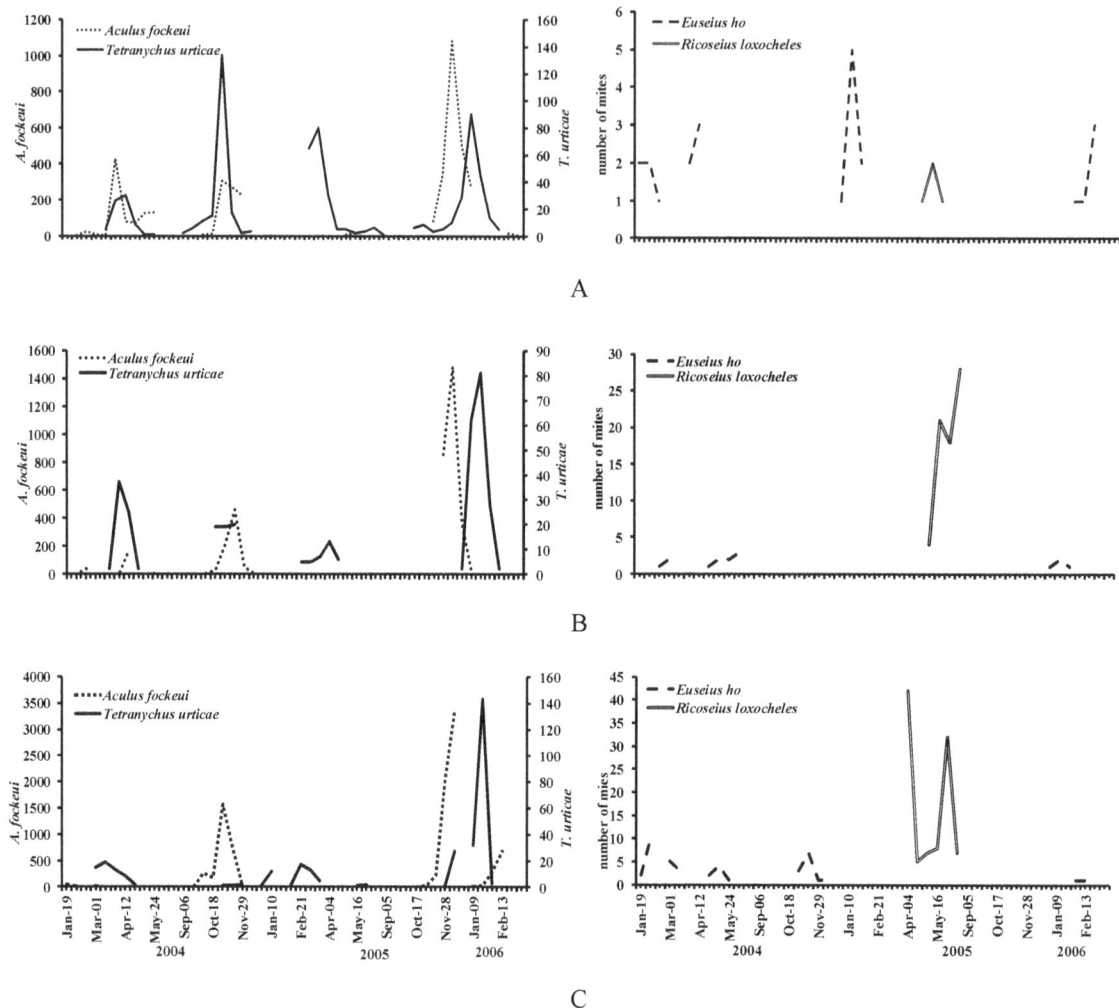

Figure 3. Population dynamics of mites *Aculus fockeui, Tetranychus urticae, Euseius ho* and *Ricoseius loxocheles* from nectarines Colombina (A), Josefina (B) and Rubro-Sol (C) in Capão Bonito, SP, Brazil

Ricoseius loxocheles was found in Colombina only in 9.7 % of the samples between April and May 2005, when its smallest population among the cultivars was found (Figure 3). In Josefina, *R. loxocheles* occurred from April to June 2005, when it had its population acme (Figure 3). In Rubro-Sol, *R. loxocheles* was the most abundant predator (101), occurring at the same time period as Josefina cultivar. No correlation was detected between this species and weather variables. However, the populations of this predator displayed a common pattern of occurrence among the three nectarine cultivars under study, and do not seem to have been negatively affected by the increase in rainfall, provided that the temperature was close to 20°C (Figures 2 and 3).

The eriophyid *A. fockeui* displayed 4,029 individuals in Colombina during the study. The most abundant population was registered between February and May, November 2004 and December 2005, with a population peak on 12 December, 2005 in the three cultivars (Figure 3). The Josefina cultivar displayed the smallest number of specimens of *A. fockeui* among the nectarine trees under study (3,797). The largest occurrence of *A. fockeui* was found between the last and the first months of each year, when the temperature was close to or above 20 °C (Figures 2 and 3). Rubro-Sol displayed the greatest abundance of *A. fockeui* in the study (9,583). In spite of the high number of this eriophyid, its frequency was similar to the frequency observed in Josefina cultivar (46 %). *Aculus fockeui* displayed 3,356 individuals in Rubro-Sol cultivar on 12 December, 2005 (Figure 3). No significant correlations were detected between the population of *A. fockeui* and the weather variables analyzed.

Tetranychus urticae was recovered from 25 out of the 48 samples taken. Colombina was the cultivar in which spotted mites displayed the greatest frequency and abundance, without displaying definite periods to occur. The population peaks of *T. urticae* were observed on 1 November, 2004, 21 March, 2005 and 9 January, 2006.

Rainfall and temperature observed on these dates were respectively 59 mm and 22 °C, 147 mm and 23 °C, and 81 mm and 25°C (Figures 2 and 3). In Josefina, *T. urticae* displayed 338 individuals and its frequency was the smallest among the cultivars. The occurrence of spotted mites in Josefina was limited to 4 isolated periods, with the greatest abundance on 16 January, 2006, with 81 individuals recovered (Figures 2 and 3). Rubro-Sol cultivar displayed the second greatest abundance of *T. urticae* among the cultivars (325 specimens), with its peaks also on 16 January, 2006, when there was no rainfall and the mean temperature was approximately 24 °C (Figures 2 and 3). The populations of spotted mites displayed distinct patterns among frequency and abundance in the cultivars analyzed. In Josefina, this species was more abundant, and in Rubro-sol it was less abundant (Figure 3).

No significant differences were found among the cultivars when submitted to Anova analysis. Table 2 shows the results of this analysis more clearly and determines that the means followed by the same letter do not differ statistically from each other.

Table 2. Statistical analyses (ANOVA) of the four main representative mite species among phytophagous and predator mites, recovered from leaves of nectarines 'Colombina', 'Josefina' and 'Rubro-sol'. Capão Bonito, SP, Brazil.

Species	Colombina	Josefina	Rubro-sol	F
Aculus fockeui	83.94 a	79.10 a	199.64 a	1.42 [ns]
Tetranychus urticae	14.00 a	7.04 a	6.77 a	1.68 [ns]
Euseius ho	0.52 a	0.60 a	0.89 a	0.74 [ns]
Ricoseius loxocheles	0.08 a	1.52 a	2.10 a	1.73 [ns]

Means within rows followed by the same letter were not significantly different by Tukey's test at ($p < 0.05$).

4. Discussion

The large abundance (19,297) of mites split among the different families of the mite fauna indicates that this group of arthropods deserves attention in terms of pest management in nectarine cultivars. Some species of mites can reach the status of pests, such as *A. fockeui,* which is considered a universal pest in peach, nectarine, plum and almond trees (Abou-Awad, AL-Azzazy & El-Sawi, 2010). However, the species of predatory mites can be important allies in programs of biological control of pests (Watanabe et al., 1994; Gerson, 2014) either due to strategies of inundation or conservationist release.

The high number of mite families (14) recovered in the present study reflects a relatively common result observed in the surveys of the mite fauna of peach (*Prunus persica* (L.) Batsch) in Brazil: 16 families (Montes et al., 2010), 14 families (Eichelberger, Johann, Majolo & Ferla, 2011), 17 families (Montes et al., 2011), 12 families (Montes et al., 2012).

In spite of the large diversity of families and species registered, it was found that only the frequency of *T. urticae* (52 %) was above 50 % among the three cultivars studied, whereas the second most frequent species was *A. fockeui* (48 %). The frequency of the other species was below 28 %. The other species occurred less often, and in some cases only one individual was found. Eichelberger et al. (2011) also found species of mites that are not very constant in the south of Brazil and different from the present study. *B. phoenicis*, *Tetranychus ludeni* Zacher and *Typhlodromalus aripo* De Leon, were predominant.

Although *R. loxocheles* was registered as the most abundant predator in the present study, this Phytoseiidae did not showed constancy and neither correlation with phytophagous mites. It is possible that this mite is not an efficient predator of other mites.

Ricoseius loxocheles has no preference for a host plant in particular. This phytoseiidae is known from vegetable materials collected in Brazil, in Florida (USA), Colombia, Costa Rica, Guadalupe, Honduras, Martinique (Flechtmann, 1967; Denmark & Muma, 1973; Flechtmann, 1976; Moraes, Mesa & Braun, 1991; Denmark, Evans, Aguilar, Vargas & Ochoa, 1999; Moraes, Kreiter & Lofego, 1999). It is believed that its eating habits contributed to its wide distribution, since this predator does not need to have a specific host or food to survive.

Euseius ho was the most frequent phytoseiid mite, with a mean of 27.08 % of occurrence among the 3 cultivars. With this, the hypothesis arises that this species may be a potential agent of biological control of phytophagous mites in Rosaceae. Moreover, in the present study, this predator displayed a positive correlation with rainfall in Colombina cultivar. This suggests that its population increased with high rainfall. This fact may also interfere in future programs of biological control.

In peach trees in Brazil, Montes et al. (2010) found that *E. citrifolius* and *E. concordis* were the most abundant and frequent predators in Presidente Prudente (SP). Several species of the *Euseius* genera were registered in the different agroecosystems where studies of the fauna of mites of *Prunus* species were made (Castagnoli & Nannelli, 1987; Montes et al., 2010; 2011; 2012; Sonoda et al., 2012).

Aculus fockeui has been frequently reported as the important species of phytophagous mites in stone Rosaceae (Ferreira & Carmona, 1997; Kondo & Hiramatsu, 1999; Montes et al., 2010; 2011). The results on population fluctuation of the key species of mites observed in the present work coincide with the data from studies conducted by Montes et al. (2010; 2012), in which *A. fockeui* was the most abundant pest mite in all cultivars assessed in the western region of the State of São Paulo. In these studies, the greatest frequency of *A. fockeui* also took place in December, when the mean temperature was close to 20 °C.

According to Abou-Awad et al. (2010), the appropriate temperature in the orchards, better eating conditions and absence of a natural predator would have led to quick growth of *A. fockeui*, which is considered a disaster to nectarine and peach trees. These authors studied the biology of *A. fockeui* in Egypt and concluded that this eriophyid mite was able to develop successfully from egg to adult at temperatures between 21 and 32 °C and 70 % relative humidity. The environmental parameters such as temperature and relative humidity are intrinsic to the development of these phytophagous mites, and can explain the significant differences in population fluctuation of this species observed in this study.

Another factor that would explain the pattern of cyclic occurrence of *A. fockeui* is the hibernation strategy of this mite, which coincides with the fall of the leaves of Rosaceae of temperate climate. Females in diapause hide in buds or cracks in branches and barks, and resume their activity when plant sprouts start to expand in spring (Beers & Oldfield, 1993). Mites in diapause were found more often in sprouts of host plants with deciduous leaves (De Lilo & Shoracka, 2010). In Brazil, there are no information about the hibernation behavior of *A. fockeui*.

Tetranychus urticae was the second most numerous phytophagous mites in this study. Montes et al. (2010) also observed this behavior in peach trees of the region of Presidente Prudente (SP). Ferreira and Carmona (1997), in peach trees in Portugal there was abundance of 41.6 % of *A. fockeui and* 20 % of *Tetranychus cinnabarinus* (Boisduval). According to Moraes and Flechtmann (2008), *T. urticae* is the most important species in Brazil among the 6 species of tetranychid mite with economic significance since they have a large number of hosts and can cause severe damage to many crops. However, more studies are required to know the potential for damage that this species can cause in nectarine trees.

In this study the mite communities of Colombina and Josefina cultivars were similar. Rubro-Sol was distinct as to diversity and abundance in comparison to the other nectarine cultivars. These results suggest that mites have a preference for a certain cultivar rather than others. According to Gerson et al. (2003), the chemical aspects and the physical structure of host plants may affect the growth rate of pest mites.

The three cultivars studied had characteristics of early growth, which excludes the possibility that mites prefer distinct growing seasons. However, differences in plant structure such as shape of the tree top (which favors control of humidity inside the tree), the presence of villosities in branches used as hiding place, or even morphological aspects of leaves such as chemical composition, water contents or even thickness can be key characteristics for mites to prefer certain cultivars.

New studies oriented to determine the key factors are involved on attractiveness and dynamics of mite fauna in Rosaceae. These studies will help to support future programs for genetic breeding and integrated pest management in nectarine trees.

References

Abou-Awad, B. A., AL-Azzazy, M. M., & El-Sawi, A. (2010). The life – history of the peach silver mite, *Aculus fockeui* (Acari: Eriophyidae) in Egypt. *Arch. Phytopathol.,* *43*(4), 384-389. http://dx.doi.org/10.1080/03235400701806427

Ayres, M., Ayres Jr, M., Ayres D. L., & Santos, A. A. S. (2007). BioEstat 5.0. - *Aplicações estatísticas nas áreas das Ciências Biomédicas. Sociedade Civil Mamirauá:* Belém, Pará-Brasil. p. 324.

Barbosa, W., Ojima, M., Campo-DallOrto, F. A., & Martins, F. P. (1993). *Melhoramento do pessegueiro para regiões de clima subtropical-temperado: realização do Instituto Agronômico no período de 1950-1990.* Instituto Agronômico, Campinas. p. 30.

Beers, E. H., & Oldfield, G. N. (1993). (Acari: Eriophyidae) Plum nursery mite Plum rust mite Cherry rust mite *Aculus (Phyllocoptes) fockeui* (Nalepa and Trouessart) Peach silver mite *Aculus cornutus* (Banks). *Washington State University.* Retrieved from http://www.tfrec.wsu.edu/pages/opm/PRU

Castagnoli, M., & Nannelli, R. (1987). Further observations on populations trend of mites in an experimental peach meadow orchard in central Italy. Firenze. *Redia, 70,*121-133.

Castagnoli, M., & Oldfield, G. N. (1996). Other fruit trees and nut trees. In: Lindquist, E. E., Sabelis, M. W. and Bruin, J., (Eds.) *Eriophyoid mites—their biology, natural enemies and control* (pp. 543-560). Amsterdam. Elsevier. http://dx.doi.org/10.1016/S1572-4379(96)80034-0

De Lillo, E., & Skoracka, A. (2010). What's "cool" on eriophyoid mites?. *Exp. Appl. Acarol., 51,* 3-30. http://dx.doi.org/10.1007/s10493-009-9297-4

Denmark, H. A., Evans, G. A., Aguilar, H., Vargas, C., & Ochoa, R. (1999). *Phytoseiidae of Central America* (Acari: Mesostigmata). Indira Publishing House, West Bloomfield. p.125.

Denmark, H. A., & Muma, M. H. (1973). Phytoseiid mites of Brazil (Acarina: Phytoseiidae*). Rev. Brasil. Biol. 33,* 235-276.

Eichelberger, C. R., Johann, L., Majolo, F., & Ferla, N. J. (2011). Mites fluctuation population on peach tree (*Prunus persica* (L.) Batsch) and in associated plants. *Rev. Bras. Frutic., 33,* 765-773. http://dx.doi.org/10.1590/S0100-29452011005000102

Ferreira, M. A., & Carmona, M. M. (1997). Acarofauna do pessegueiro em Portugal. *Bol. San. Veg. Plagas., 23,* 473-478.

Flechtmann, C. H. W. (1967). Ácaros de plantas frutíferas. *Escola Superior de Agricultura Luiz de Queiroz/Universidade de São Paulo, Piracicaba, Brasil, Boletim Técnico-Científico.* No. 30. p.24. http://dx.doi.org/10.1590/S0071-12761967000100028

Flechtmann, C. H. W. (1976). Observações sobre dois ácaros (Mesostigmata, Acari) de vida livre. *An. Soc. Entomol. Brasil., 5,* 95-96.

Flechtmann, C. H. W. (1979). *Ácaros de importância agrícola.* São Paulo: Livraria Nobel, p.189.

Gallo, D., Nakano, O., Silveira Neto, S., Carvalho, R. P. L., Batista, G. C. de, Berti Filho, E., ... Omoto, C. (2002). *Entomologia agrícola* (10th Ed.). FEALQ, Piracicaba, p. 920.

Gerson, U. (2014). Pest control by mites (acari): present and future. *Acarologia, 54,* 371-394. http://dx.doi.org/10.1051/acarologia/20142144

Gerson, U., Smiley, R. L., & Uchoa, R. (2003). *Mites (Acari) for pest control.* Blackwell Science, Oxford. P. 539. http://dx.doi.org/10.1002/9780470750995

Hammer, Ø., Harper, D. A. T., & Ryan, P. D. (2001). PAST: Paleontological statistics software package for education and data analysis. *Palaeontol. Electron., 4,* 9.

Jeppson, L. R., Keifer, H. H., & Baker, E. W. (1975). *Mites injurious to economic plants.* Berkeley: University of California Press. p. 614.

Keifer, H. H., Baker, E.W., Kono, T., Delfinado, M., & Styer, W. E. (1982). *An illustrated guide to plant abnormalities caused by eriophyid mites in North America.* United States Department of Agriculture, Washington. p. 178.

Kondo, A. E., & Hiramatsu, T. (1999). Analysis of peach tree damage caused by each silver mite, *Aculus fockeui* (Nalepa e Trouessart). *Jpn. J. Appl. Entomol. Zool., 43,* 189-193. http://dx.doi.org/10.1303/jjaez.43.189

Montes, S. M. N. M., Raga, A., Boliani, A. C., Mineiro, J. L. C., & Santos, P. C. dos. (2010). Composição acarina em diferentes cultivares de pessegueiros [*Prunus persica* (L.)] em Presidente Prudente, Estado de São Paulo. *Rev. Bras. Frut., 32,* 414-422. http://dx.doi.org/10.1590/S0100-29452010005000050

Montes, S. M. N. M., Raga, A., Boliani, A. C., Mineiro, J. L. C., & Santos, P. C. dos (2011). Effect of fungicides on the mite fauna of *Prunus persica* L. cultivars in Presidente Prudente, SP, Brazil. *J. Plant Prot. Res., 5,* 285-293. http://dx.doi.org/10.2478/v10045-011-0047-3

Montes, S. M. N. M., Raga, A., Boliani, A. C., Mineiro, J. L. C., & Santos, P. C. Dos. (2012). Mite Fauna (Arachnida: Acari) on Peach Cultivars in Presidente Prudente, São Paulo, Brazil. *J. Plant Stud. 1,* 173-179.

Moraes, G. J., Mesa, N. C., & Braun, A. (1991). Some phytoseiid mites of Latin America (Acari: phytoseiidae). *Int. J. Acarol., 17,* 117-139. http://dx.doi.org/10.1080/01647959108683892

Moraes, G. J., Kreiter, S., & Lofego, A. C. (1999). Plant mites (Acari) of the French Antilles. 3. Phytoseiidae (Gamasida). *Acarologia, 40,* 237-264.

Moraes, G. J., & Flechtmann, C. W. H. (2008). *Manual de acarologia: acarologia básica e ácaros de plantas cultivadas no Brasil*. Ribeirão Preto: Holos, p. 288.

Ojima, M., Campo Dall'orto, F. A., Barbosa, W., Tombolato, A. F. C., Martins, F. P., & Rigitano, O. (1986). Josefina: nova nectarina de polpa branca. In: Congresso Brasileiro de Fruticultura, 8., Brasília, 1986. *Anais. Brasília: SBF., 2*, 417-419.

Raseira, M. Do, & Centellas-Quezada, C. B. E. A. (2003). *Pêssego*: Produção. Embrapa Clima Temperado (Pelotas, RS). – Brasília: Embrapa Informação Tecnológica. p. 162.

Rigitano, O., Ojima, M. and Campo Dall'orto, F. A. (1975). *Comportamento de novas seleções de pêssegos introduzidos da Flórida*. Campinas: Instituto Agronômico, (Circular, 46). p. 12.

Silva, F. A. S., & Azevedo, C. A. V. (2006). A New Version of The Assistat-Statistical Assistance Software. In: World Congress on Computers in Agriculture, 4, Orlando-FL-USA: Anais... Orlando: *ASABE*. 393-396.

Sonoda, S., Kohara, Y., Siqingerile, Toyoshima, S., Kishimoto, H., & Hinomoto, N. (2012). Phytoseiid mite species composition in Japanese peach orchards estimated using quantitative sequencing. *Exp. Appl. Acarol., 56*, 9-22. http://dx.doi.org/10.1007/s10493-011-9485-x

Watanabe, M. A., Moraes, G. J. de, Gastaldo Jr, I., & Nicolella, G. (1994). Controle biológico do acaro rajado com ácaros predadores fitoseídeos (Acari: Tetranychldae, Phytoseiidae) em culturas de pepino e morango. *Sci. Agric., 51*, 75-81. http://dx.doi.org/10.1590/S0103-90161994000100012

17

Development of a Method to Measure Mechanical Properties of Single Elongated Callose Fibers in Protoplast Cultures of *Larix Leptolepis* and *Betula Platyphylla*

Tomoya Oyanagi[1], Asami Kurita[1], Toshihiko Shiraishi[1,2], & Hamako Sasamoto[2,3]

[1]Graduate School of Environment and Information Sciences, Yokohama National University, Yokohama 240-8501, Japan

[2]Faculty of Environment and Information Sciences, Yokohama National University, Yokohama 240-8501, Japan

[3]Research Institute for Integrated Science, Kanagawa University, Hiratsuka, Kanagawa 259-1293, Japan

Correspondence: Hamako Sasamoto, Research Institute for Integrated Science, Kanagawa University, Hiratsuka, Kanagawa 259-1293, Japan. E-mail: sasamoto@ynu.ac.jp

Abstract

We developed a method to measure mechanical properties of single fibers of callose in liquid protoplast cultures of *Larix leptolepis* and *Betula platyphylla,* which were formed in media containing 50 mM of $MgCl_2$ or 100 mM of $CaCl_2$, respectively. Tensile test was performed using two micromanipulators loading micropipettes under an inverted microscope. Spring constant of the pipette used was first calibrated and calculated from using a microbalance. The callose fiber was wired between the two micropipettes. The Young's modulus of single fibers for *Larix* and *Betula* was 7-9 kPa (1.4-1.9×10^4 N/m^2) though the diameters of the fiber varied from 10 μm for *Larix* and 22-26 μm for *Betula*. No big difference was found between experiments with and without medium containing high concentrations of salts. Tensile strength at break was 1.1-1.8 kPa (2.3-3.6×10^3 N/m^2). The values are compared to other materials including cellulose containing plant cell wall, cell membranes, and amorphous callose. The value of the Young's modulus observed was discussed.

Keywords: callose, fiber, mechanical property, protoplast cultures, Young's modulus

1. Introduction

We reported novel long and large spiral callose fibers developed in plant protoplast cultures of a fast-growing coniferous tree, *Larix leptolepis* and a fast-growing broadleaved tree, *Betula platyphylla* by treatment with high concentrations of $MgCl_2$ or $CaCl_2$ in the liquid media (Sasamoto et al., 2003). Spiral structures with diameters of 10 to 30 μm and up to 2 mm in length were observed under an inverted microscope. The fibers were proven to be composed of callose by using specific degrading enzymes for beta-1,3-glucan and beta-1,4-glucan. And sub-structures of a *Larix* fiber, bundles of fibrils (0.7 μm diameter) and sub-fibrils (0.17 μm diameter), were also clarified by atomic force microscopy (Fukumoto et al., 2005). Callose is a natural component of plant cell wall and it is abundant in pollen tubes (Cresti & van Went, 1976). In some situation, its formation can be regarded as a stress response (Chen & Kim, 2009). Our reported fiber structure of callose is unique. It elongates from a single point on the surface of a protoplast, and it is easily separated from a protoplast (Sasamoto et al., 2003; Fukumoto et al., 2005).

Measurement of mechanical properties of unique callose fiber is an interesting topic, which might contribute to the development of an alkali-labile nano-machine at a cellular level (Oyanagi et al., 2014). Mechanical properties of single tomato suspension cell (Blewett et al., 2000) and large sized (several centimeters) discs of cell wall components including cellulose, were measured by compression test (Chanliaud et al., 2002). Parre and Geitmann (2005) found that the amorphous polymer, callose, is related to the tension stress in pollen tubes, in which cell wall is composed of not only callose but also pectin and cellulose. The mechanical properties of plasma membrane of plant protoplasts were measured by aspiration with micropipettes (Wolf & Steponkus, 1983). By improvement of the holding of single cells, Onishi et al. (2006) reported measurement of Young's modulus of an animal cell adhered on glass plate by micropipette aspiration using micromanipulators. Simple tensile testing of a whole protoplast fiber is important to our understanding of the fundamental mechanical properties of the callose fibers. However, since the size of protoplast fibers is of the order of μm diameter and mm in length and they also float in a liquid medium, the reported methods in the literature could not be applied directly. In this report, we developed a novel method to measure mechanical properties of single fibers of callose using micropipettes mounted on micromanipulators under an inverted microscope.

2. Method

2.1 Culture of Protoplasts

Fiber formation from protoplast cultures of *Larix leptolepis* and *Betula platyphylla* was performed as described previously (Sasamoto et al., 2003). Briefly, *Larix* protoplasts were isolated by 1% Cellulase RS (Yakult) and 0.25% Pectolyase Y-23 (Seishin) in 0.4 M mannitol solution from embryogenic cells, which were sub-cultured in mCD medium containing 7 μM 2,4-dichlorophenoxyacetic acid (2,4-D), 3 μM of benzyladenine (BA), and 3% sucrose. Protoplasts were cultured in NH_4NO_3-free Murashige and Skoog's (MS, Murashige and Skoog 1962) basal medium containing 6% sucrose, 10 μM each of 2,4-D and BA and 50 mM $MgCl_2$ in a 96-well (No.3075, Falcon) or a 24-well (No. 3047, Falcon) culture plate. *Betula* protoplasts were isolated from leaves of shoot culture, by 1% each of Cellulase R-10 (Yakult) and Driselase 20 (Kyowa Hakko Kogyo) in 0.6 M mannitol solution for 20 hrs. The leaves were floated on enzyme solution without cutting. Protoplasts were cultured in 1/2 strength MS basal medium containing 3% sucrose, 1 μM each of naphthaleneacetic acid and *N*-(2-chloro-4-pyridyl)-*N'*-phenylurea (CPPU, Sigma) and 100 mM $CaCl_2$. Fibers were used after 1-2 week of culture for *Larix* or 2 month of culture for *Betula*. They were incubated in a humid incubator (CO_2-incubator without the supply of CO_2-gas, APC-30D, CL-30, ASTEC Co. Ltd.) at 28°C.

2.2 Fabrication of Micropipettes

Micropipettes were prepared by a pipette puller (PB-7, Narishige) from 10 μL calibrated pipettes (Drummond). Picking up-pipettes were made as described previously (Sasamoto et al., 2000). Briefly, after one-step heating procedure (dial 600), the tips were removed by using a sandpaper (fine, No. 2000) to make inner diameter of 50 to 200 μm and washed with water. For tensile testing, tip of 10 μm diameter and 7 mm long was made by a two-step heating procedure (dial No. 1: 70, No. 2: 600). And their tips were cut into different lengths by using a microforge (MF-900, Narishige), or were smoothened by burning. Shape of micropipettes was modified by a micro-burner (Prince) as previously described (Ogita et al., 1999).

2.3 Calibration of Micropipettes

Calibration of micropipettes was performed before tensile testing of fibers for measuring the load of the order of μN (Onishi et al., 2006). The wide end of a micropipette was connected to a microstage for fine motion in the vertical direction. The micropipette was moved downwards and pressed against a microbalance (AB104, Mettler Toledo). The pressing load caused the deflection of the tip of the micropipette like a cantilever beam. The load and deflection of the tip were measured by the microbalance and a microscope, respectively.

The spring constant of the micropipette was obtained from the slope of the load versus deflection plot. In the linear range, its spring constant k_p is given by Equation (1), where δ_p is the deflection of a micropipette, F_p is the load applied to the micropipette.

$$k_p = \frac{F_p}{\delta_p} \tag{1}$$

After calibrating micropipette *A* with a tip length $l_{p,A}$, micropipette *B*, with a different tip length $l_{p,B}$ was calibrated using the relationship between the spring constant and the tip length of micropipettes as cantilever beams (Timoshenko, 1955) by Equation (2) where $k_{p,A}$ and $k_{p,B}$ are the spring constants of micropipettes *A* and *B*, respectively.

$$k_{P,B} = k_{P,A} \frac{l_{p,A}^3}{l_{p,B}^3} \tag{2}$$

2.4 Tensile Testing of Fibers

Protoplast fibers of about 10 μm diameter for *Larix* and of about 30 μm diameter for *Betula* were selected under an inverted microscope. They were picked up and transferred to a medium or pure water in a well of a 4-well plastic dish (Nunc) or a 60 mm center-well organ culture dish (BD Falcon) by using a micromanipulator as previously described (Oyanagi et al., 2014; Sasamoto et al., 2003). The medium for *Larix* contained 50 mM

MgCl$_2$, while that for *Betula* contained 100 mM CaCl$_2$ or pure water to study the effect of a medium containing high concentrations of Ca^{2+} ions. The dish was set on the microscope stage of the micro-tensile testing system which we have designed (Figure 1a). A single fiber in the dish was held with the fine ends of two micropipettes prepared as in sections *2.1* and *2.2*. The wider ends of two micropipettes were connected to two manual-type micromanipulators (MM188, Narishige-Nikon; MO202, Narishige) and microinjectors (IM-188, Narishige) set to an inverted microscope (IX-71, Olympus). One deflectable micropipette was slowly moved in the horizontal direction to stretch the fiber. The fiber stretching process was observed and recorded with a CMOS camera combined with a video recorder (ivis HF M52, Canon), which was connected to a microscope with an attachment (NY-VS (811276), Microscope Network).

Two methods to hold a single fiber by two micropipettes have been newly developed. In one method for *Larix* fibers, a single fiber was wired between two micropipettes to make a loop of the fiber and stretched (Figure 1b). In the other method for *Betula* fibers, both ends of a single fiber was pushed down onto the bottom of the dish by one rigid micropipette of which the tip was smoothened by burning and was wider than the other deflectable micropipette to make a loop of the fiber (Figure 1c). The fiber loop was hooked on the deflectable micropipette and stretched.

2.5 Evaluation of Mechanical Properties of Fibers

The load of a single fiber F_f is half the load of a micropipette F_p because a loop of the fiber was made between micropipettes during tensile testing (Figure 1b, c). The fiber load F_f was obtained using Equation (1) with the deflection δ_p and spring constant k_p of the calibrated micropipette as in the Equation (3).

$$F_f = \frac{1}{2}F_p = \frac{1}{2}k_p\delta_p \tag{3}$$

Young's modulus of the fiber E_f was obtained by dividing the stress by the strain as in the Equation (4),

$$E_f = \frac{\dfrac{F_f}{A_f}}{\dfrac{\Delta l_f}{l_f}} \tag{4}$$

where A_f is the cross-sectional area of the fiber obtained from its diameter d_f by assuming the cross section as circular, l_f and Δl_f are the initial fiber lengths and its elongation between the points holding the fiber by two micropipettes, respectively. δ_p, d_f, l_f, and Δl_f were measured under the microscope of the tensile test system. When F_f was maximum, the tensile strength of the fiber σ_B was given by Equation (5).

$$\sigma_B = \frac{F_f}{A_f} \tag{5}$$

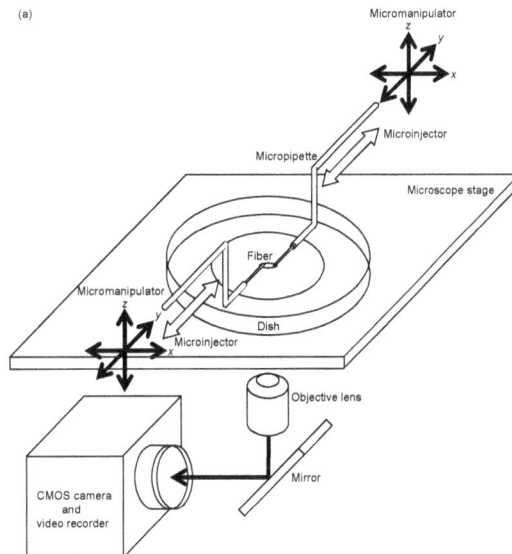

(a)

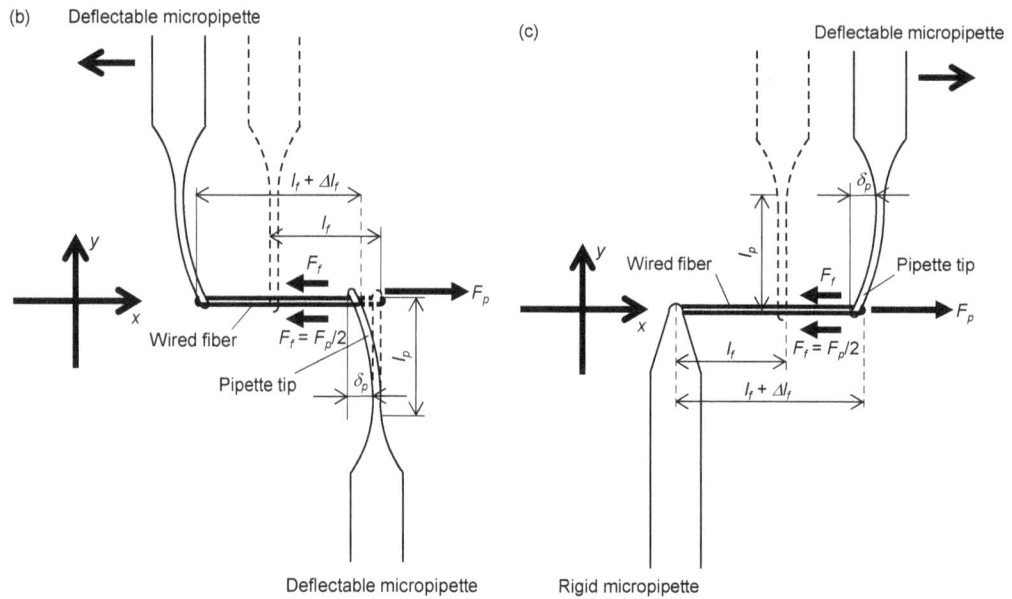

Figure 1. Schematic diagram of the micro tensile test system (a) and the detail of its test section for a *Larix* fiber (b) and a *Betula* fiber (c)

3. Results

3.1 Calibration of Micropipettes

The calibration of a micropipette (tip length 3.97 mm) was performed. The relationship between the load and deflection was almost linear (Figure 2). The spring constant of the micropipette was calculated as 7.28 mN/m from Equation (1). The spring constant of another micropipette (tip length 3.45 mm) was determined as 11.1 mN/m from Equation (2). During tensile testing, the latter micropipette was used for measuring the load of fibers.

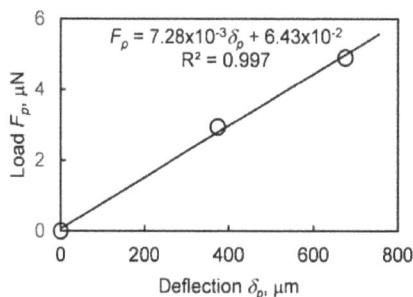

Figure 2. Load versus deflection of a micropipette for calibration

3.2 Tensile Testing of Fibers

3.2.1 *Larix Fiber*

An example of the time-lapse micrographs of the deformation of a *Larix* fiber in a medium containing 50 mM $MgCl_2$ during tensile testing is shown in Figure 3. When the fiber was firstly twisted, it was successfully wired between two deflectable micropipettes to make a loop of the fiber (Figure 3a). When the left micropipette was slowly moved to the left by a micromanipulator, the fiber was untwisted and made straight without the deflection of the right micropipette, the fiber length was 230 μm and defined as the initial fiber length l_f (Figure 3b). As the left micropipette continued to move, the fiber was stretched (Figure 3c) and finally fractured (Figure 3d).

Just before fiber fracture, the deflection of the fixed right micropipette reached a maximum and its deflection δ_p was measured as 21.6 μm using the captured image of the micropipette deformed from its initial shape (Figure 4, dotted line). The maximum load applied to the *Larix* fiber F_f was determined as 0.120 μN using Equation (3) and at that time, the maximum fiber elongation Δl_f was measured as 38.5 μm. Since the diameter of the fiber d_f was 10.2 μm from the captured image, the cross-sectional area A_f was 81.7 μm². Using Equation (4), Young's modulus of the *Larix* fiber E_f was determined as 8.75 kPa (1.75 x 10⁴ N/m²). Using Equation (5), the tensile strength of the fiber σ_B was obtained as 1.47 kPa at the strain $\Delta l_f / l_f$ of 0.170 at the breaking point.

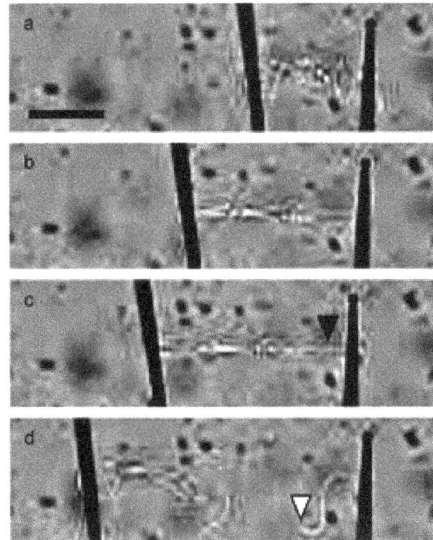

Figure 3. Time-lapse micrographs of deformation of a *Larix* fiber during tensile testing. The twisted fiber was wired between micropipettes in initial state to make a loop of the fiber (a). While the left micropipette was moved to left, the fiber was made straight (b) and elongated (c). The fiber fracture finally occurred (d). Black and white arrowheads indicate the fiber in elongated and fractured state, respectively. Scale bar, 100 μm

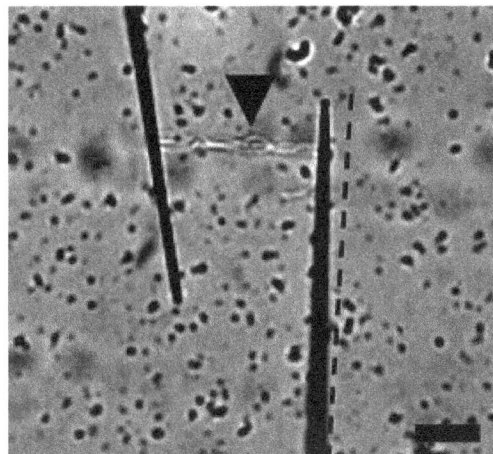

Figure 4. Micrograph of deflection of micropipettes in tensile testing for a *Larix* fiber. The arrowhead indicates the fiber wired between two micropipettes. While the left micropipette was slowly moved to left, the fixed right micropipette was deflected from its initial state (dotted line) and the fiber was elongated. The shape of the right micropipette shows the maximum deflection just before the fiber failure. Scale bar, 100 μm

3.2 Betula Fibers

In the tensile testing of a *Betula* fiber in a medium containing 100 mM $CaCl_2$, it was also successfully held with rigid and deflectable micropipettes (Figure 5). In the same manner as the *Larix* fiber, the diameter d_f, Young's modulus E_f, tensile strength σ_B, and strain at the breaking point $\Delta l_f / l_f$ of the *Betula* fiber were obtained as 25.5 μm, 9.30 kPa (1.86×10^4 N/m^2), 1.13 kPa, and 0.122 using the initial fiber length l_f of 1123 μm, the maximum micropipette deflection δ_p of 104 μm, the maximum fiber load F_f of 0.579 μN, and the maximum fiber elongation Δl_f of 137 μm, respectively.

In the tensile testing of a *Betula* fiber in pure water, the diameter d_f, Young's modulus E_f, tensile strength σ_B, and strain at the breaking point $\Delta l_f / l_f$ of the *Betula* fiber were obtained as 21.5 μm, 6.95 kPa (1.39×10^4 N/m^2), 1.78 kPa, and 0.256 using the initial fiber length l_f of 404 μm, the maximum micropipette deflection δ_p of 116 μm, the maximum fiber load F_f of 0.643 μN, and the maximum fiber elongation Δl_f of 103 μm, respectively.

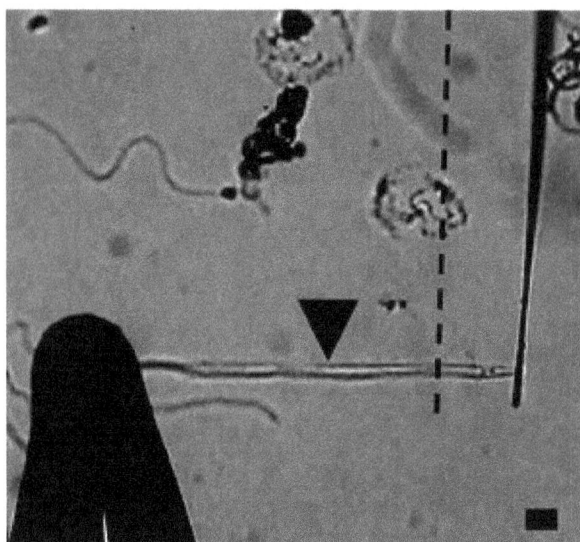

Figure 5. Micrograph of deflection of micropipettes in tensile testing for a *Betula* fiber. The arrowhead indicates the fiber wired between two micropipettes. While the right micropipette was slowly moved from the initial position (dotted line) to right and the rigid left one was fixed, the fiber was elongated and the right micropipette was deflected from its straight position. The shape of the right micropipette shows the maximum deflection just before the fiber failure. Scale bar, 100 μm

The diameter d_f, Young's modulus E_f, tensile strength σ_B, and strain at the breaking point $\Delta l_f / l_f$ in each tensile testing are summarized in Table 1.

Table 1. Mechanical properties of single fibers of callose developed in plant protoplast cultures.

Fiber type	Diameter d_f, μm	Young's modulus E_f, kPa	Tensile strength σ_B, kPa	Strain at the breaking point $\Delta l_f / l_f$
Larix in medium	10.2	8.75	1.47	0.17
Betula in medium	25.5	9.3	1.13	0.122
Betula in pure water	21.5	6.95	1.78	0.256

4. Discussion

When mechanical properties of single suspension-cultured tomato cells were investigated by micromanipulation probe on glass surface, the Young's modulus was 2.3 GPa in an uni-axial compression testing (Blewett et al., 2000; Wang et al., 2004). Chanliaud et al. (2002) reported Young's modulus (0.2-0.5 GPa) of pure cellulose by compression testing using cm-sized sheet materials with ~1 mm thickness and reduction of values by incorporation of pectin and xyloglucan, which are the components of plant cell wall. They also reviewed the Young's modulus (0.1-4 GPa) of plant cell walls and cells of different species. Cellulose is a natural fiber-forming glucan with beta-1,4-linkage composed of microfibril, which size is much smaller (1/ 1000) than the callose sub-fibrils (Fukumoto et al., 2005). Compression testing for sheet type materials of plant cell wall composites might result in an overestimation of the Young's modulus. Therefore, it is difficult to apply such a method on small protoplast fiber composed of callose. A value of 1.25 MPa was obtained in an uni-axial tensile testing, in which the size of material was 30x3x1 mm (Whitney et al., 1999). A low value, 0.25 MPa was obtained when xyloglucan was added to cellulose. Parre and Geitmann (2005) found that amorphous polymer, callose, is related to the tension stress in pollen tubes, which are composed of callose, pectin and cellulose. The modulus value, 4-5 μN / μm was reported using microindentation test. This value corresponds to 3.4 MPa. Mechanical properties of the plasma membrane of plant protoplast (Wolf & Steponkus, 1983) and animal cells (Onishi et al., 2006) were measured by aspiration using micromanipulator. The former value was 230 mN /m and the latter was 100 μN / m. These values correspond to 0.3 kPa and 4.0 kPa, respectively. As the protoplast fibers are thin and elongated, aspiration method could not be applied. It is needed to measure the mechanical property of whole fiber, but not of localized portion.

In this report, unique tensile test was developed using micromanipulators-loading micropipettes. New methods for wiring of a *Larix* fiber between two micropipette and the pushing down of *Betula* fibers for testing were successful. Improvements of holding methods of fibers was a prerequisite for the measurement of mechanical property of unique callose fiber. Before the development of the method of wiring of a fiber between two micropipettes by friction force between a fiber and glass surface, several glues were tried in vain to attach fibers to the tip of micropipettes (data not shown). The Young's modulus for the callose fiber was of the order of 10^4 [N/m^2]. This is the first report that the mechanical properties of a single callose fiber were determined. The value is two-fold larger than that of animal cells, and much smaller than those of plant cell, protoplast membrane and cell wall components. Tensile strain at break ($\Delta l_f / l_f$, 0.12-0.26) was similar in value as cellulose (Chanliaud, 2002).

Though, high concentration of Ca^{2+} (100-200 mM) in the protoplast culture medium is a prerequisite for *Betula* fiber formation, no big difference was obtained for Young's modulus from the tensile test in the culture medium and in pure water. Under tensile testing, high concentrations of Ca^{2+} ions are not necessary at the surface of the fiber structures. Furthermore, the value of Young's modulus of *Larix* fibers was similar to those of *Betula* fibers, which was described as area base. The differences between *Larix* and *Betula* are that large *Betula* fibers (30 μm diameter) take a long period (2 months), while *Larix* fibers (10-20 μm diameter), take 1-2 weeks to form (Sasamoto et al., 2003; Fukumoto et al., 2005). As similar Young's modulus values were obtained for both *Larix* and *Betula* fibers, similar substructures of callose fibers can be considered.

Recent reports indicate that protoplast cultures of salt-tolerant mangrove plants can serve as excellent callose fiber materials as fiber formation (4 μm diameter) takes only a few days of culture and without the need of additional divalent cations (Kurita et al., 2008, 2009; Oyanagi et al., 2012). In our laboratory, similar spiral fiber structures were also observed in a protoplast culture of *Arabidopsis* recently (Sasamoto et al., unpublished data) which might offer information on molecular biology in callose fiber formation. The method developed in this paper for measuring mechanical properties of callose fibers could be applied to different protoplast fibers. The callose fibers, which are developed in liquid protoplast cultures of several plant species, are easy to separate from the original protoplasts, when they were transferred to pure water or in medium containing fixation chemical, *e. g.* glutaraldehyde (Oyanagi et al., 2014). There is no deposition of callose on membrane of fiber-forming protoplast, as aniline blue dye stains only the fiber itself but not the surface of protoplast (Fukumoto et al., 2005). Difference in mechanical properties between callose fibers and the protoplast surface might be considered.

Callose is soluble in alkaline conditions which is a unique characteristic among cell wall polymer components. Clarifying the mechanical properties of callose fiber might help for possible use as spiral microfilament material in micro-world for specific medicinal purpose (Oyanagi et al., 2014).

Acknowledgments

The author H. S. thanks to Prof. E. C. Yeung of the University of Calgary for his valuable suggestions on this manuscript.

References

Blewett, J., Burrows, K., & Thomas, C. (2000). A micromanipulation method to measure the mechanical properties of single tomato suspension cells. *Biotechnology Letters, 22*, 1877-1883. http://dx.doi.org/10.1023/A:1005635125829

Chen, X. Y., & Kim, J. Y. (2009). Callose synthesis in higher plants. *Plant Signaling & Behavior, 4*, 489-492. http://dx.doi.org/10.4161/psb.4.6.8359

Chanliaud, E., Burrows, K. M., Jeronimidis, G., & Gidley, M. J. (2002). Mechanical properties of primary plant cell wall analogues. *Planta, 215*, 989-996. http://dx.doi.org/10.1007/s00425-002-0783-8

Cresti, M., & van Went, J. L. (1976). Callose deposition and plug formation in Petunia pollen tubes in situ. *Planta, 133*, 35-40. http://dx.doi.org/10.1007/BF00386003

Fukumoto, T., Hayashi, N., & Sasamoto, H. (2005). Atomic force microscopy and laser confocal scanning microscopy analysis of callose fibers developed from protoplasts of embryogenic cells of a conifer. *Planta, 223*, 40-45. http://dx.doi.org/10.1007/s00425-005-0065-3

Kurita, A., Hasegawa, A., Hayashi, N., & Sasamoto, H. (2008). Novel Callose fiber formation developed in protoplast cultures of a mangrove plant. *Proceedings of the 72th Annual Meeting of the Botanical Society of Japan*, P. 233.

Kurita, A., Fukumoto, T., Hayashi, N., Shiraishi, T., & Sasamoto, H. (2009). Characteristics of novel elongated callose fiber and regulatory mechanisms of the fiber formation in tree protoplast cultures: Mechanical properties; AFM; LCSM; TEM; GFP gene transformation. *Proceedings of the 59th Annual Meeting of the Japan Wood Research Society*, p. 96.

Murashige, T., & Skoog, F. (1962). A revised medium for rapid growth and bioassays with tobacco tissue cultures. *Physiologia Plantarum, 15,* 473-497. http://dx.doi.org/10.1111/j.1399-3054.1962.tb08052.x

Ogita, S., Sasamoto, H., & Kubo, T. (1999). Selection and microculture of single embryogenic cell clusters in Japanese conifers: *Picea jezoensis, Larix leptolepis* and *Cryptomeria japonica. In Vitro Cellular and Developmental Biology Plant, 35,* 428-431. http://dx.doi.org/10.1007/s11627-999-0061-6

Onishi, T., Shiraishi, T., & Morishita, S. (2006). Mechanical Properties of a Cultured Osteoblast. *Journal of advanced Science, 18,* 95-98. http://dx.doi.org/10.2978/jsas.18.95

Oyanagi, T., Hasegawa, A., Sasamoto, H., Hayashi, N., & Fukumoto, T. (2012). Cell divisions and callose fiber formation in cultures of protoplasts isolated from suspension cells of a mangrove plant, *Sonneratia caseolaris. Proceedings of the 62nd Annual Meetings of the Japan Wood Research Society*, p. 116.

Oyanagi, T., Kurita, A., Fukumoto, T., Hayashi, N., & Sasamoto, H. (2014). Laser Confocal Scanning Microscopy and Transmission Electron Microscopy to Visualize the Site of Callose Fiber Elongation on a Single Conifer Protoplast Selected With a Micromanipulator. *Journal of Plant Studies, 3,* 23-29. http://dx.doi.org/10.5539/jps.v3n2p23

Parre, E., & Geitmann, A. (2005). More than a leak sealant. The mechanical properties of callose in pollen tubes. *Plant Physiology, 137,* 274-286. http://dx.doi.org/10.1104/pp.104.050773

Sasamoto, H., Wakita, Y., Yokota, S., & Yoshizawa, N. (2000). Large electro-fused protoplasts of *Populus alba* selected by a micromanipulator: Techniques and some characteristics of cells and their regenerants. *Journal of Forest Research*, 5, 265-270. http://dx.doi.org/10.1007/BF02767120

Sasamoto, H., Ogita, S., Hayashi, N., Wakita, Y., Yokota, S., & Yoshizawa, N. (2003). Development of novel elongated fiber-structure in protoplast cultures of *Betula platyphylla* and *Larix leptolepis. In Vitro Cellular and Developmental Biology Plant, 39,* 223-228. http://dx.doi.org/10.1079/IVP2002388

Timoshenko, S. (1955). *Strength of Materials*. Van Nostrand, New York.

Wang, C. X., Wang, L., & Thomas, C. R. (2004). Modelling the mechanical properties of single suspension-cultured tomato cells. *Annals of Botany, 93,* 443-453. http://dx.doi.org/10.1093/aob/mch062

Wolfe, J., & Steponkus, P. L. (1983). Mechanical properties of the plasma membrane of isolated plant protoplasts. Mechanism of hyperosmotic and extracellular freezing injury. *Plant Physiology, 71,* 276-285. http://dx.doi.org/10.1104/pp.71.2.276

18

Genetic Analysis of Needle Morphological and Anatomical Traits among Naturc Populations of *Pinus Tabuliformis*

Mei Zhang[1], Jing-Xiang Meng[1], Zi-Jie Zhang[1], Song-Lin Zhu[2] & Yue Li[1]

[1]National Engineering Laboratory for Forest Tree Breeding, Key Laboratory for Genetics and Breeding of Forest Trees and Ornamental Plants of Ministry of Education, College of Biological Sciences and Technology, Beijing Forestry University, Beijing 100083, China

[2]The Forestry Bureau of Xixian, China

Correspondence: Mei Zhang, College of Biological Sciences and Technology, Beijing Forestry University, Beijing 100083, China. E-mail: liyue@bjfu.edu.cn

Abstract

The morphological and anatomical traits of needles are important to evaluate geographic variation and population dynamics of conifer species. Variations of morphological and anatomical needle traits in coniferous species are considered to be the consequence of genetic evolution, and be used in geographic variation and ecological studies, etc. *Pinus tabuliformis* is a particular native coniferous species in northern and central China. For understanding its adaptive evolution in needle traits, the needle samplings of 10 geographic populations were collected from a 30yr provenience common garden trail that might eliminate site environment effect and show genetic variation among populations and 20 needle morphological and anatomical traits were involved. The results showed that variations among and within populations were significantly different over all the measured traits and the variance components within population were generally higher than that among populations in the most measured needle traits. Population heritabilities in all measured traits were higher than 0.7 in common garden sampling among populations. Needle traits were more significantly correlated with longitude than other factors. First five principal components accounted for 81.6% of the variation with eigenvalues greater than 1; the differences among populations were mainly dependent on needle width, stomatal density, section areas of vascular bundle, total resin canals, and mesophyll, as well as area ratio traits. Ten populations were divided into two categories by Euclidean distance. Variations in needle traits among the populations have shown systematic microevolution in terms of geographic impact on *P. tabuliformis*. This study would provide empirical data to characterize adaptation and genetic variation of *P. tabuliformis*, which should be more available for ecological studies.

Keywords: genetic structure, needle, morphological and anatomical traits, geographic population, *Pinus tabuliformis*

1. Introduction

Pinus tabuliformis is a native conifer species in northern and central China. It survived from the quaternary glacial with complex genetic components and climate factors widely changed in its distributions (Hewitt, 2000; Chen, 2007; Guo, 2008). Significantly variations among nature populations in *P. tabuliformis* have been reported on growth, wood property, physiology and propagation traits (Mao *et al.*, 2009, Mao *et al.*, 2011; Xu *et al.*, 1991; Yuan *et al.*, 2014; Niu *et al.*, 2013; Yang *et al.*, 2015). Evidences from molecular have indicated the genetic differentiation among populations bounded up with the geographic distance (Chen, 2007; Wang, 2010; Gao, 2009). Like other plant in temperate arid steppe zone, *P. tabuliformis* have formed phenotypic characters to resist lower temperature, water depletion and light conditions (Zhang, 2010; Liu, 2012).

Needle is the most vigorous assimilation organs for conifer species, and the morphological and anatomical traits of needle are important references to plant taxonomy (Anna K. *et al.*, 2013; Xing, 2014; Huang, 2016). Needle traits are closely associated with physiological and functional attributes of plants including photosynthesis, respiration, water metabolism, nutritional status, as well as stress resistance (Oleksyn *et al.*, 1997; Eguchi *et al.*, 2004; Wu, 2007; Mao *et al.*, 2012). Phenotypic variations of needle morphological and anatomical characteristics were considered the results in physiological and adaptive genetic evolution (Anna K. *et al.*, 2013;

Balkrishna *et al.*, 2014), and have been widely used as available indicators in geographic variation, phylogenesis and evolutionary studies (Cole *et al.*, 2007; Xing, 2014; Melville, 2002, B. Nikolić *et al.*, 2013; Androsiuk Piotr *et al.*, 2011; K. Boratynska *et al.*, 2009).

However, phenotypic variations of needle traits should be influenced by both environment and genetic impacts (Li, 2009; Xu,1991; Cole *et al.*, 2008) in studies with samples from natural forests within species (Legoshchina *et al.*, 2013; Michael *et al.*, 2012), hardly tell the amount of genetic or environment contributions and limit in explaining the real effects on changes among populations. Variations of needle traits from population samples based on common garden trail could provide the genetic contributions among populations. The research and analysis of *P. tabuliformis* mainly involve the natural distribution characters of needles, and results that might be limited because of environmental variations at different population sites (Nikolić *et al.*, 2013; Xing *et al.*, 2014). In this study, we collected the needle samples from a 30yr provenience common garden trail with ten populations in *P. tabuliformis* in order to eliminate environment effects on needle traits among populations. The objectives of this study were to (1) reveal the variations in needle morphological and anatomical traits among populations; (2) illustrate the variations in needle traits among individuals within population; (3) evaluate the genetic and environment impacts on phenotypic variation for each needle trait; and (4) clarify the phylogenetic relationships among populations. This study provides theoretical and methodological reference for the conservation biology of populations in the coniferous morphological variation, ecological adaptability, system evolution and population genetics. Provide the basis for the use of coniferous morphological traits in each study area.

2. Materials and Methods

2.1 Sample Populations

Ten geographic populations, located in typical habitat of *P. tabuliformis*, were involved (Fig. 1) in our experiment. Ten populations are as follows: Heilihe, Inner Mongolia(HLH); Dongling, Heibei(DL); Lingkongshan, Shanxi(LKS); Guandishan, Shanxi(GDS); Shangzhuang, Shanxi(SZ); Nanyang, Henan(NY); Luonan, Shannxi(LN); Shuanglong, Shannxi(SL); Xiaolongshan, Gansu(XLS); Huzhu, Qinghai(HZ).

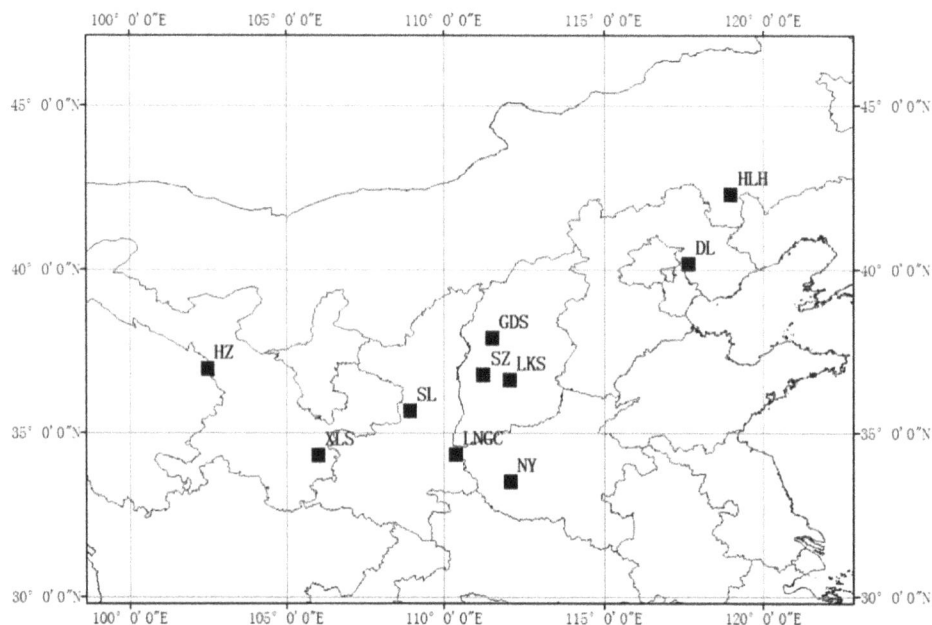

Figure 1. Location of ten *P. tabuliformis* populations sampled in North China

The environmental information of sampled populations is shown in Table 1. The annual mean temperatures and precipitations in each geographic location of populations were obtained from www.worldclim.com (Hijmans *et al.*, 2005).

Table 1. Sampling populations of *P. tabuliformis*

Population	Sample size	Longitude(°E)	Latitude (°N)	Altitude(m)	Annual mean temperature(°C)	Annual precipitation (mm)
HZ	29	102°28'	36°58'	2300	2.4	443
XLS	21	106°00'	34°20'	1630	8.5	663
NY	14	112°03'	33°32'	810	12.6	776
LNGC	12	110°21'	34°21'	1220	10.0	720
HLH	25	118°58'	42°17'	1300	7.6	360
GDS	19	111°29'	37°54'	1500	1.5	561
SL	25	108°56'	35°41'	1650	10.2	543
LKS	13	112°02'	36°37'	1665	5.6	604
SZ	13	111°12'	36°46'	1660	6.2	545
DL	14	117°38'	40°11'	200	10.6	587

2.2 Field Experimental Design

The common garden test of provenience in *P. tabuliformis* was conducted in Xixian, Shanxi province (111°10'E, 36°48'N), the distribution central of the species and represented the general environmental conditions of the species. The experiment was established in 1981 with 2a seedlings from ten nature populations from the central to the edge of geographic distribution regions. A Randomized Complete Block Design (RCBD) with 6 blocks (replication), 24-individuals rectangular plot was set for the trail, 2m×2m in planting space. However, the remaining plants for each population were varied after 30yr nature selection. Needle samples from the provenience test were collected in September on 30yr plants in 2011, 11-30 individual trees for each population were involved from randomized sample trees. For each plant, 3 needles would be randomly selected for analysis. We examined a total of 20 traits involving morphology, anatomy and ratio of needle structures (Table 2). All the traits were adaptive- related and have been reported in previous researches (Huang *et al.*, 2016; Xing *et al.*, 2014; Zhao *et al.*, 2008).

2.3 Measurement of Needle Traits

Twenty needle traits were measured according to Xing's work (Xing *et al.*, 2014), including 16 directly measured morphological and anatomical traits and four area-ratio indices (the ratios between needle organizational structures in needle cross sections) (Table 2).

(1) Morphological traits (Fig. 2A): seven morphological traits were measured, including needle length (NL, 1 cm); the width and thickness in the middle part of the needle (NW and NT, 0.1 mm); the number of stomatal rows on the front and back sides of the needle (CSRN and FSRN, measured by stereomicroscope);the number of stomata rows(NSR, calculated as NSR=(FSRN+CSRN)); the mean number of stomata in 2 mm sections (SR2N, counted using Photoshop CS5 (Adobe Systems, Mountain View, CA) after photographing); the stomatal density on the convex surface of needle(CSD, calculated as CSD=(SR2N×CSRN)/ (2×NW))); and the mean stomatal density of the needle (MSD, calculated as MSD=(SR2N×(CSRN+FSRN))/ (2×2×NW)).

(2) Anatomical traits: The needles were sliced into cross sections in the middle part by hand (Xu and Tao 2006) and then observed under a microscope (BA2100; Motic, Xiamen, China). Vascular bundles, resin canals, and mesophyll were the measured tissues to assess the abilities of substance transduction, stress tolerance, and organic synthesis of needles. A total of 11 anatomical traits (Fig. 2B) were measured by using Motic Images Plus 2.0 software, including vascular bundle width (VBW, 0.001 mm); vascular bundle thickness (VBT, 0.001 mm); vascular bundle area (VBA, 0.001 mm^2); resin canal number (RCN); total resin canal area (RCA, 0.001 mm^2); needle section area (NSA, 0.01 mm^2); and mesophyll area (MA, 0.01 mm^2, calculated as MA=NSA-VBA-RCA).

(3) Area ratio traits (indices): four ratio traits of comparative area between different organizational structures in needle cross sections were measured, including the ratio of mesophyll area/vascular bundle area (MA/VBA); the ratio of mesophyll area/resin canal area (MA/RCA); the ratio of vascular bundle area/resin canal area (VBA/RCA); and the ratio of mesophyll area/(resin canals and vascular bundles) area [MA/(VBA+RCA)] (Table 2).

A   B

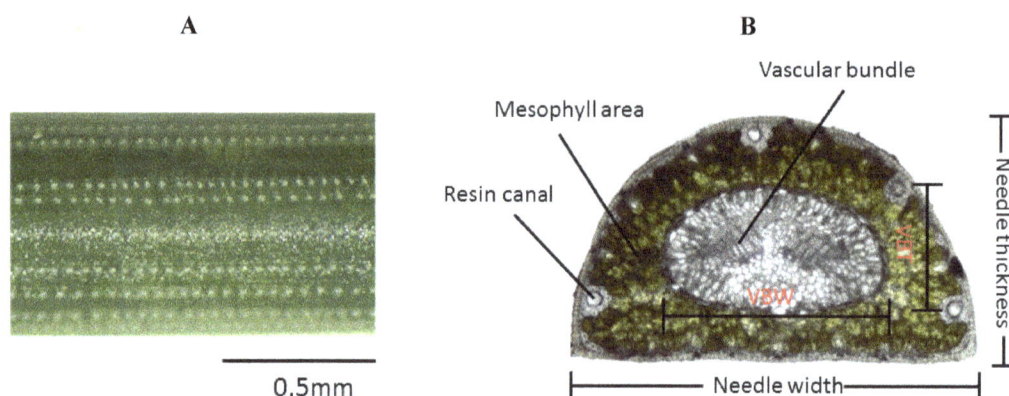

Figure 2. (**A**) stomatal rows on convex side of needle; (**B**) cross-section of the needle by unarmed slice

Table 2. List of morphological and anatomical traits of needles for *P. tabuliformis*

Abbreviation	Unit	Traits
NL	cm	Needle length
NW	mm	Needle width
NT	mm	Needle thickness
CSRN	No.	Number of stomatal rows on convex side of needle
FSRN	No.	Number of stomatal rows on flat side of needle
SR2N	No.	Mean number of stomata in a 2-mm-long section of needle
NSR	No.	Mean Number of stomata rows
CSD	No./mm2	Stomatal density on the convex surface of needle
MSD	No./mm^2	Mean stomatal density of needle
VBW	mm	Vascular bundles width
VBT	mm	Vascular bundles thickness
VBA	mm	Vascular bundle area
RCN	No.	Resin canals number
RCA	mm^2	Total resin canals area (sum of areas for total resin canals on needle section)
NSA	mm^2	Needle section area
MA	mm^2	Mesophyll area
MA/VBA	-	Mesophyll area/Vascular bundle area
MA/RCA	-	Mesophyll area/Resin canals area
VBA/RCA	-	Vascular bundle area/Resin canals area
MA/(VBA +RCA)	-	Mesophyll area/(Resin canals area and Vascular bundle area)

2.4 Statistical Analyses

Means, variation coefficients (CVs) and standard deviations for each trait among the populations were estimated. Data from each single needle measurement were used to conduct variance analysis. The analysis of variance (ANOVAs) for each morphological and anatomical trait of needle was carried out by using a similar nested-linear model and estimates of variance components:

$$y_{ijk} = \mu + p_i + T_{j(i)} + e_{ijk}$$

In this analysis, *yijk* is observation value of sample needle; μ is the mean of experiment; p_i is the effect of populations (random); $T_{j(i)}$ is the effect (random) of individual trees within populations; and e_{ijk} is the residual among sampled needles (random). All analyses were conducted in the SPSS 20.0 (SPSS Inc., Chicago, IL) and R (R, The University of Auckland, New Zealand) software. Population heritability (H^2) of each trait was estimated as described in previous author in situ and common garden test respectively (Becker, 1992).

Principal components analysis (PCA) was applied to scale data and evaluate the underlying dimensionality of the variables. We used PCA with a standardized matrix containing data on measured traits. The PCA scores of the five components were used to calculate the species distance. All the multivariate statistical analyses were performed in R.

Phenotypic correlation was analyzed at the level of individual and population. Morphological and anatomical traits of needles with a correlation coefficient of < 0.8 were used to do further analyses in order to minimize over-fitting of niche models by highly correlated traits and improve the interpretability of niche axes in the multivariate analyses.

Clustering analysis among populations of *P. tabuliformis* was performed using the Ward method in R, and the Euclidean distance among populations was calculated from z-cores. The average values of each needle trait were used for cluster analysis. Fifteen correlational needle phenotypic traits [NL, NW, SR2N, CSRN, FSRN, MSD, RCN, RCA, VBA, NSA, MA, MA/VBA, MA/RCA, VBA/RCA, MA/ (VBA+RCA)] were chosen to conduct cluster analysis.

3. Results

It was common for *P. tabuliformis* to appear in two-needle fascicle. Cross-sections of needles were fan-shaped. Resin canals were marginal or near the vascular bundle. The number of resin canal almost is 7~10.

3.1 Variations in Needle Morphological and Anatomical Traits among Populations

The mean and CV values of 20 needle traits among each population were listed in Table 3. The variation coefficients of 20 needle traits among populations were between 1.99% (SR2N) and 19.26% (RCA). The CV of each trait was different, which showed different genetic differences among populations. The values of variation coefficient in section -area traits (RCA) and four ratio traits (MA/RCA, MA/VBA, MA/(VBA+RCA) and VBA/RCA) were higher than 10%, meaning that there were larger phenotype variations among populations on those traits, while the morphological traits (NL, NT, NW,SR2N) and VBT were less than 5%. This means that there were smaller phenotype variations among populations on these traits, and anatomic traits were in moderate phenotype variation in general.

Table 3. Mean values and variability coefficient (%) analysis of needle characters in each *P. tabuliformis* population

Traits[a]	HZ		XLS		NY		LNGC		HLH	
	Mean	CV[b]	Mean	CV	Mean	CV	Mean	CV	Mean	CV
NL	10.87	16.63	11.68	15.30	10.40	13.51	10.53	12.77	10.76	13.52
NT	0.67	15.09	0.64	16.85	0.65	11.92	0.66	16.52	0.74	17.32
NW	1.27	12.83	1.25	12.67	1.31	9.81	1.28	10.29	1.37	13.59
FSRN	6.86	24.86	7.13	23.97	7.55	27.17	7.39	24.05	8.25	23.41
CSRN	7.64	23.60	7.89	24.00	8.33	26.06	8.86	26.59	8.99	22.66
NSR	14.51	16.88	15.02	14.38	15.88	20.64	16.25	13.75	17.24	15.75
SR2N	20.53	7.86	20.29	8.23	21.36	9.18	19.97	8.67	20.77	6.65
CSD	62.94	30.52	65.38	30.70	68.91	30.44	70.30	31.56	69.33	28.36
MSD	59.60	24.57	62.04	22.10	65.67	26.18	64.27	19.63	66.56	23.32
RCN	9.80	18.72	8.65	20.67	8.26	19.29	8.53	13.29	7.63	23.79
RCA	0.001	39.52	0.001	43.57	0.001	32.89	0.002	46.43	0.002	37.99
VBT	0.31	20.45	0.32	20.39	0.31	17.19	0.32	16.21	0.36	19.70
VBW	0.65	17.83	0.64	14.25	0.66	14.23	0.67	12.27	0.73	14.97
VBA	0.17	34.50	0.16	30.96	0.17	29.66	0.17	25.05	0.21	34.51
NSA	0.68	25.78	0.64	27.52	0.68	20.62	0.67	25.61	0.81	31.15
MA	0.51	26.22	0.48	28.82	0.51	21.39	0.49	28.50	0.60	32.22
MA/VBA	3.37	34.89	3.10	26.72	3.28	29.95	2.95	25.03	2.96	23.21
MA/RCA	410.74	44.33	507.64	44.99	397.16	43.50	385.91	66.77	354.47	52.61
VBA/RCA	129.27	46.96	175.99	56.45	125.28	39.95	135.26	75.52	123.27	55.40
MA/(VBA+RCA)	3.34	34.61	3.07	26.49	3.24	29.81	2.92	24.97	2.93	23.14

Traits[a]	GDS		SL		LKS		SZ		DL		Among populations	
	Mean	CV	Mean	CV	Mean	CV	Mean	CV	Mean	CV	Mean	CV
NL	10.71	14.72	11.16	14.28	11.12	13.25	11.30	15.70	10.53	22.50	10.91	3.72
NT	0.66	17.29	0.67	14.60	0.69	17.43	0.73	17.26	0.69	19.90	0.68	4.54
NW	1.31	12.35	1.30	8.68	1.34	11.46	1.36	15.45	1.33	13.66	1.31	3.00
FSRN	8.49	26.16	8.48	27.03	7.97	23.81	7.90	27.53	8.50	21.15	7.85	7.61
CSRN	9.30	23.96	9.37	26.53	8.62	23.00	8.74	24.57	8.88	20.08	8.66	6.50
NSR	17.79	16.28	17.85	19.37	16.59	11.63	16.64	18.00	17.38	14.49	16.51	6.84
SR2N	20.42	7.96	21.01	10.25	20.26	6.28	20.74	8.69	20.33	9.82	20.57	1.99
CSD	73.07	25.11	75.96	30.08	65.38	24.27	67.74	26.55	68.92	24.46	68.79	5.54
MSD	70.16	20.36	72.62	25.09	63.00	14.84	64.60	20.93	67.31	19.04	65.58	5.84
RCN	9.07	20.82	8.39	22.43	8.00	14.62	7.79	26.52	8.14	15.72	8.43	7.61
RCA	0.002	35.19	0.002	46.03	0.002	42.36	0.002	39.55	0.002	45.64	0.002	19.26
VBT	0.34	18.96	0.33	17.55	0.34	19.01	0.31	21.91	0.33	28.48	0.33	4.69
VBW	0.71	13.54	0.68	21.36	0.75	15.64	0.67	19.54	0.67	23.20	0.68	5.15
VBA	0.19	27.62	0.18	33.95	0.20	32.52	0.17	40.16	0.18	50.78	0.18	9.38
NSA	0.69	26.73	0.69	21.05	0.74	28.73	0.79	30.99	0.73	32.15	0.71	7.70
MA	0.50	29.01	0.51	20.14	0.53	30.02	0.62	31.05	0.55	31.16	0.53	8.77
MA/VBA	2.69	22.55	3.21	47.82	2.78	38.93	3.98	25.82	3.47	36.07	3.18	11.87
MA/RCA	327.29	50.45	340.12	42.90	275.61	43.36	461.62	48.96	324.27	41.73	378.48	18.31
VBA/RCA	123.27	43.24	114.87	42.24	108.71	52.41	120.84	51.51	102.49	44.36	125.92	15.91
MA/(VBA+RCA)	2.66	22.38	3.17	47.31	2.74	38.08	3.94	25.71	3.42	35.85	3.15	11.87

[a] See Table 2 for definitions of the traits; [b] Variation coefficients(%).

Results of ANOVA showed that variations in all 20 needle traits were significant within and among populations. It showed that the variation of needle traits was controlled by significant genetic effects, and the genetic diversity and abundance variation among populations in *P. tabuliformis*. The values of variance components among populations were changed from 3.65% to 16.47%. The variance components of all traits were less than 17%, meaning that the genetic contribution of population on phenotype of each trait was different among them. In each measured trait, the population heritability among populations of the morphological traits was greater than 0.70, the anatomical traits were greater than 0.86, and the ratio traits were greater than 0.88.

3.2 Variation in Needle Traits among Individuals within Population

The average CV of each measured needle trait was from 22.81%~40.89% in individuals within population. And the SZ population was the largest (40.02%), which showed that the variation of the needle phenotype was abundant, and the diversity was high, while the NY population was the least (23.67%).

All the measured needle traits in sampling have shown significant variations of individuals within population by analyzing each sample data respectively as well (Table 4). Results from ANOVA showed that variance components of needle traits among individuals within population were generally higher than those among populations. Needle traits variation is influenced by individual genetic.

Table 4. ANOVA for morphological and anatomical needle traits among and within populations of *P. tabuliformis*

Traits[a]	Variance component /%			H^2_p	CV_p
	P (9)[b]	T (175)[c]	R (370)[d]		
NL	4.64**	64.45**	30.91	0.84	3.72
NT	6.82**	69.82**	23.36	0.92	4.54
NW	5.66**	68.29**	26.05	0.89	3.00
CSRN	8.43**	35.09**	56.48	0.83	7.61
FSRN	7.80**	39.95*	52.25	0.83	6.50
NSR	15.70**	68.57**	15.73	0.98	6.84
CSD	4.17**	45.72**	50.11	0.71	1.99
MSD	7.32**	67.32**	25.35	0.92	5.54
SR2N	3.65**	52.16**	44.19	0.70	5.84
RCA	16.47**	52.25**	31.28	0.95	19.26
RCN	13.94**	62.21**	23.85	0.96	7.61
VBT	5.19**	64.03**	30.78	0.86	4.69
VBW	7.81**	68.25**	23.94	0.93	5.15
VBA	6.73**	68.52**	24.76	0.91	9.38
NSA	7.10**	70.33**	22.58	0.92	7.70
MA	7.84**	67.48**	24.68	0.92	8.77
MA/RCA	10.88**	48.91**	40.21	0.91	11.87
MA/VBA	8.84**	55.16**	36.00	0.90	11.87
VBA/RCA	7.84**	52.33**	39.83	0.88	18.31
MA/(VBA+RCA)	8.98**	55.12**	35.9	0.90	15.91

*Statistical significance at P = 0.05; **Statistical significance at P = 0.01; [a] See Table 2 for definitions of the traits; [b] df = 9 for populations; [c] df = 175 for individuals within populations; [d] df = 370 for error; We abbreviate individual and populations as ind and pop, respectively; H^2_p is population heritability; CV_P is coefficient of variation among populations.

3.3 The Correlations between Needle Traits

Based on analysis by population, NL, SR2N, MSD and MA in morphological traits have not shown significant correlation with other traits (Table 5), while NW was positively correlated with RCA, VBA, NSA and MA significantly, and CSRN was positively correlated with MSD only. VBA and RCD were positively correlated with NSA and MA significantly. These results showed that NL and the stomatal characteristics were independent from other traits in population level.

By individual trees, there were more correlation coefficients significantly between measured traits. The trends of correlation were similar for both population and individual tree level, and meant that the same genetic relation between measured needle traits existed in both genetic level.

Table 5. Correlation coefficients between needle traits of P. tabuliformis

Traits[a]	NL	NW	FSRN	CSRN	SR2N	MSD	RCN	RCA	VBA	NSA	MA	MA/VBA	MA/RCA	VBA/RCA	MA/(VBA+RCA)
NL	1.00	-.18	-.19	-.26	-.13	-.21	.03	.31	.16	.03	.03	.16	.48	.51	.17
NW	-.07	1.00	.64*	.51	.28	.29	-.73*	.74*	.69*	.95**	.86**	.14	.47	-.69*	.13
FSRN	-.11**	.14**	1.00	.89**	.17	.85**	-.49	.69*	.65*	.51	.36	-.13	-.66*	-.66*	-.14
CSRN	-.08	.12*	-.01	1.00	.07	.88**	.45	.59	.55	.38	.24	-.23	-.58	-.54	-.23
SR2N	-.01	.01	.11*	.06	1.00	.35	-.21	-.08	-.09	.19	.26	.34	.07	-.21	.33
MSD	-.08	-.41**	.51**	.53**	.46**	1.00	-.27	.36	.35	.15	.04	-.16	-.45	-.42	-.16
RCN	-.06	.3**	-.04	-.04	.01	-.21**	1.00	-.47	-.43	-.69*	-.65*	-.16	.16	.30	-.16
RCA	-.04	.33**	.18**	.15**	-.03	-.03	-.07	1.00	.87**	.61	.41	-.29	-.91**	-.82**	-.31
VBA	-.09*	.73**	.18**	.15**	.01	-.22**	.26**	.33**	1.00	.60	.34	-.56	-.77**	-.51	-.56
NSA	-.06	.92**	.12**	.14**	-.01	-.36**	.26**	.32**	.80**	1.00	.96**	.29	-.27	-.55	.28
MA	-.04	.90**	.09**	.12*	-.01	-.38**	.23**	.28**	.63**	.97**	1.00	.54	-.03	-.46	.54
MA/VBA	.10*	-.07	-.14**	-.07	-.04	-.09	-.07	-.16**	-.59**	-.08	.14**	1.00	.49	-.12	.99**
MA/RCA	.02	.19**	-.13**	-.08*	.00	-.21**	.2**	-.68**	.05	.23**	.29**	.22**	1.00	.79**	.24**
VBA/RCA	-.01	.19**	-.06	-.05	.03	-.15**	.26**	-.57**	.37**	.24**	.17**	-.29**	.82**	1.00	-.11
MA/(VBA+RCA)	.09*	-.07	-.14**	-.07	-.04	-.09*	-.07	-.17**	-.58**	-.08	.14**	.97**	.24**	-.29**	1.00

[a] See Table 2 for definitions of the traits; Above diagonal line is the correlation coefficients among populations; Below diagonal line is the correlation coefficients within individuals; *Statistical significance at P = 0.05; **Statistical significance at P = 0.01.

3.4 The Correlations between Needle Traits and Environmental Factors

Results of correlation analysis between geological, meteorological factors and some major needle traits of populations in *P. tabuliformis* showed that there was more significant correlation of needle traits with longitude than other factors (Table 6). Among major needle traits, longitude was significant associated with NW, CSRN, FSRN, RCN, RCA, VBA, NSA and MA; and latitude promoted traits in NW, FSRN, RCA, VBA, NSA and MA for positively significant correlation. Elevation was negatively correlated with the stomata (FSRN, CSRN, SR2N, MSD) and area traits (RCA, VBA, NSA, MA). The annual average temperature and the annual precipitation had no significant correlation with the measured traits. The higher or moderate correlation coefficients estimated between geological, meteorological factors and population needle traits might reveal microevolution of needle traits among populations by environmental impact.

Table 6. Correlation coefficients between geological, meteorological factors and some major needle traits of population in *P. tabuliformis*

Traits[a]	Longitude	Latitude	Altitude	Amt[b]	Ap[c]
NL	-0.443	-0.203	0.575[#]	-0.218	-0.153
NW	0.764*	0.664*	-0.234	-0.051	-0.453
CSRN	0.596[#]	0.345	-0.341	0.111	-0.118
FSRN	0.707*	0.554[#]	-0.441	0.063	-0.252
SR2N	0.094	-0.078	-0.066	0.315	-0.046
MSD	0.427	0.207	-0.352	0.218	-0.052
RCN	-0.774**	-0.307	0.478	-0.512	-0.019
RCA	0.644*	0.578[#]	-0.233	-0.069	-0.337
VBA	0.646*	0.695*	-0.097	-0.227	-0.497
NSA	0.672*	0.739*	-0.114	-0.104	-0.615[#]
MA	0.553[#]	0.615[#]	-0.098	-0.040	-0.542
MA/VBA	-0.083	-0.033	-0.062	0.199	-0.062
MA/RCA	-0.486	-0.411	0.264	0.088	0.169
VBA/RCA	-0.518	-0.448	0.319	0.009	0.250
MA/(VBA+RCA)	-0.089	-0.038	-0.058	0.199	-0.059

[a] See Table 2 for definitions of the traits; [b] Annual Mean Temperature; [c] Annual Precipitation; [#] Statistical significance at P=0.1; *Statistical significance at P = 0.05; **Statistical significance at P = 0.01.

3.5 Principal Component and Cluster Analysis on Needle Traits

Principal component analysis (PCA) on needle traits revealed five principal components with eigenvalues >1. The five principal components accounted for 81.659% of the variation. Individually, the following traits NW, NT, NSA, VBW, VBT, VBA, and MA had high loadings in the first component, more than |0.8| (Table 7). CSRN, NSR, CSD, MSD had high loadings in the second component, more than |0.6|. MA, MA/VBA, and MA/ (VBA+RCA) had high loadings in the third component, more than |0.5|. RCA, MA/RCA, and VBA/RCA were the main loading in the fourth component, more than |0.7|. FSRN, CSRN were the main loading in the fifth component, more than |0.5|. The main loadings in first, second, third and fourth component were different. Ten populations were not clearly distinguished from each other. All of them had broad distributions that covered almost all of the each other.

Ten populations were divided into two categories by cluster analysis (Fig. 3). Populations in HZ and XLS (in the west of populations) were clustered into one group. Populations in GDS, SL, NY, LNGC, SZ, LKS (the central populations) and HLH, DL (in the northeast populations), were clustered into another group.

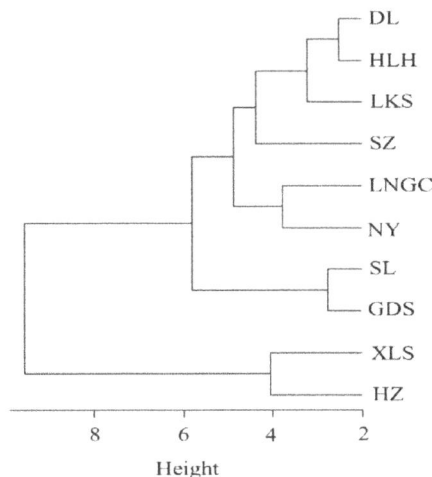

Figure 3. Cluster analysis based on needle traits in ten populations of *P. tabuliformis*

Table 7. Loadings on the first five components of needle traits in PCA in *P. tabuliformis*

Traits[a]	Comp1	Comp2	Comp3	Comp4	Comp5
NL	-0.100	-0.154	0.024	-0.033	-0.180
NT	0.891	-0.093	0.296	0.031	-0.047
NW	0.874	-0.125	0.317	-0.044	0.020
FSRN	0.171	0.480	0.116	-0.014	0.822
CSRN	0.108	0.684	0.344	0.282	-0.513
NSR	0.197	0.830	0.331	0.196	0.192
SR2N	-0.027	0.313	0.069	0.255	0.141
CSD	-0.302	0.754	0.187	0.352	-0.410
MSD	-0.339	0.847	0.121	0.290	0.196
RCN	0.332	-0.190	-0.068	0.197	-0.016
RCA	0.340	0.342	0.200	-0.761	-0.080
VBT	0.894	0.131	-0.186	0.006	-0.047
VBW	0.894	0.180	-0.235	-0.053	-0.033
VBA	0.947	0.151	-0.189	-0.008	-0.041
NSA	0.929	-0.110	0.334	0.013	-0.015
MA	0.816	-0.207	0.511	0.023	-0.002
MA/VBA	-0.414	-0.415	0.794	0.055	0.051
MA/RCA	0.161	-0.477	0.071	0.820	0.078
VBA/RCA	0.363	-0.247	-0.375	0.775	0.048
MA/(VBA+RCA)	-0.411	-0.420	0.794	0.064	0.052

[a] See Table 2 for definitions of the traits.

4. Discussion

4.1 Variation Patterns and Genetic Structures of Needle Morphological and Anatomical Traits in Populations of P. Tabuliformis

Needle variations were considered the visual indicators to investigation of genetic variability and adaptive divergence (Wu, 1965; Balkrishna *et al.*, 2014). All organizations have a special function of needles in conifer growth and metabolism physiological protection. The needle trait of mesophyll tissue is the main place of photosynthesis and fixed carbon assimilation of plant, vascular tissue is the main area of needles to divert water, inorganic salts and organic compounds, stomatal regulation of water transpiration and gas exchange with the environment of the channel, resin only with resistance. Populations of different environments in order to adapt to the local environment, the adaptability traits of its needle will have varying degrees of variation (Pensa M. et al., 2004). In order to adapt to the local environment, the adaptability traits of the needle will have different degree of variation. Understanding the stability of needle traits would be helpful for employing them in ecological and evolutionary studies (Xing, 2014). In this study, to test natural populations, *P. tabuliformis* is planted in the same environmental conditions so that variations of needle traits should be mainly impacted by genetic effects. The needles of population phenotypic variation are extremely rich, populations and individuals in a population were significant difference (Table 3 and 4). Variation components of needle traits within population were higher than that among general populations, which means the more genetic variations of needle traits existed in population in general. The morphological and anatomical needles traits within population diversity were higher than among populations, this is a kind of adaptability to complicated environment of *P. tabuliformis*. The needle traits were also significant variation among populations, reflected the extent to adapt to different environments. Different original provenances may lead the needle traits had higher population genetic variations. And this phenomenon was showed lately in seedling period of *P. yunnanensis* (Huang *et al.*, 2016; Woo *et al.*, 2002). It showed that the effect of environmental effects on plants can be inherited, and its distribution is closely related to the geographical environment, reproductive isolation and human activities. The morphological and anatomical needle traits were affected by the habitat change and natural selection in the distribution area, which reflected the adaptive strategy of population in the process of evolution. The morphological and anatomical characters of needles were affected by the habitat change and natural selection in the distribution area, which reflected the adaptive strategy of population in the process of evolution. The measured traits were much lower population contributions than individuals within population, and populations and individuals within population were significant difference. Phenotypic variation was determined by the effects of genetic variation, environmental variation, and the interactions between genetic and environmental factors (Rehfeldt, 1991).

Plant respiration, transpiration, and photosynthesis take place in the stomata (Batos *et al.*, 2010). In this study, variance components of three stomatal traits (CSRN, FSRN, and SR2N) among populations were less than 10%, and correlations between stomatal traits and other traits were low (Table 5). What this showed is that variation in stomatal traits among populations were relatively stable and independent. Variations in stomatal traits among populations were significant, probably due to the respiratory and photosynthetic pathway of the leaves, which are controlled by genetic factors(Zhang *et al.*, 2012) and HIC CO_2 genes (Gray *et al.*, 2000). In the process of needle formation, the CO_2 level in the atmosphere was relatively stable, and the source of the variation was reduced. The excessive stomatal variation is not good for the growth and development of plants. Clearly, the environmental factors of provenance had a certain impact on genetic changes of stomatal density in population (Tiwari *et al.*, 2013). In addition, needles length (NL) variation among geographical populations of pine is considered the result of ecology selection (Xiao, 2003). The length of the needles (NL) with little differences between conifer samples from different sources, indicating that needle length is mostly genetic-determined. Geographic populations at high-latitude and high-altitude usually have shorter needles, which should benefit coniferous plants resist snow pressure (López, 2010). Needles morphology and anatomy characteristics should be genetic determined (Eguchi, 2004; Li, 2013; Xing, 2014), and these indicators could be used as a reliable and efficient fast method to explore the genetic variation among geographical populations.

4.2 The Relationship between Geographical Variation and Ecological Adaptability of Morphological and Anatomical Traits of Needles in Populations of P. Tabuliformis

Differences in leaf traits reflect the adaptability of species in complex habitats, as well as the evolutionary history of species (Eo and Hyun, 2013). Genetic differentiation and environment divergence promote the phenotypic variation within species (Li, 2013; Kitajima, 2012). The influence of environment on different geographical populations is very significant. In our test, variations of needle traits in common garden sampling should represent the genetic differentiation among and within populations. The most variation components were observed within population, demonstrating that the differentiation among individuals is a major source of variation in *P. Tabuliformis*. Results from correlation analysis show that lots of needle traits are correlated with multiple environmental factors in different provenances (Table 6). MA/VBA presented a trend of decrease with increasing altitude, and increased with increasing temperature, which reflected the responses of needles to coldness in high altitude habitats. RCN was positively correlated with altitude, but negatively correlated with longitude, annual mean temperature and precipitation. Resin canals were considered to be the first line of defense stress (Eo and Hyun, 2013). Increasing the number of resin canals could help increase stress resistance in habitats with higher altitude. Likewise, increasing the number of resin canals could reduce the influence of torridity and drought. All the tested traits (except MA/RCA, VBA/RCA) were negatively correlated with annual precipitation. Correlation between traits and precipitation was generally higher than temperature, which meant precipitation was the limitation factor for the needle growth. Correlation between needle traits and longitude generally higher, indicating needle traits are relatively sensitive to longitude factor comprehensive effect, in other words phenotypic variation of *P. tabuliformis* showed variation pattern with longitude based on spatial distribution. Correlation analysis showed that most needle indicators increased from southwest to northeast, which is accordance with the distribution of water and heat in winter conditions. For previous research, temperature and precipitation gradual changed with longitude and latitude should have a selection on the genetic variation of *P. tabuliformis* (Liang, 2008; Liu, 2012).

The higher values of estimated population heritability in all measured needle traits in common garden sampling indicated that significant genetic differences are existence between populations in *P. tabuliformis* (Table 4). Habitat divergence among geographical populations should promote the genetic differentiation of plant species. *P. tabuliformis* is widely distributed in the northern China, the same natural population habitat area difference, directly affected by local habitat, thus forming a local ecotype. These variations could reflect the geographic and ecological distribution trend of species to a certain extent.

4.3 Impacts of Environmental Effects on the Evolution of P. Tabuliformis Populations

The ability of a species or population to adapt to the environment depends on the level of genetic variation and the distribution pattern of genetic variation (Nobis, 2012). Genetic differentiation and environmental variation will promote the phenotypic variation of species (Yuan, 2014; Li, 2008).Common garden test is one of the most detect method to exam genetic differences among geographical populations (Xiao 2003). In previous studies, most of them were sampled under the nature forest, which could not be used to represent the genetic difference. Environmental factors in the original areas had important impacts on the growth of plants (Körner, 2007; Lavadinović *et al.*, 2011). Schlichting (1986) found that the changes of phenotypes were in accordance with the changes in environmental gradients.

In this study, populations in the closer area were gathered into the same groups (Fig 3). Due to heterogeneity and independence of the habitats, directional selection played a main role during the process of differentiation in individual habitats(Sękiewicz *et al.*, 2013). There were differences between populations in the east and west of the distribution center of *P. tabuliformis.* Previous studies with molecular markers divided the *P. tabuliformis* into distinct groups (Chen, 2009), while our results with the test samples got the similar results. Molecular studies have shown that the western populations maybe recolonize from the southern population after the Quaternary glaciation (Wang, 2009; Chen, 2009), while the eastern populations may have originated from a refuge (Guo, 2008). Populations in the northeastern (DL, HLH, LKS), central west (SL, GDS), southeastern (NY, LNGC) probably were originated in the late Quaternary glacial refuge different (Guo, 2008; Liang, 2006).

Different correlation coefficients of needle traits and habitat factors are observed. Correlation analysis showed that measurement values of the most needle traits increased from southwest to northeast, which is accordance with the distribution tendency of water and heat of population location. For previous research, temperature and precipitation gradual changed with longitude and latitude should have a selection on the genetic variation of *P. tabuliformis* (Liang, 2008; Liu, 2012). Environment effected should be dissected when we compared the results of correlation analysis. Elevation has a positive impact on NL and resin ducts indicator (RCN), but a negative effect on the vascular bundles and stomatal density. Average annual temperature has a negative impact on the morphological traits, resin ducts indicator and vascular bundle. Annual precipitation was negative effect on most needle traits except MA/RCA, VBA/RCA. In earlier research, needle variations are always supposed to be the ecological consequence, but in this study, a complex reason of mutation, genetic drift and evolutionary history should be used explained the different direction of environment effected and genetic tendency of *P. tabuliformis*.

5. Conclusion

In conclusion, all measured morphological and anatomical needle traits were significantly different in both population and individual genetic level, and genetic contributions of individuals within population were higher than among populations in most of traits. Phenotypic variation of needle traits is considered to be the consequence of genetic evolution and higher population heritability in all measured traits were estimated. The traits related with needle size in both morphology and anatomy decreased with annual precipitation. Variations of needle traits among populations have shown systematic microevolution trends by geographic impact on *P. tabuliformis*, which provide genetic basis for the utilization of needle traits in various research fields and help us to understand the genetic pattern of needle in future.

Acknowledgements

This research was funded by the Fundamental Research Funds for the Central Universities (NO.

2015ZCQ-SW-02).

We would like to express our gratitude to the laboratory of forest tree breeding, Xixian Forestry Bureau,, Yu-Jie Huang, for their kind help.

References

Androsiuk, P., Kaczmarek, Z., & Urbaniak, L. (2011). The morphological traits of needles as markers of geographical differentiation in European *Pinus sylvestris* populations. *Dendrobiology, 65*, 3-16.

Anna, K., Jasinska, & Krystyna, Boratynska et al. (2013). Relationships among Cedrus libani, C. brevifolia and C. atlantica has revealed by the morphological and anatomical needle characters. *Plant Syst Evol., 299*, 35-48. https://doi.org/10.1007/s00606-012-0700-y

Balkrishna, G., Chunghee, L., & Kweon, H. (2014) Leaf anatomy and its implications for phylogenetic relationships in Taxaceaes. *Journal Plant Research, 127*, 373-388. https://doi.org/10.1007/s10265-014-0625-3

Batos, B., Vilotić, D., Orlović, S., & Miljković D. (2010). Inter and intra-population variation of leaf stomatal traits of *Quercus robur* L. in Northern Serbia. *Archives of Biological Sciences, 62*, 1125-1136. https://doi.org/10.2298/ABS1004125B

Becker, W. A. (1992). Manual of quantitative genetics. 5th ed. Academic Enterprises, Washington.

Boratynska, K., & Bobowicz, M. A. (2001). *Pinus uncinata Ramond* taxonomy based on needle characters. Plant Syst. *Evol., 227*, 183-194. https://doi.org/10.1007/s006060170047

CHEN, K., et al. (2008). Phylogeography of *Pinus tabuliformis Carr.* (*Pinaceae*), a dominant species of coniferous forest in northern China. *Molecular Ecology, 17*(19), 4276-4288. https://doi.org/10.1111/j.1365-294X.2008.03911.x

Cole, K. L., Fisher, J., & Arundel, S. T. (2008). Geographical and climatic limits of needle types of one- and two-needled pinyon pines. *Journal of Biogeography, 35*(2), 257-269.

Eguchi, N., Fukatsu, E., & Funada, R. (2004). Changes in morphology, anatomy, and photosynthetic capacity of needles of Japanese larch (*Larix kaempferi*) seedlings grown in high CO_2 concentrations. *Photosynthetica, 42*(2), 173-178. https://doi.org/10.1023/B:PHOT.0000040587.99518.a8

Eo, J. K., & Hyun, J. O. (2013). Comparative anatomy of the needles of *Abies koreana* and its related species. *Turkish Journal of Botany, 37*, 553-560.

Gao, M. W. Æ. F. Genetic Variation in Chinese Pine (*Pinus tabuliformis*), a Woody Species Endemic to China. Biochem Genet, 2009.

Gray, J. E., Holroyd, G. H., Bahrami, A. R., Sijmons, P. C., & Woodward, F. I. (2000). The HIC signalling pathway links CO_2 perception to stomatal development. *Nature, 408*, 713-716. https://doi.org/10.1038/35042663

Guo, L., Huang, G., & Wang, Y. (2008). Seasonal and Tissue Age Influences on Endophytic Fungi of *Pinus tabuliformis* (*Pinaceae*) in the Dongling Mountains, Beijing. *Journal of Integrative Plant Biology, 50*(8), 997-1003. https://doi.org/10.1111/j.1744-7909.2008.00394.x

Hewitt, G. M. (2000). The genetic legacy of the Quaternary ice ages. *Nature, 405*, 907-913. https://doi.org/10.1038/35016000

Hijmans, R. J., Cameron, S. E., Parra, J. L., Jones, P. G., & Jarvis, A. (2005). Very high resolution interpolated climate surfaces for global land areas. *International Journal of Climatology, 25*, 1965-1978. https://doi.org/10.1002/joc.1276

Huang, Y. J., Mao, J. F., & Chen, Z. Q. (2016). Genetic structure of needle morphological and anatomical traits of *Pinus yunnanensis*. *Journal of Forestry Research, 27*(1), 13-25. https://doi.org/10.1007/s11676-015-0133-x

Körner, C. (2007). The use of 'altitude' in ecological research. *Trends in Ecology & Evolution, 22*, 569-574. https://doi.org/10.1016/j.tree.2007.09.006

Lavadinović, V., Miletić, Z., Lavadinović, V., & Isajev, V. (2011). Variability in Magnesium Concentration in Needles of Different Douglas-Fir Provenances. *Forestry, 17*, 74-79.

Legoshchina, O., Neverova, O., & Bykov, A. (2013). Variability of the anatomical structure of *Picea obovata*

Ledeb. Needles under the influence of emissions from the industrial zone of Kemerovo. *Contemporary Problems of Ecology, 6,* 555-560. https://doi.org/10.1134/S1995425513050065

Li, B., & Gu, W. C. (2003). Review on genetic diversity in *Pinus. Hereditas, 25* (6), 740-748.

Li, C., Chai, B., & Wang, M. (2008). Population genetic structure of *Pinus tabuliformis* in Shanxi Plateau, China[J]. *Russian Journal of Ecology, 39*(1), 34-40. https://doi.org/10.1134/S1067413608010062

Liang, E., Eckstein, D., & Liu, H. (2008). Climate-growth relationships of relict *Pinus tabuliformis* at the northern limit of its natural distribution in northern China. *Journal of Vegetation Science, 19*(3), 393-406. https://doi.org/10.3170/2008-8-18379

Liu, J., Sun, Y., & Ge, X. (2012). Phylogeographic studies of plants in China: Advances in the past and directions in the future. *Journal of Systematics and Evolutio, 50*(4), 267-275. https://doi.org/10.1111/j.1759-6831.2012.00214.x

López, R., Climent, J., & Gil, L. (2010). Intraspecific variation and plasticity in growth and foliar morphology along a climate gradient in the Canary Island pine. *Trees, 24,* 343-350. https://doi.org/10.1007/s00468-009-0404-2

Mao, J. F., & Wang, X. R. (2011). Distinct niche divergence characterizes the homoploid hybrid speciation of *Pinus densata* on the Tibetan Plateau. *The American Naturalist, 177,* 424-439. https://doi.org/10.1086/658905

Mao, J. F., Li, Y., & Wang, X. R. (2009). Empirical assessment of the reproductive fitness components of the hybrid pine *Pinus densata* on the Tibetan. *Evolutionary Ecology, 23*(3), 447-462. https://doi.org/10.1007/s10682-008-9244-6

Mao, Q. Z., Watanabe, M., & Imori, M. (2012). Photosynthesis and nitrogen allocation in needles in the sun and shade crowns of hybrid larch saplings: effect of nitrogen application. *Photosynthetica,* 1-7. https://doi.org/10.1007/s11099-012-0049-z

Meiners, J., & Winkelmann, T. (2011). Morphological and Genetic Analyses of Hellebore Leaf Spot Disease Isolates from Different Geographic Origins Show Low Variability and Reveal Molecular Evidence for Reclassification into Didymellaceae. *Journal of Phytopathology, 159*(10), 665-675. https://doi.org/10.1111/j.1439-0434.2011.01823.x

Michael, P., Nobisa, Christopher, T., & Anita, R. (2012). Latitudinal variation in morphological traits of the genus *Pinus* and its relation to environmental and phylogenetic signals. *Plant Ecology & Diversity, 5,* 1-11. https://doi.org/10.1080/17550874.2012.687501

Nikolić, B., Bojović, S., & Marin, P. (2013). Variability of morpho-anatomical characteristics of the needles of *Picea omorika* from natural populations in Serbia. *Plant Biosystems, 10,* 1080-1126.

Nikolić, B., Bojović, S., & Marin, P. D. (2013). Variability of morpho-anatomical characteristics of the needles of Picea omorika from natural populations in Serbia. *Plant Biosystems, 10,* 1080-1126.

Niu, S., W. L. (2013). Open Pollinated Progeny Test and Stability Analysis of Seedlot from Clonal Seed Orchard of *Pinus tabuliformis. Journal of Northwest Forestry University, 2,* 013.

Nobis, M. P., Traiser, C., & Roth-Nebelsick, A. (2012). Latitudinal variation in morphological traits of the genus *Pinus* and its relation to environmental and phylogenetic signals. *Plant Ecology & Diversity, 5*(1), 1-11. https://doi.org/10.1080/17550874.2012.687501

Oleksyn, J., Modrzyński, J., & Tjoelker, M. G. (1998). Growth and physiology of *Picea abies* populations from elevational transects: common garden evidence for altitudinal ecotypes and cold adaptation. *Functional Ecology, 12*(4), 573-590. https://doi.org/10.1046/j.1365-2435.1998.00236.x

Pensa, M., Aalto, T., & Jalkanen, R. (2004). Variation in needle-trace diameter in respect of needle morphology in five conifer species. *Trees-Structure and Function, 18*(3), 307-311. https://doi.org/10.1007/s00468-003-0307-6

Rehfeldt, J. (1991). A model of genetic variation for applications in gene resource management. *Can J For Res., 2,* 1491–1500. https://doi.org/10.1139/x91-209

Schlichting, C. D. (1986). The evolution of phenotypic plasticity in plants. *Annual review of ecology and systematics, 17,* 667-693. https://doi.org/10.1146/annurev.es.17.110186.003315

Sękiewicz, K., Sękiewicz, M., Jasińska, A., Boratyńska, K., & Boratyński, A. (2013). Morphological diversity and structure of West Mediterranean Abies species. *Plant Biosystems, 147*, 125-134. https://doi.org/10.1080/11263504.2012.753130

Tiwari, S. P., Kumar, P., Yadav, D., & Chauhan, D. K. (2013). Comparative morphological, epidermal, and anatomical studies of *Pinus roxburghii* needles at different altitudes in the North-West Indian Himalayas. *Turkish Journal of Botany, 37*, 65-73.

Wang, M. B., & Hao, Z. Z. (2010). Rangewide genetic diversity in natural populations of Chinese pine (Pinus tabuliformis). *Biochem Genet, 48*(7-8), 590-602. https://doi.org/10.1007/s10528-010-9341-4

Woo, K. S., Fins, L., McDonald, G. I., Wenny, D. L., & Eramian, A. (2002). Effects of nursery environment on needle morphology of *Pinus monticola Dougl.* and implications for tree improvement programs. *New Forests, 24*, 113-129. https://doi.org/10.1023/A:1021230304530

Wu, Y. S., & Xiao, J. Y. (1991). A preliminary study on vegetation and climate changes in DianChi Lake area in the last 40000 years. *Acta Botanica Sinica, 33*, 450-458.

Xiao, Y. (2003). Variation in needle longevity of *Pinus tabuliformis* forests at different geographic scales. *Tree Physiology, 23*, 463–471. https://doi.org/10.1093/treephys/23.7.463

Xing, F. Q., Mao, J. F., & Meng, J. X. (2014). Needle morphological evidence of the homoploid hybrid origin of Pinus densata based on analysis of artificial hybrids and the putative parents, *Pinus tabuliformis* and *Pinus yunnanensis*. *Ecology and Evolution, 4*(10), 1890-1902. https://doi.org/10.1002/ece3.1062

Xu, B., & Tao, W. (2006). Application of a hand-held slicing method for wood species identification. *China Wood Ind, 20*, 41-43.

Xu, H. C. (1991). Geographical Variation and Selection of Provinces of *Pinus tabuliformis*. *China Forestry Publishing House, Beijing*, 21-30.

Yang, L., & Liu, Z. (2015). Genetic structure of *Pinus henryi* and *Pinus tabuliformis*: Natural landscapes as significant barriers to gene flow among populations. *Biochemical Systematics and Ecology, 61*, 124-132. https://doi.org/10.1016/j.bse.2015.06.003

Yuan, H., & Li, Z. (2014). Variation and Stability in Female Strobili Production of a First-Generation Clonal Seed Orchard of Chinese Pine (*Pinus tabuliformis*). *Silvae Genetica, 63*(1-2), 41-47.

Zhang, S. B., Guan, Z. J., Sun, M., Zhang, J. J., Cao, K. F., & Hu, H. (2012). Evolutionary association of stomatal traits with leaf vein density in *Paphiopedilum, Orchidaceae*. *Plos One, 7*, e40080. https://doi.org/10.1371/journal.pone.0040080

Zhao, C. M., Chen, L. T., Ma, F., Yao, B. Q., & Liu, J. Q. (2008). Altitudinal differences in the leaf fitness of juvenile and mature alpine spruce trees (*Picea crassifolia*). *Tree Physiology, 28*, 133-141. https://doi.org/10.1093/treephys/28.1.133

19

Impact of Different Environmental Stimuli on the Release of 1-MCP from Boron-MCP Complexes

Tasnuva Shahrin[1], Majher I. Sarker[1], Yanli Zheng[2], Peggy Tomasula[1], & LinShu Liu[1]

[1]Dairy and Functional Foods Research Unit, Eastern Regional Research Center, Agricultural Research Service, US Department of Agriculture, 600 E Mermaid Lane, Wyndmoor, PA 19038, USA

[2]Laboratory of Food Nutrition and Safety, Tianjin University of Science and Technology, Tianjin, 300457, China

Correspondence: LinShu Liu, Eastern Regional Research Center, Agricultural Research Service, US Department of Agriculture, 600 E Mermaid Lane, Wyndmoor, PA 19038, USA. E-mail: linshu.liu@ars.usda.gov

Abstract

In our previous report, boron derivatives of methylene cyclopropane complexes (B-MCP) were developed to stabilize the gaseous 1-MCP (1-methylcyclopropene), a commercial plant growth regulator, for eventual release in open crop fields when under humid conditions or in contact with water. To meet the requirements of various end-use applications, B-MCP compounds that can release 1-MCP at different rates for different time periods are demanded. In this paper, we examined the impact of different environmental stimuli such as humidity, surface area, water pH and water volume on the release kinetics of 1-MCP from B-MCP compounds of various chemical structures. The results showed that the release of 1-MCP from B-MCP compounds can be tailored by altering the architectures and chemical natures of the two short-chain moieties attached to the boron atom and thus either the electronic affinity or the hydrophobic or hydrophilic properties or both.

Keywords: boron derivatives of methylene cyclopropane, B-MCP, open field application, plant growth regulator

1. Introduction

Physiological studies of plants reveal that diffusion of ethylene into and out of plant tissue is responsible for regulating various growth and developmental activities such as ripening of fruits (Yang, 1995; Alexander & Grierson, 2002; Atta-Aly & Brecht, 2000; Klee, 2002; Fan, Blankenship & Mattheis 1999; Fan & Mattheis, 2000), opening of flowers, and senescence (Abeles, Morgan & Saltveit, 1992). However, ethylene also is responsible for the factors that cause defects in product quality or premature death such as loss of cellular turgor and chlorophyll, pigment degradation, irregular opening, yellowing or shedding of flowers and leaves (Serek, Woltering, Sisler, Frello & Sriskandarajah, 2006). 1-Methylcyclopropene (1-MCP), a cyclic olefin used commercially as a synthetic plant growth regulator, is applied onto post-harvest agricultural products such as fruits (including climacteric and non-climacteric), flowers and vegetables to retard their ripening by working as a preservative and thus prolong their shelf-lives (Sisler & Serek, 2003; Blankenship & Dole, 2003). Because of structural similarity to the natural plant hormone, 1-MCP works as a nontoxic antagonist of ethylene, interacting with and then covalently binding to ethylene receptors. (Sisler & Serek, 1997; Blankenship & Dole, 2003; Watkins, 2006; Sisler, Grichko & Serek, 2006). The affinity of 1-MCP to the ethylene receptors is ten times greater than that of ethylene (Watkins, 2006), which means a low concentration of 1-MCP is sufficient for application to horticultural products. 1-MCP completely blocks the absorption of ethylene and its hormonal action (Serek, Tamari, Sisler & Borochov, 1995; Sisler & Serek, 2003; Blankenship & Dole, 2003) for a certain time (Figure 1) until all 1-MCP diffuses from the binding sites or new ethylene receptors develop (Bayer, 1976a, 1976b, 1978; Veen, 1983).

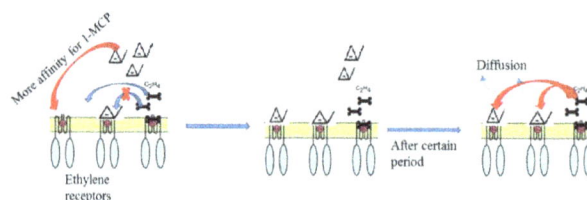

Figure 1. Competition between 1-MCP and Ethylene in blocking ethylene receptors and reopening of receptors for ethylene after diffusion of 1-MCP

Under environmental stresses, plants tend to produce more ethylene, resulting in premature opening or early ripening (Chow & McCourt, 2006; De Paepe & Van der Straeten, 2005). By this mechanism, extreme weather, such as drought, can cause yield losses of grain crops, such as corn, wheat, and rice, etc. In this case, application of 1-MCP in fields is expected to play a role in yield protection as a pre-harvest treatment (Kawakami, Oosterhuis & Snider, 2010). The use of 1-MCP in an open field indicates a huge potential, since about 45% of the total world's agricultural lands are subjected to continuous or frequent drought conditions (Bot, Nachtergaele & Young, 2000).

1-MCP is an odorless gas at ambient conditions (b.p. 9-12 °C), which makes it difficult to handle in an open environment. At present, there is no technology available for application of 1-MCP in open fields conveniently and effectively. However, the reported B-MCP analog complexes (Sarker, Fan & Liu, 2015; Sarker, Tomasula & Liu, 2016) (Figure 2) have shown significant potential to be directly used in open fields by releasing 1-MCP over a long time period depending on their individual structures. The effectiveness of these complexes to inhibit tomato ripening was demonstrated in an open, laboratory environment and reported in our earlier publication (Sarker et al., 2016).

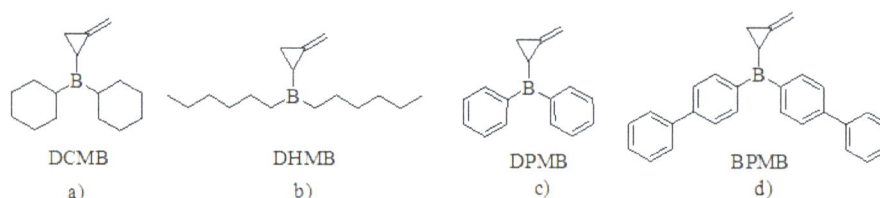

Figure 2. B-MCP derivatives: a) **DCMB**, Dicyclohexyl-(2-methylene cyclopropyl)-borane; b) **DHMB**, Dihexyl-(2-methylene cyclopropyl)-borane; c) **DPMB**, Diphenyl-(2-methylene cyclopropyl)-borane; d) **BPMB**, Bis-biphenyl-4-yl-(2-methylene cyclopropyl)-borane

The chemistry involved in release of 1-MCP from a B-MCP analog is controlled by hydrolysis reaction (Sarker et al., 2015) and the rate of this reaction mainly depends on the chemical and physical properties of the moieties attached to the central boron atom. Therefore, 1-MCP is found to be released at different rates from different complexes which are useful depending on the application. For practical application in an open field, the flexibility to tailor the structure of B-MCP complex is important for having the 1-MCP release at the rate as demanded under different environmental conditions, such as temperature, humidity, wetted area, and pH; e.g. To have a better understanding of the interdependency of the release of 1-MCP on environmental stimuli and the structure of B-MCP, in this paper, we compared the release pattern of 1-MCP of two newly synthesized boron compounds, **BNMB**, Bis-naphthalen-1-yl-(2-methylene cyclopropyl)-borane and **BPNMB**, Bis-phenanthren-9-yl-(2-methylene cyclopropyl)-borane (Figure 3) with **DCMB**, Dicyclohexyl-(2-methylene cyclopropyl)-borane and **BPMB**, Bis-biphenyl-4-yl-(2-methylene cyclopropyl)-borane (Figure 2) that was reported previously (Sarker et al., 2016).

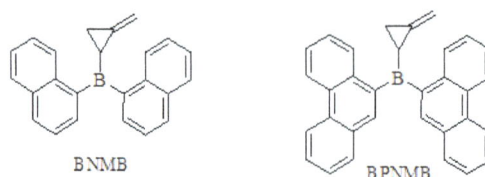

Figure 3. B-MCP derivatives: **BNMB**, Bis-naphthalen-1-yl-(2-methylene cyclopropyl)-borane and **BPNMB**, Bis-phenanthren-9-yl-(2-methylene cyclopropyl)-borane

2. Materials and Methods

2.1 Chemicals

All the chemicals and solvents used for synthetic purposes were purchased from Sigma-Aldrich unless otherwise stated. All solvents used were of HPLC grade and moisture dry.

2.2 Syntheses of Complexes *BNMB* (13) and *BPNMB* (14) (Figure 4)

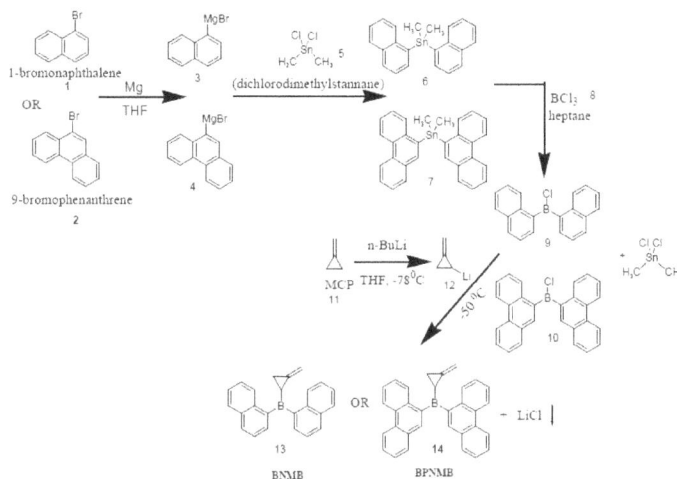

Figure 4. Synthetic scheme of complexes **BNMB** and **BPNMB**

BNMB and **BPNMB** were prepared by the reaction of bis-naphthalen-1-yl-chloro-borane (9) and bis-phenanthren-9-yl-chloro-borane (10) respectively with lithiated methylenecyclopropane (12), whereas, bis-naphthalen-1-yl-chloro-borane and bis-phenanthren-9-yl-chloro-borane were previously made from the reaction of bis-naphthalen-1-yl-dimethyl-stannane (6) and bis-phenanthren-9-yl-dimethyl-stannane (7) respectively with boron trichloride (8) in heptane. Bis-naphthalen-1-yl-dimethyl-stannane and Bis-phenanthren-9-yl-dimethyl-stannane were synthesized by the reaction of dichlorodimethylstannane (5) with Grignard reagents of 1-bromonaphthalene (1) and 9-bromophenanthrene respectively (2). Methylenecyclopropane (MCP) (11) was prepared (not shown in Figure 4) from the reaction between Potassium-bis(trimethylsilyl)-amide and methallylchloride (Binger, Brinkmann & Wedemann, 2002).

2.2.1 Synthesis of **BNMB** Complex (13)

2.2.1a Synthetic Procedure of Bis-naphthalen-1-yl-dimethyl-stannane (6)

An oven-dried 200 mL Schlenk flask cooled under argon was charged with 2.4 g (100 mmol) Mg turnings and 60 ml of anhydrous THF. To the mixture, 14.7 g (9.93 mL, 71 mmol) of 1-bromonaphthalene (1) was added slowly. The reaction mixture was stirred at room temperature for 30 min before it was refluxed for 2 h then cooled at room temperature through Argon flow. The freshly prepared Grignard reagent (3) was transferred via cannula into a 200 mL 2-neck flask equipped with a condenser and cooled at 0 °C. A solution of dichlorodimethylstannane (5) (7.5 g, 34.1 mmol) in 15 mL of dry THF was added into the 2-neck flask via a cannula. The reaction mixture was refluxed for 3 h then stirred at room temperature for 16 h. After the reaction the solution was cooled at 0 °C, treated little by little with a total 15 mL of saturated NH_4Cl solution and extracted with dichloromethane in a separatory funnel. The organic layer was washed three times with a total of 250 mL water followed by quenching with brine solution, then dried with Na_2SO_4, concentrated in vacuo to give crude product mixed with naphthalene. The crude product was separated with 100% hexane through flash silica column chromatography. Two fractions were collected: mostly naphthalene in the 1st fraction and pure product in the 2nd fraction. The 2nd fraction was concentrated in vacuo and dried under vacuum to obtain 11.13 g (27.6 mmol) white solid of Bis-naphthalen-1-yl-dimethyl-stannane (6). Product yield was found to be 80.94%. ^{1}H-NMR (400 MHz, $CDCl_3$): δ 0.82 (3H, s), 7.36 - 7.55 (3H, m), 7.71 (1H, d, J = 6.42 Hz), 7.89 (3H, d, J = 7.55 Hz); ^{13}C NMR (100 MHz, $CDCl_3$) δ - 7.7, 125.4, 125.5, 125.9, 128.87, 129.04, 129.87, 133.65, 135.5, 138.43, 141.16.

2.2.1b Synthetic Procedure of Bis-naphthalen-1-yl-chloro-borane (9)

An oven dried 250 mL thick-walled flask with Teflon screw cap cooled under argon was charged with 5 g (12.4

mmol) of Bis-naphthalen-1-yl-dimethyl-stannane (6), 40 mL of anhydrous heptane and 13 mL of 1M boron trichloride solution (8) in heptane (13 mmol) in a nitrogen saturated glove box. The mixture was stirred for 30 minutes at room temperature and then heated at 120 °C for 96 h. After the reaction, the solution was cooled and subjected to vacuum filtration under nitrogen in the glove box to remove dichlorodimethylstannane (5) as a solid. The solid was washed with the least amount of anhydrous heptane and the collected heptane solution was concentrated in a rotary evaporator. The rest of dichlorodimethylstanane was removed by a sublimation technique at least 3 times and dried under vacuum to obtain pure 3.57 g (11.88 mmol) ash colored solid material of Bis-naphthalen-1-yl-chloro-borane (9). Product yield was 95.8%. ^{1}H NMR (400 MHz, CDCl$_3$): δ 7.34 (1H, t, J = 7.60 Hz), 7.48 (1H, t, J = 7.59 Hz), 7.54 (1H, t, J = 7.59 Hz), 7.91 (1H, d, J = 7.94 Hz), 7.98 (1H, d, J = 6.90 Hz), 8.06 (1H, d, J = 7.94 Hz), 8.10 (1H, d, J = 8.29 Hz); ^{13}C NMR (100 MHz, CDCl$_3$) δ 125.85, 127.16, 127.48, 127.66, 128.93, 130.47, 132.72, 135.69, 136.51, 137.71; ^{11}B NMR (128 MHz, CDCl$_3$): δ 64.9.

2.2.1c Synthetic Procedure of Bis-naphthalen-1-yl-(2-methylenecyclopropyl)-borane (13), **BNMB**

An oven-dried 100 mL Schlenk flask cooled at -78 °C was charged with 1 mL MCP (11) (0.82 g, 15.2 mmol) in 15 mL of anhydrous THF. To the solution, 5.5 mL of 2.5M n-BuLi in hexane (13.75 mmol) was added slowly and stirred at room temperature for 3 h. The mixture was cooled at -50 °C and 4.54 g (15.1 mmol) of bis-naphthalen-1-yl-chloro-borane (9) dissolved in 15 mL of dry THF was added slowly, over 15 min, via a cannula. The reaction mixture was stirred at room temperature for 24 h. The solution was filtered under Argon to remove salt and concentrated in vacuo. The solid was dissolved in 20 mL of anhydrous dichloromethane and filtered with a syringe filter to get a clear solution. The solution was concentrated in vacuo and dried under vacuum to obtain 3.1 g (9.74 mmol) solid of **BNMB** (13). Product yield 64.5%. ^{11}B NMR (128 MHz, CDCl$_3$): δ 48.3.

2.2.2 Synthesis of **BPNMB** Complex

2.2.2a Synthetic Procedure of Bis-phenanthren-9-yl-dimethyl-stannane (7)

An oven-dried 200 mL Schlenk flask cooled under argon was charged with 1.9 g Mg turnings (78.2 mmol) and 55 mL of anhydrous THF. To the mixture, 12.6 g (49 mmol) of 9-bromphenanthrene (2) dissolved in 15 mL of anhydrous THF was added slowly via cannula. The reaction mixture was stirred at room temperature for 30 min before it was refluxed for 2 h then cooled at room temperature through argon flow. The freshly prepared Grignard reagent (4) was transferred via cannula into a 200 mL 2-neck flask equipped with a condenser over an ice bath. A solution of dichlorodimethylstannane (5) (5 g, 22.76 mmol) in 10 mL of dry THF was added into the 2-neck flask via a cannula. The mixture was refluxed for 3 h then stirred at room temperature for 16 h. After the reaction, the solution was cooled at 0 °C, treated little by little with a total of 10 mL saturated NH$_4$Cl solution and extracted with dichloromethane in a separatory funnel. The organic layer was washed three times with a total of 150 mL water followed by quenching with brine solution, then dried with Na$_2$SO$_4$, concentrated in vacuo to give a crude product mixed with phenanthrene. The crude product dissolved in least amount (~30 mL) of CH$_2$Cl$_2$ was added 50 mL of hexane to get precipitation. The precipitate was filtered under vacuum and checked for the purity with thin layer chromatography (TLC) in 100% hexane as mobile phase. For any further impurity, the solid was heated at 80 °C with 30 mL of hexane and filtered immediately to obtain pure 7.4 g (14.68 mmol) solid of Bis-phenanthren-9-yl-dimethyl-stannane (7). Product yield 64.5%. ^{1}H NMR (400 MHz, CDCl$_3$): δ 0.87 (3H, s), 7.49 (1H, t, J = 7.52 Hz), 7.54 – 7.73 (3H, m), 7.80 (1H, d, J = 7.52 Hz), 7.96 (1H, d, J = 7.82 Hz), 8.01 (1H, s), 8.72 (1H, d, J = 8.12 Hz), 8.76 (1H, d, J = 8.12 Hz).

2.2.2b Synthetic Procedure of Bis-phenanthren-9-yl-chloro-borane (10)

An oven dried 250 mL thick-walled flask with Teflon screw cap cooled under argon was charged with 6.5 g (12.9 mmol) of Bis-phenanthren-9-yl-dimethyl-stannane (7), 15 mL of anhydrous heptane and 13.5 mL (13.5 mmol) of 1M boron trichloride solution in heptane (8) in a nitrogen saturated glove box. The mixture was stirred for 30 minutes at room temperature and then heated at 120 °C for 96 h. After the reaction, the solution was cooled, 40 mL of anhydrous dichloromethane was added and then subjected to vacuum filtration under nitrogen in the glove box. The solid was washed three times with a total of 15 mL of additional dichloromethane to remove dichlorodimethylstannane (5). The rest of the dichlorodimethylstanane was removed by a sublimation technique to obtain pure 3.85 g (9.6 mmol) of Bis-phenanthren-9-yl-chloro-borane (10). Product yield 74.4%. ^{1}H NMR (400 MHz, CDCl$_3$): δ 7.57 (2H, q, J = 7.48 Hz), 7.65 – 7.73 (2H, m), 7.81 (1H, d, J = 7.70 Hz), 8.09 (1H, s), 8.49 (1H, d, J = 8.11 Hz), 8.74 (1H, d, J = 8.11 Hz), 8.80 (1H, d, J = 8.11 Hz); ^{13}C NMR (50 MHz, CDCl$_3$) δ 122.79, 123.32, 126.73, 126.92, 127.13, 128.07, 128.75, 129.49, 129.62, 130.43, 131.25, 131.82, 134.23, 137.35; ^{11}B NMR (128 MHz, CDCl$_3$): δ 64.63.

2.2.2c Synthetic Procedure of bis-phenanthren-9-yl-(2-methylenecyclopropyl)-borane (14), BPNMB

An oven-dried 100 mL Schlenk flask cooled at -78 °C was charged with 0.544 g (0.66 mL, 10.1 mmol) of MCP (11) in 15 mL of anhydrous THF. To the solution, 3.9 mL of 2.5M n-BuLi (9.75 mmol) in hexane was added slowly and stirred at room temperature for 2 h. The mixture was cooled at -50 °C and 3.85 g (9.6 mmol) of bis-phenanthren-9-yl-chloro-borane (10) dissolved in 20 mL of dry THF was added slowly over 15 min via cannula. The reaction mixture was stirred at room temperature for 24 h. The solution was filtered to remove salt and concentrated in vacuo. The solid was dissolved in 25 mL of anhydrous dichloromethane and filtered with a syringe filter to get a clear solution. The solution was concentrated in vacuo and dried under vacuum to obtain 1.73 g (4.14 mmol) of the final product, BPNMB (14). Product yield 43.1%. ^{11}B NMR (128 MHz, $CDCl_3$): δ 48.79.

2.3 GC Analysis

For all GC analysis, a Hewlett-Packard 5890 GC with capillary column (30 m x 0.25 mm i.d.) coated with a 0.25 μm film of 5% phenyl methyl silicon and a flame ionization detector was used. The temperature of oven was programmed at 30 °C isothermal with an injection point temperature of 50 °C. The detector was operated at 235 °C and sample was injected under split less condition. Helium was used as carrier gas with a 1.5 mL/min column flow.

2.4 Sample Preparation of BNMB and BPNMB for Controlled Release Analysis of 1-MCP

Two separate GC analyses were conducted to study the controlled release capability of 1-MCP. For BNMB, 272 mg (0.85 mmol) of sample and for BPNMB, 321 mg (0.77 mmol) of sample, were each mixed with 0.8 mL of H_2O in a 1.5 mL air tight vial at room temperature (22±1°C). The head space vapors collected from the solutions were injected into the GC.

2.5 Comparative Study of Release of 1-MCP from BPMB, BNMB and BPNMB at Different Time Intervals

Three air tight 4 mL vials were separately charged with equivalent amounts (0.79 mmol) of BPMB (291 mg), BNMB (254 mg) and BPNMB (329 mg) respectively and 0.8 mL of water in each vial. All the samples were prepared at room temperature (22±1 °C). The mixtures in each vial were vigorously stirred throughout the 99 h experiment. The vapors collected from the head spaces of the vials were injected into a GC to quantify the accumulated 1-MCP for four different time segments, 0-21 h, 22-46 h, 47-76 h and 77-99 h. After each segment the flasks were kept under vacuum for 10 minutes and then flushed with nitrogen.

2.6 Experimental Procedure for the Impact of Environmental Stimuli on the Release Rate of 1-MCP from B-MCP Complexes

Two B-MCP complexes (DCMB, BNMB) were chosen to analyze the impact of humidity, surface area, water pH and amount of water on the release rate of 1-MCP. Two different humidities (95% and 44%), two surface areas (7.06 cm^2 and 10.75 cm^2), three pH (3.9, 6.8, 10.6) and four different volumes of water (0.1 mL, 0.4 mL, 0.7 mL 1.0 mL) were used for a specific amount of complexes in each type of analysis. All the samples for GC analyses were prepared at room temperature (22±1 °C).

2.6.1 Impact of Humidity on B-MCP Complexes in Release of 1-MCP

A total of eight 250 mL air tight glass jars with screw caps and rubber septa were used. Four of the jars were loaded with a supersaturated solution of $Na_2HPO_4.12H_2O$ (30 g salt + 30 mL water) and four were loaded with K_2CO_3 (30 g salt + 15 mL water) salts to provide 95% and 44% humidity conditions, respectively, inside the jars. A tripod stand was placed in each jar supporting a petri dish of surface area 10.75 cm^2 (Figure 5). Four of the petri dishes were loaded with 150 mg (0.65 mmol) of DCMB and the other four dishes with 100 mg (0.31 mmol) of BNMB. The samples (DCMB, BNMB) were dissolved in the least amount of anhydrous dichloromethane and then equally distributed in the petri dishes to obtain a homogeneous layer of sample. Loaded petri dishes were dried under vacuum before placing them in the humidity chambers. Two petri dishes from each group were then placed in the four jars with 95% humidity and the other two dishes from each group were placed in the four jars with 44% humidity to generate duplicate analyses. GC readings for accumulated 1-MCP were measured from the headspace of the jars at different time intervals.

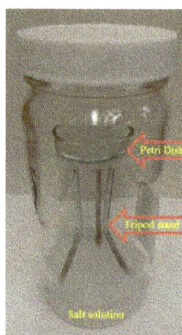

Figure 5. Humidity chamber

2.6.2 Impact of Surface Area on B-MCP Complexes at Specific Humidity (95%)

Eight 95% humidity chambers were designed as mentioned above. Four of them were used for the **DCMB** and the other four chambers were for the **BNMB** complex. Two different sizes of petri dishes (surface area: 7.06 cm^2 and 10.75 cm^2, respectively) were placed in each two **DCMB** chambers containing same amount (150 mg, 0.65 mmol) of **DCMB**. The samples were dissolved in the least amount of anhydrous dichloromethane and then equally distributed in the petri dishes to get a homogeneous layer. Loaded petri dishes were dried under vacuum before placing them in the humidity chambers. Same design of analysis was used for **BNMB** complex containing 100 mg (0.31 mmol) of sample in each petri dish. GC readings for accumulated 1-MCP were measured from the headspace of the jars at different time intervals.

2.6.3 Impact of Water pH on Release Rate of 1-MCP from B-MCP Complexes

Six 2 mL airtight vials were charged with 56 mg (0.24 mmol) of **DCMB** individually where, each two vials were added 0.4 mL water of pH 3.9, 6.8 and 10.6 respectively. Quantification of the accumulated 1-MCP from the head space of the vials at the respective pH were made by GC analyses at different time intervals. Between the time intervals the mixtures were continuously stirred. The same procedure was followed in the case of **BNMB** analysis, charging each vial with 51.8 mg (0.16 mmol) of sample.

2.6.4 Effect of Water Volume on the Rate of Hydrolysis Reactions to Release 1-MCP

Eight 2 mL air tight vials with screw caps and septa were charged with 57.2 mg (0.25 mmol) of **DCMB** in which, 0.1, 0.4, 0.7 or 1.0 mL of water, respectively, was added in two of the vials with continuous stirring. The released 1-MCP was collected from the head space of each vial and quantified with GC analysis at specific intervals. After each GC injection, the vials were opened and flushed with nitrogen flow to make sure there was no 1-MCP left in the vials which was released during the previous intervals. For the analysis with **BNMB**, the same procedure was followed charging each vial with 52.2 mg (0.16 mmol) of sample.

2.7 Contact angle Measurements for Hydrophilicity/Hydrophobicity Characteristics

The contact angle (θ) is the angle formed between the surface of a liquid and the outline of the contact surface to a solid where a liquid-vapor interface meets a solid surface at a specific temperature and pressure and quantifies the wettability of a solid by a liquid by Young's equation (Kwok & Neumann, 1999). In this study with water as the wetting liquid, the solid surface can be classified as hydrophilic or hydrophobic (Förch, Schönherr & Jenkins, 2009). If a water drop spreads out over a surface of a solid, the attraction between the solid and liquid is strong and the contact angle is 0°. For contact angles between 0° and 90°, the solid is considered as hydrophilic and above 90° it is considered as hydrophobic.

To measure the contact angle of water droplets on the B-MCP complexes, glass slides were prepared with a thin homogenous layer of each B-MCP complex (**BPMB**, **BNMB** and **BPNMB**). The thin films of B-MCP complexes were made by placing 10 mg of each complex dissolved in 0.5 mL of dichloromethane on individual slide. A contact angle Automated Goniometer (Ramé-hart Instrument Co., Succasunna, NJ, USA) equipped with a high resolution camera (Model 590 F4 Series) was used to determine the hydrophilicity and wettability of the complexes. Experiments were carried out in air through static sessile drop measurements using ultrapure water as the liquid phase (11 ± 0.6 μL drops). The distance between needle-tip and compound surface was kept constant to ensure consistency between the different measurements. The reported angles were recorded after ≤10 s to avoid as much as possible penetration of water into the surface layer of the complexes. Determination of the contact angles of the droplets were performed using Dropimage Advanced v2.5 software. The results were the average of five to ten water drops deposited on each compound layer surface.

3. Results and Discussion

There is no existing technology available for use of 1-MCP in an open environment as it is a gas at ambient conditions, although it has been shown that 1-MCP can potentially protect crops from environmental stresses such as drought to reduce yield loss (Kawakami & Oosterhuis, 2006; Kawakami et al., 2010). Our invented B-MCP complexes (Figure 2) have shown enormous potential for open field applications as they are stable under ambient conditions owing to higher boiling points than 1-MCP and capable of releasing 1-MCP for a long time period from hours to days depending on their structural differences and availability of moisture. In our previous article (Sarker et al., 2016) the effectiveness of two B-MCP complexes (**DCMB**, **BPMB**) in an open environment was reported in which they were found active in delaying the ripening process of matured green tomatoes for more than 7 days. These encouraging results obtained in the laboratory have driven us to develop new B-MCP complexes and to check their effectiveness under environmental stimuli such as humidity, water pH, water volume and surface area. In synthesizing new B-MCP complexes, polycyclic aromatic hydrocarbons (PAH) were chosen to provide the branch chains of the complexes (Figure 3, **BNMB** and **BPNMB**) as they are readily available, cheap, possess pesticidal properties and are biodegradable. A four-step synthetic scheme (Figure 4) for **BNMB** (13) and **BPNMB** (14) resulted in good yields at every step. An important part of this synthetic route is the recovery of dichlorodimethylstannane (5) in the following step that can be reused thus reducing the synthetic costs.

The B-MCP complexes with PAH (**BNMB**, **BPNMB**) as side chains could be used for dual purposes such as in controlled release of 1-MCP and as a pesticide for crop fields. For example, after complete release of 1-MCP from **BNMB** and **BPNMB**, naphthalene and phenanthrene, which are well known pesticides, will be deposited, respectively, due to uncontrolled and complete hydrolysis of B-MCP complexes. Naphthalene has been used as a household fumigant. Other fumigant uses of naphthalene include use in soil as a fumigant pesticide. Naphthalene is broken down by soil bacteria to naphthalene diol, salicylic acid, and catechol. Some bacteria may utilize naphthalene as their sole carbon source (Cerniglia, 1984). Bacterial oxidation pathways include five metabolites of naphthalene along the degradation pathway: cis-1,2-dihydroxy-1,2-dihydronaphthalene, 1,2-dihydroxynaphthalene, cis-o-hydroxybenzalpyruvic acid, salicylic acid, and catechol, which was subsequently subject to ring cleavage (Bouwer, J., McCarty, Bouwer, H. & Rice, 1984). Fungal degradation of naphthalene produces naphthalene-1,2-oxide via cytochrome P-450 oxidation. The oxide can be subsequently hydrolyzed to trans-1,2-dihydroxy-1,2-dihydronaphthalene, or alternatively conjugate with glucuronide or sulfate to break down first to 1-naphthol and 2-naphthol and subsequently to 4-hydroxy-1-tetralone (Jury, Spencer & Farmer, 1984). Other reported metabolites include naphthalene trans-1,2-dihydrodiol, 1,2-naphthoquinone, and 1,4-naphthoquinone (Garon, Krivobok, Seigle-Murandi, 2000). Phenanthrene could be an alternative option as it is used as pesticides and also biodegradable (Shetty et al., 2015).

*3.1 Controlled Release of 1-MCP from **BNMB** (13) and **BPNMB** (14)*

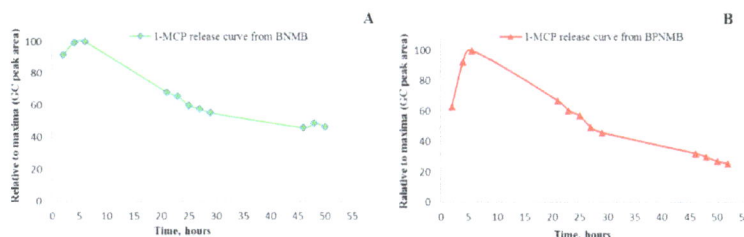

Figure 6. Controlled release of 1-MCP from **BNMB** and **BPNMB** when in contact with water

As shown in Figure 6, **BNMB** (13) and **BPNMB** (14) are capable of gradual release of 1-MCP when in contact with water. Both **BNMB** and **BPNMB** require close to 6 h to reach their highest point of release and after that both descending curves indicate either completion of 1-MCP release from the complexes or much slower release rate accumulating less amount than the amount of 1-MCP required to be pulled out from the airtight vials to obtain an ascending curve in GC analysis. The curves (Figure 6) also show **BNMB** and **BPNMB** are more reactive toward water than **DCMB** (Figure 2a) (Sarker et al., 2015) in which the highest point was reached in 21 h. The faster release of 1-MCP makes **BNMB** and **BPNMB** better choices for applications which require relatively high concentrations of 1-MCP for a short period of time. The differences in release rate of 1-MCP from the B-MCP analog is controlled by the mechanism of the hydrolysis reaction (Figure 7). The driving force of these particular hydrolysis reactions are generated differently due to the structural differences of the attached moieties of the B-MCP complexes. Following the mechanism (Figure 7), it is unquestionable that the more electron deficient the central boron atom is, the faster the nucleophilic attack by the hydroxyl group of the water molecule will be, leading to the cleavage of the carbon-boron bond to produce 1-MCP. In this regard, the central boron atom in **DCMB** is more electron rich than that in **BNMB** or **BPNMB** because of two electron releasing cyclohexyl groups present in the DCMB structure. Alkyl groups are known to be electron donating groups because of inductive effects and also hyperconjugation. As a result of being electron rich, the boron atom in **DCMB** is more resistant toward nucleophilic attack than in **BNMB** or **BPNMB** offering better control in release of 1-MCP.

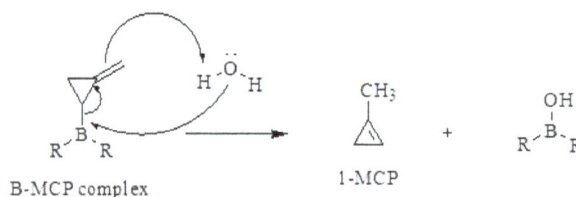

Figure 7. 1-MCP releasing mechanism from B-MCP complexes; R= alkyl or aryl group

3.2 Comparative Study of 1-MCP release from *BPMB*, *BNMB* and *BPNMB*

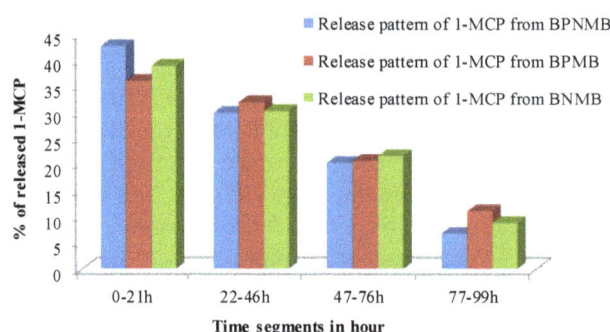

Figure 8. Comparative release pattern of 1-MCP from **BPNMB**, **BPMB** and **BNMB** when in contact with water

From this comparative study in Figure 8, it is shown that **BPNMB**, **BNMB** and **BPMB** release 1-MCP at a similar rate when in contact with water. In first 21 h, they release 42%, 39% and 36% of total released 1-MCP in 99 h, respectively. These slight differences in releasing 1-MCP in the first 21 h could be resulted from the differences in their resonance stabilities. The B-MCP complexes **BPMB**, **BNMB** and **BPNMB** have polycyclic aromatic hydrocarbons (PAH) in their structures in which internal resonance is possible (Figure 9) making these molecules stable and at the same time, the central boron atom becomes electron deficient due to the delocalization of π-electrons and thus vulnerable to nucleophilic attack by the hydroxyl groups of H_2O. The attached moieties of **BPNMB** (phenanthrenyl) have the highest number of cyclic aromatic rings creating stronger internal resonance than in **BNMB** and **BPMB** and thus make the central boron atom the most electron deficient ($B^{+\delta}$), which favors faster hydrolysis. The branches of **BNMB** (naphthalenyl) and **BPMB** (biphenyl) have the same number of cyclic aromatic rings but in the case of **BPMB**, the phenyl rings connected with a single σ-bond inhibit to provide the full impact of resonance on the central boron atom than in **BNMB** which permits hydroxyl group interaction as a slower attack on **BPMB**.

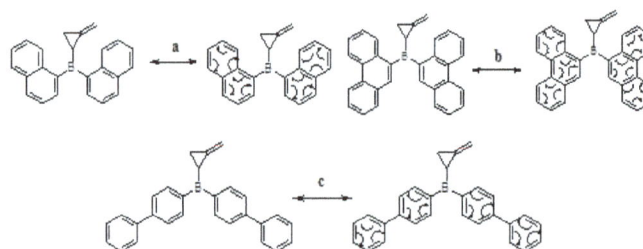

Figure 9. Resonance structures of a) **BNMB**, b) **BPNMB** and c) **BPMB**

The differences in the release rate of 1-MCP from B-MCP complexes may also be rationalized by their hydrophobic and hydrophilic characteristics. Hydrophobicity and hydrophilicity of a compound were determined by contact angle measurement analysis. The contact angles of **BPMB**, **BNMB** and **BPNMB** with a water droplet were measured as 99.5 ± 3°, 76.5 ± 3.5° and 67.0 ± 2°, respectively, suggesting that **BPMB** has hydrophobic character whereas **BNMB** and **BPNMB** have hydrophilic characteristics. **BNMB** and **BPNMB** with hydrophilic character attract water molecules more strongly than **BPMB** leading to a faster hydrolysis reaction. Similarly, **BPNMB** is more hydrophilic than **BNMB** allowing increased contact with the water molecules and a faster hydrolysis reaction. It is also important to mention that all three B-MCP complexes are capable of releasing 1-MCP even after 77 hours with at least 5% or more released. This variation in release rate will help to select the right complex for specific applications (Figure 8).

3.3 Effect of Environmental Stimuli on Releasing 1-MCP from B-MCP Complexes

3.3.1 Effect of Humidity

DCMB is found to release 1-MCP at a faster rate at 95% humidity than at 44% humidity as expected (Figure 10A). After two days, **DCMB** released 22% of its maxima at 95% humidity whereas only 6% of the maxima was released at 44% humidity. Although the maximum release rate of 1-MCP for both humidities were reached at 16 days, in the case of 95% humidity, the release rate was faster than at lower humidity on its way to maxima. It might have something more to do with the dew point. At 44% humidity and 22 °C and in atmospheric pressure, the dew point is about 9 °C. This means that there is little water to trigger 1-MCP release. At 95% and 22 °C and in atmospheric pressure, the dew point is about 21 °C, and there is more moisture causing release of the 1-MCP. On the other hand, **BNMB** (Figure 10B) did not show any significant difference in release rate of 1-MCP between 95% and 44 % humidity conditions because of its high reactivity toward water. It was found to reach 80% of the maxima in one day and maximum in three days in both humid conditions. Therefore, the releasing capability of 1-MCP under humidity will make these B-MCP complexes useful in crop field without rain or water sources.

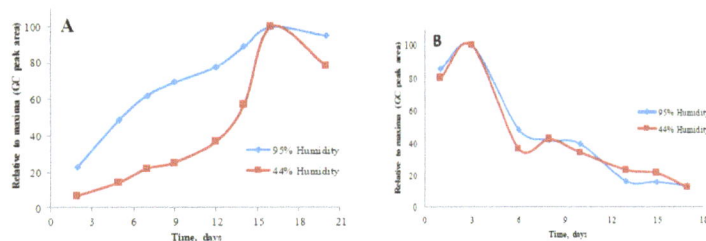

Figure 10. The effect of humidity at 22±1 °C on release rate of 1-MCP from A) **DCMB** and B) **BNMB**

3.3.2 Effect of Surface Area

At 95% humidity, **DCMB** was found to release 1-MCP at a faster rate with the larger surface area (Figure 11a). In both cases 20% of the maxima was released in the first two days and the maxima was reached at 16 days. Again in the case of **BNMB** (Figure 11b), no significant differences were observed in release rate of 1-MCP from two different surface areas due to its high reactivity. Release rate of 1-MCP can be controlled by changing the exposed area of B-MCP complexes which will provide an additional tool to manipulate this technology in open field applications.

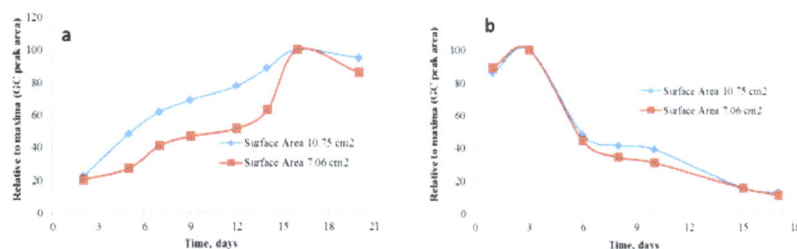

Figure 11. Effect of Surface Area on release rate of 1-MCP from a) **DCMB** b) **BNMB** at 95% humidity.

3.3.3 Impact of Water pH

In the case of **DCMB**, the release rate of 1-MCP was found to vary with the pH of water (Figure 12a). At basic conditions (pH 10.6), the rate of hydrolysis was increased resulting in a faster release rate of 1-MCP than at pH

6.8 or 3.9. Although it took about 27 h to reach the maxima at pH 6.8 or 10.6, the sudden drop of the release curve for pH 10.6 indicates the fastest rate as it released most of the 1-MCP within that time. On the other hand, at pH 3.9, the maximum was reached in 74 h. However, there was no significant difference observed in the release of 1-MCP from **BNMB** (Figure 12b) at pH 3.9 or 6.8, but similar to **DCMB** at pH 10.6, the release rate became faster by releasing the most part of 1-MCP on its way to reach the maximum point of 28 h. The independency on release rate of 1-MCP from B-MCP complexes by water pH will expand their application in different geographical locations with pH differences in rain water.

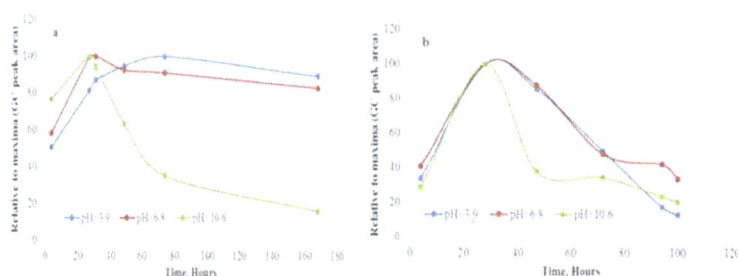

Figure 12. Effect of water pH on release rate of 1-MCP from a) **DCMB** b) **BNMB**

3.3.4 Effect of Water Volume

From Figure 13a, it is clear that, the reaction rate of hydrolysis of **DCMB** is changed up to the sufficient amount of water (0.1 mL) and after that the rate becomes independent to more water content. That is why 63% of total 1-MCP was released in first 48 hours in 0.1 mL of water where 76 % (±2) remained unchanged in 0.4-1.0 mL of water. In case of **BNMB**, very little or no change in 1-MCP release were observed due to the amount of water (Figure 13b) which is consistent with the humidity experiments where **BNMB** was found to release 1-MCP at similar rates under two humidity conditions (95% and 44%).

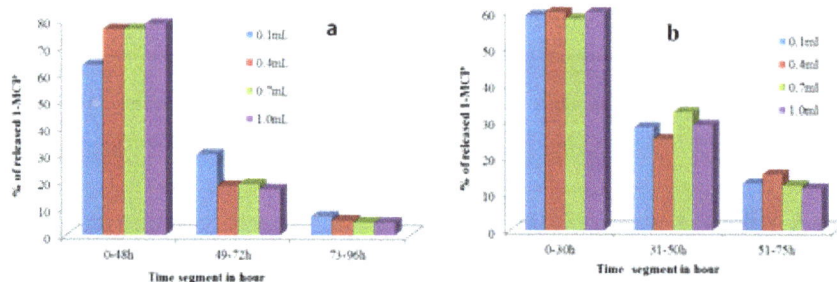

Figure 13. Effect of water volume on release rate of 1-MCP from a) **DCMB** b) **BNMB**

4. Conclusion

Two new B-MCP complexes have been synthesized which are capable of releasing 1-MCP for up to 99 h when in contact with a sufficient amount of water. It was found that they release 1-MCP at a faster rate in the beginning than the previously reported complexes, **DCMB** and **BPMB**. This difference in release rate is due to either electronic effects resulting from the internal resonance of the PAH groups or the hydrophilic characteristics of the B-MCP complexes or both. It is proven that B-MCP complexes are also capable of releasing 1-MCP not only in contact with water but also under humidity. The other factors like surface area, water pH which are of importance when considering use in open field applications. The change in environmental conditions impacts the release rate of 1-MCP from B-MCP complexes unless they are too sensitive to water molecules (hydrolysis). Overall, these B-MCP analog complexes have shown enormous potential as a means to provide 1-MCP in open environments over a long period to protect crops in hostile environmental conditions. The control of 1-MCP release from B-MCP complexes by water pH is an important finding to utilize this technology.

Acknowledgment

We thank Dr. Gary Strahan for his help with acquiring and processing the data for NMR analyses of the compounds. We also thank Mr. Raymond Kwoczak for his assistance in acquiring the contact angles of the B-MCP complexes.

References

Abeles, F. B., Morgan, P. W., & Saltveit, M. E. (1992). *Ethylene in Plant Biology* (2nd ed.). New York: Academic Press. http://dx.doi.org/10.1016/B978-0-08-091628-6.50011-4

Alexander, L., & Grierson, D. (2002). Ethylene biosynthesis and action in tomato: A model for climacteric fruit repining. *Journal of Experimental Botany, 53*, 2039-2055. http://dx.doi.org/10.1093/jxb/erf072

Atta-Aly, M., & Brecht, J. K. (2000). Ethylene feedback mechanisms in tomato and strawberry fruit tissues in relation to fruit ripening and climacteric patterns. *Postharvest Biology and Technology, 20*, 151-162. http://dx.doi.org/10.1016/S0925-5214(00)00124-1

Bayer, E. M. (1976a). Silver ion: a potent anti-ethylene agent in cucumber and tomato. *HortScience, 11*, 195-196.

Bayer, E. M. (1976b). A potent inhibitor of ethylene action in plants. *Plant Physiology, 58*, 268-271. http://dx.doi.org/10.1104/pp.58.3.268

Bayer, E. M. (1978). Method for overcoming the anti-ethylene effects of Ag+1. *Plant Physiology, 62*, 616-617. http://dx.doi.org/10.1104/pp.62.4.616

Binger, P., Brinkmann, A., & Wedemann, P. (2002). Highly Efficient Synthesis of Methylene cyclopropane. *Synthesis. 10*, 1344-1346. http://dx.doi.org/10.1055/s-2002-33122

Blankenship, S. M., & Dole, J. M. (2003). 1-Methylcyclopropene: a review. *Postharvest Biol. Technol. 28*, 1-25. http://dx.doi.org/10.1016/S0925-5214(02)00246-6

Bot, A. J., Nachtergaele, F. O., & Young. (2000). A Land resource potential and constraints at regional and country levels. *World Soil Resources Reports 90.* Land and Water Development Division, FAO, Rome.

Bouwer, E. J., McCarty, P. L., Bouwer, H., & Rice, R. C. (1984). Organic contaminant behavior during rapid infiltration of secondary wastewater at the Phoenix 23rd Avenue Project. *Water Res. 18* (4), 463-472. http://dx.doi.org/10.1016/0043-1354(84)90155-6

Cerniglia, C. E. (1984). Microbial metabolism of polycyclic aromatic hydrocarbons. *Adv. Appl. Microbiol. 30*, 31-71. http://dx.doi.org/10.1016/S0065-2164(08)70052-2

Chow, B., & McCourt, P. (2006). Plant hormone receptors; perception is everything. *Genes Dev. 20* (15), 1998-2008. http://dx.doi.org/10.1101/gad.1432806

De Paepe, A., & Van der Straeten, D. (2005). Ethylene biosynthesis and signaling; an overview. *Vitam Horm. 72*, 399-430. http://dx.doi.org/10.1016/S0083-6729(05)72011-2

Fan, X., Blankenship, S. M., & Mattheis, J. P. (1999). 1-Methylcyclopropene Inhibits Apple Ripening. *Journal of American Society of Horticultural Sciences. 124*, 690-695.

Fan, X., & Mattheis, J. P. (2000). Yellowing of Broccoli in Storage Is Reduced by 1-Methylcyclopropene. *HortScience. 35*, 885-887.

Förch, R., Schönherr, H., & Jenkins, A. T. A. (2009). *Surface Design: Applications in Bioscience and Nanotechnology*. Appendix C: Contact Angle Goniometry, 471-473. http://dx.doi.org/10.1002/9783527628599

Garon, D., Krivobok, S., & Seigle-Murandi, F. (2000). Fungal degradation of Fluorene. *Chemosphere. 40*, 91-97. http://dx.doi.org/10.1016/S0045-6535(99)00250-7

Jury, W. A., Spencer, W. F., & Farmer, W. J. (1984). Behavior assessment model for trace organics in soil: III. Application of screening model. *J. Environ. Qual. 13* (4), 573-479. http://dx.doi.org/10.2134/jeq1984.00472425001300040012x

Kawakami, E. M., & Oosterhuis, D. M. (2006). Effect of 1-MCP on the Physiology and Growth of Drought-Stressed Cotton Plants. *Summaries of Arkansas Cotton Research*, 62-66.

Kawakami, E. M., Oosterhuis, D. M., & Snider, J. L. (2010). Physiological effects of 1-Methylcyclopropene on well-watered and water-stressed cotton plants. *Journal of Plant Growth Regulation. 29*, 280-288. http://dx.doi.org/10.1007/s00344-009-9134-3

Klee, H. J. (2002). Control of ethylene mediated processes in tomato at the level of receptors. *Journal of*

Experimental Botany. 53, 2057-2063. http://dx.doi.org/10.1093/jxb/erf062

Kwok, D. Y., & Neumann, A. W. (1999). Contact angle measurement and contact angle interpretation. *Adv. in Colloid and Interface Sci. 81*, 167-249. http://dx.doi.org/10.1016/s0001-8686(98)00087-6

Mir, N., Canoles, M., Beaudry, R., Baldwin, E., & Mehla, C. P. (2004). Inhibiting tomato ripening with 1-methylcyclopropene. *J. Am. Soc. Hort. Sci., 129*, 112-120.

Reich, H. J., Holladay, J. E., Mason, J. D., & Sikorski, W. H. (1995). The Origin of Regioselectivity in an Allenyllithium Reagent. *J. Am. Chem. Soc., 117*(49), 12137-12150. http://dx.doi.org/10.1021/ja00154a014

Sarker, M. I., Fan, X., & Liu, L.S. (2015). Boron derivatives: As a source of 1-MCP with gradual release. *Scientia Horticulturae. 188*, 36-43. http://dx.doi.org/10.1016/j.scienta.2015.03.017

Sarker, M.I., Tomasula, P., & Liu, L. S. (2016). 1-MCP releasing complexes for open field application. *J. Plant Studies. 5*(1), 1-10. http://dx.doi.org/10.5539/jps.v5n1p1

Serek, M., Tamari, G., Sisler, E. C., & Borochov, A. (1995). Inhibition of ethylene-induced cellular senescence symptoms by 1-methylcyclopropene, a new inhibitor of ethylene action. *Physiol. Plant. 94*(2), 229-232. http://dx.doi.org/10.1111/j.1399-3054.1995.tb05305.x

Serek, M., Woltering, E. J., Sisler, E. C., Frello, S., & Sriskandarajah, S. (2006). Controlling ethylene responses in flowers at the receptor level. *Biotechnol. Adv. 24*, 368-381. http://dx.doi.org/10.1016/j.biotechadv.2006.01.007

Shetty et al. (2015). Complete genome sequence of the phenanthrene-degrading soil bacterium Delftia acidovorans Cs1-4. *Standards in Genomic Sciences, 10*, 55. http://dx.doi.org/10.1186/s40793-015-0041-x

Sisler, E. C., Dupille, E., & Serek, M. (1996). Effect of 1-methylcyclopropene and methylenecyclopropane on ethylene binding and ethylene action on cut carnations, *Plant Growth Regulation, 18*, 79-86. http://dx.doi.org/10.1007/BF00028491

Sisler, E. C., & Serek, M. (1997). Inhibition of ethylene responses in plants at the receptor level: Recent developments. *Physiol Plant, 100*, 577-582. http://dx.doi.org/10.1034/j.1399-3054.1997.1000320.x

Sisler, E. C., & Serek, M. (2003). Compounds interacting with the ethylene receptor in plants. *Plant Biol, 5* (5), 473-480. http://dx.doi.org/10.1055/s-2003-44782

Sisler, E. C., Grichko, V. P., & Serek, M. (2006). Ethylene action in plants. *Springer*. 1-6.

Thomas, J. C., & Peters, J. C. (2003). Bis(phosphino)borates: A New Family of Monoanionic Chelating Phosphine Ligands. *Inorg. Chem. 42*, 5055-5073. http://dx.doi.org/10.1021/ic034150x

Veen, H. (1983). Silver thiosulphate: an experimental tool in plant science. *Scientia Horticulturae, 20*, 211-224. http://dx.doi.org/10.1016/0304-4238(83)90001-8

Watkins, C. B. (2006). The use of 1-methylcyclopropene (1-MCP) on fruits and vegetables. *Biotechnology Advances. 24*, 389-409. http://dx.doi.org/10.1016/j.biotechadv.2006.01.005

Yang, S. F. (1995). The role of ethylene in fruit ripening. *Acta Horticulturae. 398*, 167-178.

20

Effect of Nitrogen Fertilizer and Fruit Positions on Fruit and Seed Yields of Okro (*Abelmoschus Esculentus* L. Moench)

Y. Zubairu[1], J. A. Oladiran[1], O. A. Osunde[1] & U. Ismaila[2]

[1]Department of Crop Production, Federal University of Technology, Minna, Niger State, Nigeria

[2]Department of Agricultural Biotechnology, National Biotechnology Development Agency Lugbe Abuja, Nigeria

Correspondence: U. Ismaila, Department of Agricultural Biotechnology, National Biotechnology Development Agency Lugbe Abuja, Nigeria. E-mail: ismailaumar72@yahoo.com

Abstract

Study was conducted in 2006 and 2007 cropping seasons at the experimental field of Federal University of Technology, Minna (9o 401N and 6o 301E), in the Southern Guinea Savanna of Nigeria to determine the effects of N fertilizer and the fruit positions on fruit and seed yield of okro. The treatment comprised factorial combination of five nitrogen levels (0, 30, 60, 90 and 120 kg/ha) and five fruit positions on a mother-plant (3, 5, 7, 9 and 10) which were replicated three times and laid in a Randomized Complete Block Design (RCBD). The results indicated that significant taller plants were recorded in the plots that received 120 kg N/ha while shorter plants were recorded in plots that received 0 N/ha in both years of the study. The higher N level of 120 kg/ha and lower fruit position of 3 significantly gave higher number of fruit yield while the yield decreased with decrease in N level and increase in fruit position on the mother-plant. Similarly, heavier fruits were recorded in lower fruit positions and higher N levels. The fruits formed at the lower position 3 and 5 produced more seeds with higher seed weight than those formed at the higher positions. The results also showed that significantly higher seed yield was recorded at N level of 120 kg/ha ?.

Keywords: Nitrogen fertilizer, fruit position, yield and okro

1. Introduction

Okra or okro (*Abelmoschus esculentus* (L.) Moench) known in many English-speaking countries as ladies fingers, bhindi, bamila, ochro or gumbo, is a flowering plant in the malvaceae family and it is valued for its edible green seed pods. It is widely cultivated in the tropics and sub- tropics and warm temperate regions around the world. It is a chief vegetable crop grown for its immature pod that can be used as a boiled or fried vegetable, or may be added to salad or soup (Kashif *et al* 2010). Okra plays an immense role in the human diet, providing carbohydrates, fats, proteins, vitamins and minerals (Abd El-Kader *et al.*, 2010). Okra is rich in unsaturated fatty acids, linoliec acid, that are generally deficient in the other human diet (Savello *et al.*, 1980). It is an important pan-tropical vegetable, particularly in West Africa, India, Brazil and Southern USA (Gruben, 1989).

Nitrogen, phosphorus and potassium are among the common major nutrients, which are essential for the growth and development of all plant species. Nitrogen is an essential element and important determinant in the growth and development of crop plants (Firoz 2009). It plays an important role in chlorophyll, protein, nucleic acid, hormone and vitamin synthesis and also helps in cell division and elongation. Several works have reported linear increase in green pod yield of okra with the application of N from 56 to 150 kg/ha (Hooda *et al.*, 1980; Mani and Ramanathan, 1980; Majanbu *et al.*, 1985 and Singh, 1995)

Tropical soils are generally low in organic matter and total nitrogen due to high temperature and low rainfall leading to sparse vegetation and high rate of mineralization, leaching and erosion (Okafor 1989; Rasyad 1990). Deficiency symptoms of nitrogen in soils where okra is grown has been reported to result in poor growth, thin shoot, yellowing and bristle leaves (Chonker and Sign, 1963). Olasantan (1998) reported that phosphorus and potassium if lacking would drastically reduce the yield due to their synergic interaction with nitrogen. According to Liu (2004) attainment of optimum yield where? is hardly possible without the use of fertilizer. Fertilizer recommendation for okra? varies from place to place due to environmental conditions including soil factors. Majanbu *et al.* (1986) stated in their work that 112 kg N/ha, is recommended in Trinidad while 80 kg N is the recommendation in Brazil. The South Eastern American soils have a recommendation of 120 kgN/ha. For Northern Nigeria Majanbu *et al.,* (1986), Ahmed and Tullock-Reid (1980) and FAO (1992) recommended 100 kg N/ha.

Thomas *et al.* (1979) stated that the position at which a carrot or celery seed is produced on the plant can markedly affect its size, germination characteristics and size of the ensuing seedling. In both species the lowest percentage seed germination and seedling emergence were obtained from seeds produced on primary umbels A similar result was also recorded in tomato by Dias *et al.* (2006). Fruit size is also known to vary with position on mother-plant. Ho and Hewit (1986) reported that during rapid growth in tomato, both the rates of maximum growth and of starch accumulation of proximal fruits are higher than those of distal fruits. However, when the assimilate supply is abundant, the proximal fruit could gain more weight than the distal ones.

There is dearth of information in respect of N requirement for okro seed production as; researchers appear to be more concerned with fruit production of okro for table consumption. Sajid, et al., (2012) observed that there are various ways of improving yield and seed production of okra and the best way is to provide appropriate amount of N fertilizers () and ensuring the use of appropriate fruit position on the mother-plant (Ibrahim and Oladiran, 2011) to select high yielding cultivars. In this connection, present study was conducted to study the effect of various levels of N fertilizer and fruit position on the mother-plant with the view to finding out the most suitable N level and appropriate fruit position for high seed and fruit yields of okra.

2. Materials and Methods

The study was conducted at the Teaching and Research Farm of the Federal University of Technology, Minna (Minna (latitude 6° 33[1]N and longitude 9° 45[1]E)), Niger State, Nigeria. Field and laboratory studies were conducted in in the Southern Guinea Savanna zone of Nigeria. Rainfall was 155.33mm in 2006 and 240.66mm in 2007. Wet and dry season temperatures were about 27°C and 39°C respectively /). Seeds of okra (*A. esculentus* L. Moench) variety NHAe 47-4 were sown on the flat at 75 cm inter row and 40 cm intra row spacing. Soil samples were collected at random from the entire field ? at the depth of 0-15cm, using an auger and the composite samples were subjected to physico-chemical analysis as described by Page *et al.* (1982).

The experiment consisted of fruit position on mother-plant (position 3, 5, 7, 9 and 10 the odd numbering suddenly changed to whole number 10 thus distorting the sequence for position, which should be 11? numbering from ground level) and nitrogen fertilizer levels (0, 30, 60, 90, and 120 kg/ha) with five treatments each giving a 25 treatment combination and planted on six 2 m x ? m ridges giving a plot size of 7.5 m^2 laid out in a Randomized Complete Block Design (RCBD) with three replications. Three seeds were sown per hole and following emergence seedlings were thinned to one per stand at two weeks after planting. Hoe weeding was carried out at 3, 6 and 9 weeks after planting (WAP).

Data were collected on plant height at 50% flowering, dry fruit weight, number of fruit harvested per plot, number of seeds per fruits, seed weight/fruit and seed yield/plot and converted to tha^{-1} Data collected were subjected to analysis of variance (ANOVA) using the ?. Means were separated where significant differences between treatments were obtained using which statistic? Least significant difference (LSD)? The amount of treatments would not be adequately separated with LSD, the Duncan's Multiple Range Test could suffice.

3. Results

Table 1 indicates some physico-chemical properties of the soil from the experimental site before each cropping season. The soil was sandy loam and near neutral in pH in 2006 and neutral in 2007. The organic carbon, total N, and exchangeable bases were low in 2006 and slightly higher in 2007. The effective cation exchange capacity for both years was low.

Plant height recorded at 50% flowering was observed to be highly significant influenced by different levels of nitrogen. (Table 1). Plants grown with minimum dose of N (120 kg N/ha) were found to be significantly taller (70.8 and 70.5cm in 2006 and 2007 respectively) than the plants grown with inadequate amount of N fertilizer. Plant height increased with the increasing rate of N up to 120 kg/ha. The shorter plants were recorded in the plots that received no N application in both years. The fruit position on the mother-plant did not have significant effect on the plant height in both years of study.

In 2006, the dry fruit weight was significantly influenced by only fruit position on the mother-plant and heaviest fruits were from position 3 while the lightest were from 10th position. In 2007, fruit weights were affected by fruit position, fertilizer levels and position by fertilizer interaction. The significant interaction of fruit position by fertilizer is an indication that responses to N level varied with fruit position. In 2007, significant higher fruit was recorded at position 3 when 90 kg/ha of N was applied while lower fruit weight was recorded at position 10 at the same N level.

Consistently, significantly higher numbers of fruits were recorded at lower position 3 in both years of study while position 10 gave significantly lower number of fruits. The application of different N levels had decisive effect on number of fruits and 120 kgN/ha gave significantly higher fruit numbers than the other N levels evaluated in this study though similar with 90 kg/ha at position 3 in 2006 and positions 3, 5 and 7 in 2007(Table 4). The interaction of fruit position by N levels had significant effect on number of fruits.

On seed yield, it was found that fruits at the lower positions contained significantly more seeds than those at higher positions in both years (Table 5). The different N levels were found to have insignificant effect on number of seed in both years of study. Also, a significant interaction of fruit position by fertilizer level was recorded in 2007. The fruit position on the mother-plant had no significant effect on seed weight per fruit in both years of study and the 120 and 90 kg N/ha gave significantly higher seed weight in both years while no N application gave lower seed weight per fruit. The interaction of fruit position by N levels had no significant effect on seed weight (Table 6).

Fruits closest to the base of the plant yielded more seeds than those closest to the plant tip. Fruit positions and fertilizer levels had significant effects on the seed yield per plot in both years of study (Table 7). Fruits from the lower positions significantly yielded more seeds than those at higher positions. Application of 90 and 120 kgN/ha resulted in significantly higher seed yield production compared to less? yields obtained at lower N levels.

4. Discussion

The higher dose of N might have enhanced cell division and formation of more tissues resulting in luxuriant vegetative growth giving rise to to the observed tall plants recorded in the present study (Firoz, 2009). Minimum plant height (66.2 and 62.6 cm) in zero N plots might be due to the poor nutritional status which resulted in retarded growth resulting in short plants. Muhammed et al. 2013 recorded similar result in their experiment conducted to evaluate the influence of nitrogen and phosphorous fertilizer on the phenology of okra. The results are also similar to that obtained by Sarnaik et al., (1986) who recorded maximum plant height with application of 120 kg N + 60 kg P ha[-1]. Similar results were also obtained by Gondane and Bhatia (1995) who reported that NPK fertilizer significantly increased plant height. Majanbu et al., (1985) also observed that plant height was enhanced by N fertilizer upto 100 kg N /ha. The significant progressive increase in plant height as fertilizer rate was increased in this study agrees with observations reported by Gupta (1981), Majambu et al. (1985) and Katung et al. (1996).

In both 2006 and 2007 trials the heaviest fruits were from position 3 (position closest to the plant base) while the lightest were from position 10. This result agrees with the trend reported by Alan and Eser (2007) for pepper in which fruit weight gradually declined from positions closest to the plant base to those at upper layers. The absence of significant increase in fruit weight in response to increasing addition of N in this study indicates that okro fruits at high positions are poor physiological sinks. Guillaspy, et al. (1993) reported that tomato fruits located on late trusses were smaller and lighter due to their low capacity to compete for photo-assimilates. Bertin, et al. (1998) also reported that tomato fruit size was affected by position on mother-plant under competition for assimilates. The above explanations may also hold for why significantly fewer fruits were produced at position 10 than at other lower positions 3, 5, and 7 irrespective of the fertilizer rates applied Both fruit position and the N level received generally influenced the number and seed weight per pod in this trial, following the trend reported for fruit weight by other workers. Khan and Passam (1992) reported a positive correlation between fruit size and seed number in pepper. Cucurbita pepo fruit size was also reported to correlate positively with seed number (Stephenson et al., 1988). The significant increase in seed yields at high fertilizer rates as recorded in this study is in agreement with the work of Malhi et al. (2000) which showed that the highest okra seed yield was obtained when fertilizer which type of fertilizer? Was it NP, NK or NPK? was increased from 100 to 120kg/ha though Saimhi and Padda (1990) had earlier recorded the best yield at 100 kg N/ha. Almeida (2002) stated that yield and quality became significantly superior when fertilizer level was raised from 0 to 100kgNha[-1]. Similarly, Chandler (1999) recorded significant okra seed quality improvement when N level was increased up to 120kg/ha. In a similar study, Sawan et al. (1998) reported that the increase in seed weight and better storability of cotton seeds due to high N rate might be due to the increased accumulation of metabolites which had direct impact on seed weight. A better fertilizer strategy would have to be worked out so as not only to obtaineseeds of better vigour at higher positions on mother-plants but also to improve overall yield. Nolan et al. (1999) stressed that higher yield and better seed quality will always be obtained at lower positions compared to higher positions unless a good strategy is devised to get greater N fertilization to upper positions in preference to lower ones. I disagree with this statement as the ability to produce seeds with higher weights and seed number is a genetic function and not phenotypic one to be influenced by N fertilization.

5. Conclusion

It is clearly established from this study that application of 90 – 120 kg N/ha gave better fruit and seed yield of okra. Seed and fruit yield of okra plant is better at a lower position on the mother-plant than higher position. With the result of this study, we therefore advise okra farmer in the zone to apply 90 kg N/ha and use the seed obtained from the lower position as the planting material.

Table 1. Selected physico-chemical properties of soil at the experimental site during the 2006 and 2007 cropping seasons.

Soil Properties	Values	
	2006	2007
Particle size analysis (gkg-1)		
Sand	800	800
Silt	57.0	57.0
Clay	46.0	46.0
Textural Class	Sandy loam	Sandy loam
PH (1:2 H2O)	6.90	7.00
Total N (%)	0.14	0.20
Organic Carbon (gkg-1)	10.1	12.4
Available P (mg kg-1)	1.99	2.01
Exchangeable bases (cmol.kg-1)		
K	0.22	0.30
Mg	0.12	0.14
Ca	0.30	0.33
Na	0.31	0.35
CEC	1.28	1.30
Exchangeable Acidity (cmol.kg-1)	0.5	0.7

Table 2. Effect of N fertilizer levels on plant height at 50% flowering in 2006 and 2007 cropping season

Plant height (cm)		
N fertilizer rate (kg/ha)	2006	2007
0	66.2	62.6
30	69.7	66.2
60	70.3	70.0
90	73.1	73.0
120	74.5	79.7
Mean	70.8	70.5
LSD (0.05)	0.9	1.9

Table 3. Effect of fruit positions and N fertilizer levels on dry fruit weight in 2006 and 2007

N fertilizer	Fruit Position 2006						Fruit Position 2007					
(kg/ha)	3	5	7	9	10	Mean	3	5	7	9	10	Mean
0	8.58	7.63	5.91	4.97	4.92	6.40	8.77	7.77	6.49	4.85	4.42	6.46
30	9.12	7.59	7.33	5.80	4.50	6.95	9.20	7.59	7.47	5.81	4.76	6.97
60	8.25	8.04	7.25	7.35	5.29	7.23	8.38	8.32	7.72	7.53	5.48	7.49
90	11.00	8.47	7.80	5.43	4.97	7.53	11.00	8.85	7.91	6.66	4.37	7.76
120	10.18	10.09	6.97	5.44	5.46	7.63	10.41	9.69	6.97	5.67	5.26	7.60
Mean	9.43	8.37	7.13	5.80	5.03		9.55	8.44	7.31	6.11	4.86	

LSD (0.05) Fruit position = 1.40   = 1.00

LSD (0.5) Fertilizer levels = NS   = 1.10

LSD (0.05) Fruit position x N fertilizer levels = NS   =1.20

Table 4. Effect of fruit positions and N fertilizer levels on the number of fruits harvested per lot in 2006 and 2007

N fertilizer (kg/ha)	Fruit Position 2006						Fruit Position 2007					
	3	5	7	9	10	Mean	3	5	7	9	10	Mean
0	11.38	9.00	6.34	8.33	6.00	8.20	12.00	10.00	9.00	7.00	6.00	8.80
30	11.00	8.00	7.33	7.66	6.33	8.27	11.00	10.66	9.00	7.67	7.00	9.06
60	10.67	10.33	9.00	6.67	6.00	8.41	10.67	10.33	9.00	8.00	7.00	9.00
90	12.00	10.66	8.00	6.66	6.33	9.23	12.00	11.33	10.00	9.33	7.33	10.00
120	12.00	11.00	10.00	9.00	6.60	9.73	12.00	11.00	10.00	8.00	7.67	9.73
Mean	11.40	9.80	8.13	8.07	6.26		11.53	10.67	9.40	8.00	7.00	

LSD (0.05) Fruit position = 1.51   =1.20

LSD (0.5) Fertilizer levels = 0.20   =1.20

LSD (0.05) Fruit position x N fertilizer levels = NS   =1.40

Table 5. Effect of fruit positions and N fertilizer levels on number of seed per fruit in 2006 and 2007

N fertilizer (kg/ha)	Fruit Position 2006						Fruit Position 2007					
	3	5	7	9	10	Mean	3	5	7	9	10	Mean
0	87.67	76.83	58.34	54.33	53.33	66.10	82.00	77.00	74.00	66.33	66.33	72.53
30	77.33	61.67	80.33	58.00	54.66	66.40	89.00	83.67	77.33	68.00	63.67	76.33
60	83.33	79.66	64.67	67.67	64.33	71.93	94.67	91.00	85.00	71.33	64.67	81.33
90	90.00	72.00	57.67	54.67	54.67	69.27	97.00	96.00	90.00	81.67	70.33	87.00
120	78.33	72.00	65.95	56.60	57.57	63.47	101.33	99.00	95.67	80.00	67.33	88.67
Mean	83.33	73.70	65.95	56.60	57.53		92.80	89.33	84.40	73.47	65.86	

Table 6. Effect of fruit positions and N fertilizer levels on seed weight (g) per fruit in 2006 and 2007

N fertilizer (kg/ha)	Fruit Position 2006						Fruit Position 2007					
	3	5	7	9	10	Mean	3	5	7	9	10	Mean
0	6.16	6.57	7.09	3.73	5.14	5.74	4.46	3.79	3.49	3.18	2.74	5.53
30	7.25	6.25	5.34	6.02	5.22	6.02	5.77	4.88	4.63	4.19	3.72	4.64
60	6.54	5.74	4.47	5.62	5.27	5.53	6.31	5.86	4.93	4.49	3.70	5.05
90	8.98	8.16	9.51	4.02	6.02	7.33	6.75	6.31	5.81	4.60	4.11	5.05
120	8.04	8.87	6.56	7.95	6.32	7.54	7.21	6.73	6.29	5.81	4.98	6.21
Mean	7.39	7.12	6.59	5.46	5.59		6.10	5.52	5.03	4.45	3.85	

LSD (0.05) Fruit position = NS   =1.12

LSD (0.5) Fertilizer levels =1.30   =1.10

LSD (0.05) Fruit position x N fertilizer levels = NS   =NS

Table 7. Effect of fruit positions and N fertilizer levels on seed yield (g) per plot in 2006 and 2007

N fertilizer (kg/ha)	Fruit Position 2006						Fruit Position 2007					
	3	5	7	9	10	Mean	3	5	7	9	10	Mean
0	66.85	53.00	39.25	25.28	29.80	43.45	67.85	54.00	40.25	29.29	30.80	44.45
30	68.01	43.67	44.04	41.82	29.66	45.43	69.01	44.67	45.04	42.82	30.66	45.43
60	59.81	55.93	37.76	37.84	32.17	44.51	60.81	55.93	38.76	38.84	33.17	45.51
90	106.84	67.63	58.21	32.46	31.93	59.43	107.84	68.63	59.21	33.56	32.93	60.43
120	92.75	80.34	58.94	46.70	39.62	63.67	93.75	81.34	59.94	47.70	40.62	64.67
Mean	78.85	59.92	47.65	34.00	32.63		79.85	73.92	48.50	38.42	35.00	

LSD (0.05) Fruit position = 1.20   =2.21

LSD (0.5) Fertilizer levels = 2.21   =3.20

LSD (0.05) Fruit position x N fertilizer levels = NS   = NS

References

Ahmed, N., & Tullock, L. T. (1980). Effect of fertilizer nitrogen, phosphorus and potassium on yield and nutrient content of okra (*Abelmoschus esculentus (L). Agro. J., 60*, 353-355. http://dx.doi.org/10.2134/agronj1968.00021962006000040006x

Almeida, I. (2002). Effect of fertilizer on yield and quality of okra. *Field Crop Research, 98*, 198-202.

Bertin, N., C. Gary, M. T., & Vaissiere, B. E. (1998). Influence of cultivar, fruit position and seed content on tomato fruit weight during a crop cycle under low and high competition for assimilates J. Hortic. *Sci. Biotechnical., 73*, 541- 548. http://dx.doi.org/10.1080/14620316.1998.11511012

Chandler, E. K. (1999). Comparison of Uniform and variable rate of fertilization in okro yield, 675-686 Precision Agriculture. Proc. Int. Cont. St Paul, M.N. 19-22 July 2000.

Dias, D. C. F. S., Ribeiro, F. P., Dias, L. A. S., Silva, D. J. H., & Vidigal, D. S. (2006). Tomato seed quality in relation to fruit maturation and post-harvest storage. *Seed Sci. Technol., 34*, 691-699. http://dx.doi.org/10.15258/sst.2006.34.3.15

Firoz Z. A. (2009) Impact of nitrogen and phosphorus on the growth and yield of okra *[Abelmoschus esculentus* (L.) Moench] in hill slope condition. *Bangladesh Journal of Agriculture Research. 34*(4), 713-722

Gruben, G. H. T. (1989). The cultivation of Amaranth as a leafy vegetable with special reference to South Dahomey. Communication 67 of Department of Agriculture Res. Royal Tropical Institute Amsterdam 96pp.

Guillaspy, G., Ben-Divid, H., & Gruissen, W. (1993). Fruits: A developmental perspective. *Plant cell, 5*, 1439-1451. http://dx.doi.org/10.1105/tpc.5.10.1439

Gondane, S. U., & Bhatia, G. L. (1995). Response of okra genotypes to different enivorments. Punjabrao Krishi Vidyapeeth, Marashthra, India. *PKV-Res. J., 19*(2), 143-146.

Hooda, R. S., Pandita, M. L., & Sidhu, A. S. (1980). Studies on the effect of nitrogen and phosphorus on growth and green pod yield of okra *(Abelmoschus esculentus* L. Moench). *Haryara J. Hort. Sci. 9*, 180-183.

Ho, L. C, & Hewit, J. D. (1986) Fruit Development: J. G. Atherton and J. Rudich (Eds.).The tomato crop. Chapman Hall, New York. 201-239. http://dx.doi.org/10.1007/978-94-009-3137-4_5

Ibrahim, H., & Oladiran, J. A. (2011). Effect of fruit age and position on mother-plant on fruit growth and seed quality in okra (*Abelmoschus esculentus* L. Moench). *International Journal of science and nature, 2*(3), 587-592.

Katung, M. D., Olanrewaju, J. D., & Kureh, E. (1996). Fruit and seed yield of okra *(Abelmoschus esculantus*(L) Moench), as influenced by nitrogen fertilizer. *Proceeding of 14th Annual Hortson Conference*, 173-178.

Khan, E. M., & Passam, E. C. (1992). Flowering, fruit set, and development of the fruit and seed of sweet pepper (*Capsicum annuum* L.) cultivated under conditions of high ambient temperature. *J. Horticultural Science, 67*, 251-258. http://dx.doi.org/10.1080/00221589.1992.11516245

Liu, M. L. (2004). Studies on the correlation between yield in different harvesting period at different fertilizer rates in okra. *Agricultural Research and extension station Research Bulletin, 15*(4), 46-55.

Majambu, I. S., & Ogunlela, V. B. (1986). Reponse of two okra ((*Abelmoschus esculantus(L)*Moench) varieties to fertilizer growth and nutrient concentration as influenced by nitrogen and phosphorus application. *Fertilizer Research, 8*, 297-306. http://dx.doi.org/10.1007/BF01048632

Majambu, I. S. Ogunlela, J. D., & Ahmed, M. K. (1985). Response of two okra (*Abelmoschus esculentus* L. Moench) varieties to fertilizer yield and yield component as influenced by nitrogen and phosphorus application. *Fertilizer Research, 6*, 257-267. http://dx.doi.org/10.1007/BF01048799

Malhi, S. S., Bill, E., & Robert, W. (2000).Yield in okra as influenced by fertilizer application. *American Society Horticultural Science, 77*(8), 554-557.

Mani, S., & Ramanathan, K. M. (1980). Effect of nitrogen and phosphorus on the yield of bhindi fruits. *South Indian Hort, 20*, 136-138.

Muhammad, A. K., Muhammad, S., Zahid, H., Abdur, R., Khan, B., Marwat, F., & Shahida, B. (2013). How nitrogen and phosphorus influence the phenology of okra. Pakistan. *Journal Botany, 45*(2), 479-482.

Okafor L. J. (1989). A preliminary study of the nitrogen response and yield potential of twenty barley varieties in Lake Chad Basin. *Nigeria Journal of Agricultural Technology, 9*, 20-25.

Olasantan, F. O. (1998). Effect of intercropping and population density on the growth and yield of okra (Abelmoschus esculentus(L)Moench) Beltrage Zur Tropischen Landwirtschaft and veterinarmedizin.

Page, A. L., Miller, R. H., & Keeney, D. R. (1982). Method of Soil Analysis.Volume 2,Second Edition. American Society of Agronomy, Madison Winsconsin, U.S.A.

Rasyad, D. A., Sanford, D. A., & TeKrony, D. M. (1990). Changes in seed viability and vigour during wheat seed maturation. *Seed Science and Technology, 18,* 259-267.

Saimhi, M. S., & Padda, D. S. (1990). Effect of N and Phosphate fertilization on the growth and yield of okra (*Abelmoschus esculentus (L)* Moench) *Indian Journal of Horticulture, 14,* 450-463.

Sajid M., Khan, M. A., Rab, A., Shah, S. N. M., Arif, M., Jan, I., Hussain, Z., & Mukhtiar, M. (2012). Impact of nitrogen and phosphorus on seed yield and yield components of okra cultivars. *Journal of Animal & Plant Sciences, 22*(3), 704-707.

Sarnaik, D. A., Bagael, B. S., & Singh, K. (1986). Response of Okra seed crop to major nutrients. *Research and Development Reporter. 3*(2), 1012.

Singh, I. P. (1995). Effect of various doses of nitrogen on seed yield and quality of okra (Abelmoschus esculentus (L) Moench). *Annals of Agril. Res. 16*(2), 227-229.

Thomas, J. F., Biddington, N. L., & O'Toole, D. F. (1979). Relationship between position on parent and dormancy characteristics of seeds of three cultivars of celery (Apium gradvedus). *Seed Science and Technology, 26,* 309-318.

21

Effects of Different Concentrations of Organic Waste on Selected Traits of Individuals *Capsicum Chinense* Jacq.

Francisco Orlando Holanda Costa Filho[1], Jefania Sousa Braga Amorim[1], Magnum de Sousa Pereira[1], Francisca Edineide Lima Barbosa[1], Rifandreo Monteiro Barbosa[1], Roberto Albuquerque Pontes Filho[1] & Franklin Aragão Gondim[1]

[1]Instituto Federal de Educação, Ciência e Tecnologia do Ceará (IFCE), Maracanaú Campus, Brazil

Correspondence: Franklin Aragão Gondim, IFCE, Maracanaú Campus, Brazil. E-mail: aragaofg@yahoo.com.br

Abstract

At present, excessive waste production makes it necessary to carry out research aimed to minimize the problems arising from waste generation and inadequate disposal. In this sense, this study aimed to analyze plant growth, fruit production, antioxidative enzyme activities and organic solute contents in fruits of *Capsicum chinense* Jacq. plants (BRS Moema cultivar) growing in substrates with different concentrations of municipal organic solid waste (MW) or shrimp waste (SW) under greenhouse conditions. A completely randomized design was used with seven treatments (control; MW at 50, 100 and 200% of recommendation; and SW also at 50, 100 and 200%) and five replications. The MW and SW were mixed with the soil. When plants already had fruits, 122 days after sowing (DAS), shoot height, stem diameter and number of leaves were determined, and 134 DAS, the visibly ripe fruits were collected and stored at -20 °C for subsequent biochemical analysis. The MW and SW provided a linear increase in the parameters of shoot height, stem diameter and number of leaves. The numbers of fruits in the treatments were quite variable, but it can be concluded that the MW 200 and SW 100 treatments provided higher fruit production. The best results for the activities of catalase (CAT), ascorbate peroxidase (APX) and guaiacol peroxidase (GPX) in the fruits occurred in the MW 100 and SW 200 treatments. Both provided an increase in CAT and reductions in APX and GPX, which can contribute to greater postharvest life of *C. chinense* Jacq. fruits. In relation to soluble protein and carbohydrate contents, the MW 100 and SW 100 treatments did not contribute to their increase; in the other residue concentrations, the results were variable. However, considering all parameters analysed in this study, the most suitable treatments would be MW 100 and SW 100. Therefore, the results demonstrate the susceptibility of using substrates with MW and SW in the cultivation of *C. chinense* Jacq. plants.

Keywords: *Capsicum chinense* Jacq., organic waste, antioxidative enzymes, organic solutes

1. Introduction

Excessive waste production and the difficulty to dispose of it properly have become big problems today that threaten environmental quality. In many cases, solid waste is simply deposited on the ground, in open places, without any control. This inadequate form of waste disposal causes numerous problems of an environmental and social nature that can be minimized through reuse and/or recycling of waste (Jacobi & Besen, 2011).

Organic solid waste has a high potential for use in agriculture. These residues, which are rich in organic matter and nutrients, can be used in the composition of agricultural substrates used in plant cultivation, improving the quality of the substrates and contributing to better plant growth (Abreu-Júnior, Boaretto, Muraoka, & Kiehl, 2005).

The potential use of organic waste for agriculture can be one of the solutions for the proper disposal of waste, in view of Brazilian Law 12.305 of Aug. 2, 2010, which established the National Policy on Solid Waste that aims to increase recycling and reuse of waste. Considering that more than 50% of Brazilian municipal solid waste consists of organic matter (Ministério do Meio Ambiente [MMA], 2012), the recycling of these residues would reduce significantly the volume to be disposed in the soil, therefore reducing pollution resulting from the disposal in dumps while providing increased lifespan of the landfills (Worrell & Vesilind, 2012).

In addition, these residues when applied to the soil can provide improvements in its physical, chemical and biological characteristics, which also may result in better plant development (Bot & Benites, 2005). At the same time, with the use of these residues in agriculture, there would be a decrease in the use of chemical fertilizers, reducing manuring costs and impacts caused by the use of fertilizers (Abreu-Júnior et al., 2005).

However, to have success in the use of these organic residues in the improvement of plant growth, it is extremely important to carry out studies to establish appropriate concentrations to be used, because a condition of deficiency or excess nutrients present in organic matter can compromise plant development (Maathuis & Diatloff, 2013). Additionally, when exposed to adverse conditions, such as drought, high and low temperatures, pathogen attack and deficient or excess nutrients, plants may suffer secondarily to oxidative stress (Ramakrishna & Ravishankar, 2011). In these situations, there is an imbalance between the production of reactive oxygen species (ROS) and the action capacity of the antioxidant defense system, resulting in an accumulation of ROS, which are substances generated by normal cellular metabolism of aerobic organisms. When present in excess, however, ROS can impair the metabolism of plants by harmful action to various cellular components (Hawlliwell & Gutteridge, 2007; Møller, Jensen, & Hansson, 2007; Noctor & Foyer, 1998).

The plant used in this research, *Capsicum chinense* Jacq. (BRS Moema), is largely grown in Brazil, has a height between 45 and 76 cm and produces pepper fruits ranging from 1 to 12 cm in length, which have variable shapes and colors in addition to high nutritional value and antioxidant potential (Smith & Heiser, 1957; P. Vaishnava & D. W. Wang, 2003).

To provide greater relevance, this study sought to use municipal organic solid waste (MW) and also shrimp waste (SW) in the composition of substrates. Shrimp farming is an activity that has great socioeconomic importance in Brazil, especially in the northeast region (Rocha, 2011). However, this activity is responsible for generating large quantities of waste from processing the shrimp. Nevertheless, there is no research in the literature that addresses the use of shrimp waste in substrate composition and, consequently, in improving growth of *C. chinense* Jacq. Therefore, the objective of this study was to evaluate the influence of substrates with different concentrations of MW and SW on plant growth, productivity of fruits, activities of antioxidative enzymes and organic solute contents of the fruits of *C. chinense* Jacq. growing under greenhouse conditions. The hypothesis of this study is that the use of substrates with MW and SW may provide greater plant growth, productivity of fruits and antioxidant defense system than the control treatment, which does not have MW and SW.

2. Method

2.1 The Experiment Design

The experiment was carried out under greenhouse conditions at the Laboratório Terra of the Instituto Federal de Educação, Ciência e Tecnologia do Ceará (IFCE), Maracanaú campus, located in Maracanaú, Ceará, Brazil. During the day, the average temperature inside the greenhouse was 33.3 °C and relative humidity 54%. The experiment was carried out between November 2013 and April 2014.

To set up the experiment, 16-L plastic buckets with holes drilled in the bottom ends for drainage of excess water were used. Irrigation conducted during the experiment was performed manually and aimed at maintaining the water content of the soil near field capacity. The seeds for the production of seedlings were obtained commercially and sown in Styrofoam trays with 128 cells containing vermiculite and worm humus as a substrate in a volumetric ratio of 2:1. After germination and early seedling growth, 49 days after sowing (DAS), plants were transplanted to the plastic buckets containing different treatments described in following section (2.2 Definition of the waste concentrations for the composition of the substrates).

The soil used in the experiment was obtained from subsoil excavation in an area close to the IFCE Maracanaú campus. To characterize the soil used in the preparation of substrates, samples were separated and sent to the Laboratório de Solo/Água of the Universidade Federal do Ceará (UFC), which carried out the necessary analyses (Table 1).

Table 1. Chemical analysis of the soil used in the composition of the substrates for cultivation of *Capsicum chinense* Jacq. Plants

g/kg			C/N	mg/kg					H₂O	dS/m
O.M	C	N		Assimilable phosphorus	Fe	Cu	Zn	Mn	pH	E.C
6.93	4.02	0.39	10	2	17.9	0.3	0.9	3.4	5.2	0.15

cmol$_c$/kg						S.B	C.E.C	%		
Ca^{2+}	Mg^{2+}	Na$^+$	K$^+$	Al^{3+}	H$^+$ + Al^{3+}			B.S	M	E.S.P
0.50	0.30	0.07	0.14	0.85	1.98	1.0	3.0	33	46	2

Note. O.M = organic matter; E.C = electrical conductivity; S.B = sum of bases; C.E.C = cation exchange capacity; B.S = base saturation percentage; M = aluminum saturation percentage; E.S.P = exchangeable sodium percentage.

The SW was obtained from an effluent treatment plant in the shrimp processing industry located in Aracati, Ceará, Brazil. During the treatment of shrimp processing waste, sludge is generated as a byproduct, which is put into a drying bed and later sent to a landfill. Part of this dry sludge was collected, analyzed and used in the experiment as shrimp waste (Table 2).

Table 2. Chemical analysis of the shrimp waste used in the composition of the substrates for cultivation of *Capsicum chinense* Jacq. plants.

g/kg							mg/kg			
N	P	P₂O₅	K	K₂O	Ca	Mg	Fe	Cu	Zn	Mn
38.2	8.5	19.5	2.5	3.1	8.3	4.3	5713.7	1297.9	1893.4	134.6

The MW was obtained from municipal waste processing performed by the company Kermais Ambiental & Lubbad Ambiental. The process known as EcoKer involves weighing the MW; separating all recyclable, inert and ferrous metals; shredding the residues for volume reduction and homogenization; separating non-ferrous metals; and ending the process, which lasts only 4 h in total, the MW is added in a mixer along with a cleaning agent and quicklime, resulting in the organic fertilizer used in this study and a fuel derived from waste. The results of the MW analysis are shown in Table 3.

Table 3. Chemical analysis of the municipal organic solid waste used in the composition of the substrates for cultivation of *Capsicum chinense* Jacq. plants.

g/kg							mg/kg			
N	P	P₂O₅	K	K₂O	Ca	Mg	Fe	Cu	Zn	Mn
3.4	0.9	2.1	2.5	3.1	185.9	9.1	3471.8	9.7	152.7	66.5

2.2 Definition of the Waste Concentrations for the Composition of the Substrates

The concentrations of the solid waste used in the composition of the different substrates were defined according to the recommendation suggested by the Brazilian Agricultural Research Corporation [EMBRAPA] (2007) of 60 kg of nitrogen ha^{-1} in the cultivation of pepper and considering the amounts of nitrogen present in the compositions of the soil (0.39 g kg^{-1}) and residues (SW: 38.2 g kg^{-1}; MW: 3.4 g kg^{-1}) (Tables 1, 2 and 3). Treatments were tested using concentrations calculated for each residue as recommended by EMBRAPA (100%), treatments using half (50%) and twice (200%) the concentrations of nitrogen recommended for each residue. The residues were thoroughly mixed with soil to form the different substrates, which were manually added to the plastic buckets. The substrate used as the control treatment was composed of only soil.

2.3 The Studied Taxon

The plant used in this study, *C. chinense* Jacq., is originally from Central and South America and largely grown in Brazil, has a height between 45 and 76 cm and produces pepper fruits ranging from 1 to 12 cm in length, which have variable shapes and colors (Smith & Heiser, 1957). Moreover, its fruits have high nutritional value and antioxidant potential (P. Vaishnava & D. W. Wang, 2003). The cultivar used in this research was the BRS Moema (RNC 22493), which has high productivity and is sensitive to low temperatures (Amaro, 2012).

2.4 Plant Growth Evaluations

When the plants already had fruits, 122 DAS, plant growth evaluations were performed. The following determinations were carried out: shoot height, stem diameter and number of leaves. The shoot height evaluations were performed with the aid of a ruler graduated in centimeters, considering the height of the base of the plant (substrate surface) to the end of the farthest leaf. The stem diameter, measured near the substrate surface, was evaluated with the aid of a digital caliper, and the number of leaves was counted manually.

When the fruits were visibly mature (dark red coloration), 134 DAS, harvest was carried out manually and selectively. At the end of harvest, the fruits were stored at -20 °C for subsequent biochemical analysis.

2.5 Obtaining Enzymatic Extracts for Biochemical Analysis

The activities of the antioxidant enzymes catalase (CAT), ascorbate peroxidase (APX) and guaiacol peroxidase (GPX) in pepper fruits were determined. Thus, enzymatic extracts were prepared by maceration of 1 g of pepper seeds in 4 mL of potassium phosphate buffer at 100 mM, containing 0.1 mM ethylenediaminetetraacetic acid (EDTA) (pH 7.0) at 4 °C. For APX activity, the extract was added to ascorbate at a concentration of 2 mM. The homogenate was then filtered through nylon cloth, transferred to Eppendorf tubes and centrifuged at 12,000 x g for 15 min at 4 °C, and the supernatant (enzymatic extracts) was removed.

2.6 Enzymatic Activity in Fruits of Capsicum Chinense Jacq.

The activities of the antioxidant enzymes CAT, APX and GPX were determined. The CAT activity was determined according to the method of Havir and McHale (1987) by the decrease in absorbance at 240 nm due to consumption of hydrogen peroxide; the GPX activity according to the method of Kar and Mishra (1976), with the reaction accompanied by the increase in absorbance at 470 nm due to the formation of tetraguaiacol; and the APX activity according to the method of Nakano and Asada (1981), with the oxidation of ascorbate measured by the decrease in absorbance at 290 nm. The CAT, APX and GPX enzyme activities were expressed in micromoles of hydrogen peroxide per minute per gram of fresh matter (FM). Each extract was measured in duplicate.

2.7 Determination of the Soluble Protein and Carbohydrate Contents in the Fruits

The soluble protein contents were determined according to the method described by Bradford (1976). The soluble protein standard curve used was formed from a standard solution of bovine serum albumin. The protein concentration was expressed in milligrams of protein per gram of FM.

The soluble carbohydrate contents were determined according to the method of Dubois, Gilles, Hamilton, Rebers and Smith (1956). The standard curve of carbohydrates used was formed from the standard solution of anhydrous glucose. The concentration of carbohydrates was expressed in micromoles of carbohydrates per gram of FM.

2.8 The Statistical Analyses

A completely randomized design was used with seven treatments (control; MW at 50, 100 and 200% of recommendation; and SW at 50, 100 and 200%) and five replications. The data were analyzed using a one-way analysis of variance (ANOVA). The means were compared through Tukey's test ($P \leq 0.05$).

3. Results

3.1 Plant Growth and Number of the Fruits

The results of the effects of the MW and SW on *C. chinense* Jacq. growth are shown in Figure 1. In general, the two residues promoted an increase in plant growth in the parameters analyzed in relation to the control treatment.

In relation to shoot height (Figure 1a), it was verified that the plants grown in the presence of MW in the substrate showed higher values than those found in substrates with the control treatment. It was also observed that the MW 200 treatment was the highest and was 67% higher than the control. Plants growing in substrates containing SW showed similar growth. SW 200 was the highest and was 94% and 16% higher than the control and MW 200 treatments, respectively.

The root collar diameter (Figure 1b) showed behavior similar to that of plant height (Figure 1a). It was observed that the MW 200 treatment exhibited the highest value among the treatments containing MW, being 32% higher than the control. For the treatments with SW, the behavior was similar. SW 200 exhibited the highest value, being 85% and 39% higher than the control and MW 200 treatments, respectively.

For the number of leaves parameter (Figure 1c), it was also verified that MW 200 exhibited the highest value among the treatments containing MW, being 129% higher than the control. For the substrates containing SW, it was observed that there was no difference in the SW 50 concentration compared to the control treatment. However, SW 100 and SW 200 differed from the control, with SW 200 being the highest value – 231% and 44% higher than the control and MW 200 treatments, respectively.

In relation to the numbers of fruits obtained, the values in the treatments were quite variable (Figure 1d). However, the MW 200 treatment followed the same trend that had been shown in relation to plant growth, being higher than the other treatments with MW and being 215% higher than the control. In relation to the SW treatments, there was not the same trend. The number of fruits in SW 200 did not differ from that of the control. The highest values occurred in the SW 100 treatment, which were 284% higher than those of the control.

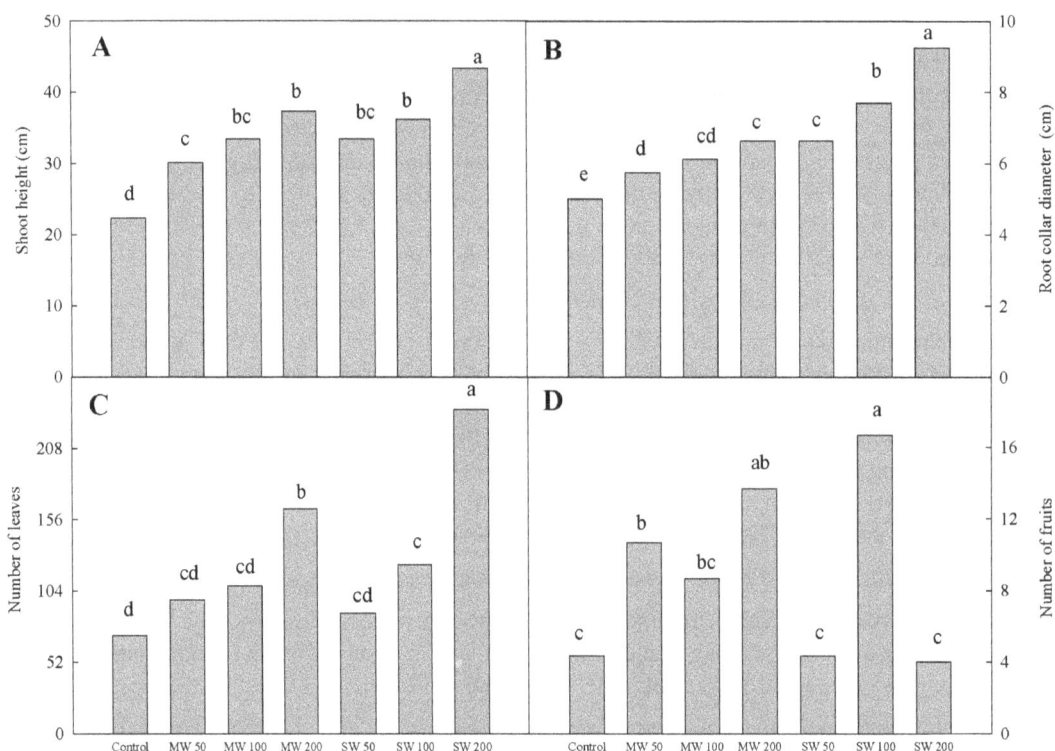

Figure 1. Shoot height (a), root collar diameter (b) and number of leaves (c) of *Capsicum chinense* Jacq. plants at 122 days after sowing under greenhouse conditions as well as number of fruits/plant (d) at 134 days after sowing. Plants were grown on substrates containing only soil (control), municipal organic solid waste (MW) (50, 100 or 200% of the fertilizer recommendation) or shrimp waste (SW) (50, 100 or 200% of the fertilizer recommendation). Bars represent the mean values of five replications. Values followed by the same letters are not statistically different according to Tukey's test ($P \leq 0.05$)

3.2 Enzymatic Activity

Figure 2 shows the CAT, APX and GPX enzyme activities in the fruits of *C. chinense* Jacq. plants growing in substrates containing different concentrations of MW and SW.

As shown in Figure 2a, the highest values for CAT activity in substrates containing MW occurred in the MW 50 and MW 100 treatments, with no difference between these treatments. The average values of MW 50 and MW 100 were 169% higher than those of the control. The results of the MW 200 treatment did not differ in relation to those of the control. In relation to the substrates containing SW, an increase in CAT activity was observed as the concentrations of this residue increased in the substrate. The SW 50 and control treatments did not differ. However, the SW 100 and SW 200 treatments increased by 57% and 192%, respectively, compared to the control. Thus, the highest values of CAT activity occurred in the MW 50, MW 100 and SW 200 treatments, which did not differ among themselves.

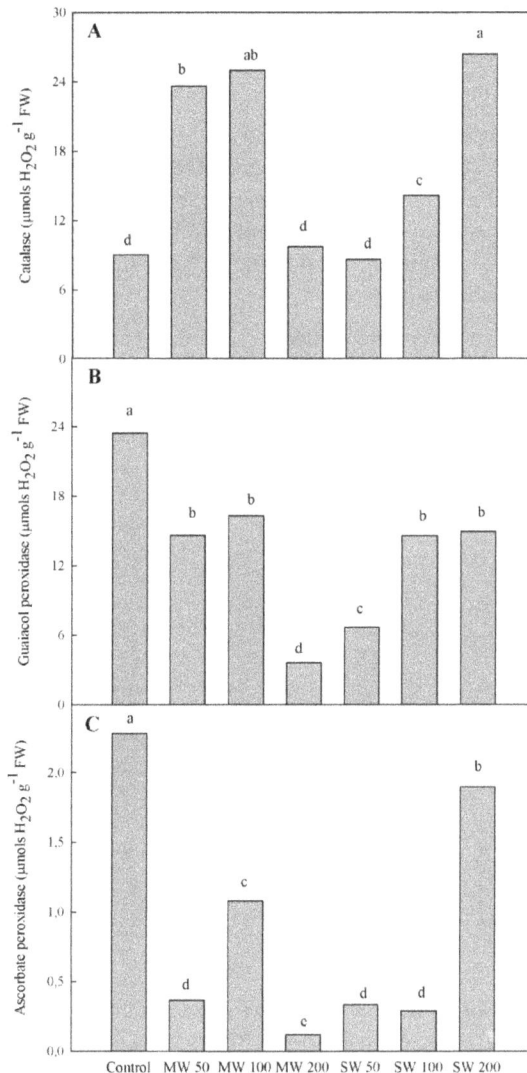

Figure 2. Activities of catalase (A), guaiacol peroxidase (B) and ascorbate peroxidase (C) in fruits of *Capsicum chinense* Jacq. plants at 134 days after sowing under greenhouse conditions. Plants were grown on substrates containing only soil (control), municipal organic solid waste (MW) (50, 100 or 200% of the fertilizer recommendation) or shrimp waste (SW) (50, 100 or 200% of the fertilizer recommendation). Bars represent the mean values of five replications. Values followed by the same letters are not statistically different according to Tukey's test ($P \leq 0.05$)

In Figure 2b, it was observed that the highest value of GPX activity occurred in the control treatment. In relation to the substrates containing MW, the reduction compared to the control treatment was more accentuated in MW 200, being 547% lower. In the substrates containing SW, the reduction was lower in the SW 100 and SW 200 treatments, which did not differ among themselves. The average value of GPX of these two treatments was 58% lower than that of the control.

In relation to APX activity (Figure 2c), the control treatment was again higher compared to other treatments, and, among the MW treatments, MW 200 showed a greater reduction, being 18 times lower than the control. In relation to the substrates containing SW, SW 200 was the closest treatment compared to the control, being 20% lower.

At high concentrations of MW (MW 200), there was a decrease in the activity of the enzymes (CAT, GPX and

APX) to values similar to or below those of the control. However, the greatest results occurred in the treatment that used the fertilizer recommendation (MW 100). In relation to the SW treatments, linearity in the results was observed such that the activities of the enzymes were higher in treatments with higher concentrations.

3.3 Organic Solutes

Figure 3 shows the effects of different concentrations of MW and SW on the soluble protein and carbohydrate contents in fruits of *C. chinense* Jacq. plants.

There was no difference in soluble protein content (Figure 3a) between the control and MW 100 treatments. However, the MW 50 and MW 200 treatments were 63% and 342% lower than the control, respectively. In relation to the treatments containing SW, there was an increase of soluble protein contents with an increase in concentrations of the residues in the substrates. The SW 50 and SW 100 treatments were 175% and 29% lower than the control, respectively. The SW 200 treatment showed the highest soluble protein content in relation to all the treatments used, being 26% higher than the control.

In the treatments with MW, the soluble carbohydrate contents (Figure 3b) exhibited a behavior similar to that of the soluble protein contents. The MW 100 treatment did not differ from the control, and the MW 50 and MW 200 treatments were 140% and 503% lower than the control, respectively. In the substrates with SW, the SW 50 and SW 100 treatments did not differ and were, on average, 82% lower than the control. The SW 200 treatment did not differ from the control.

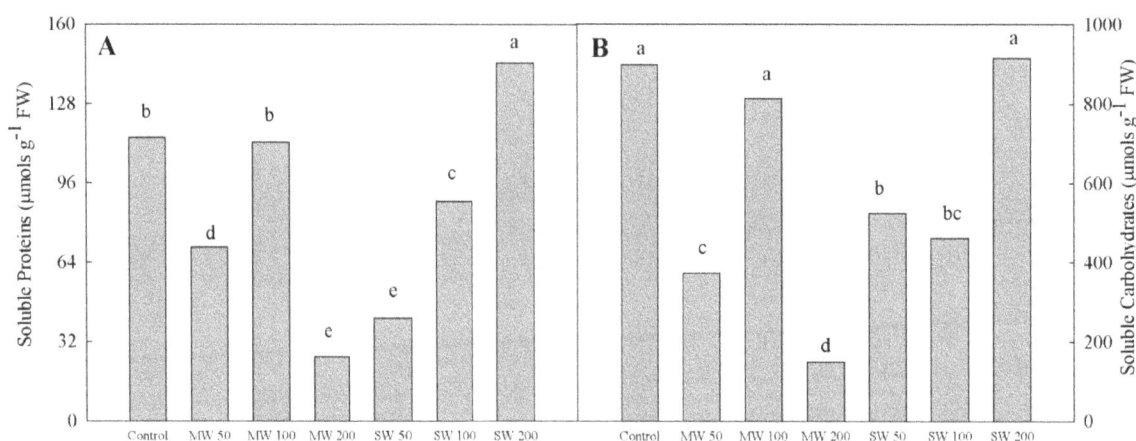

Figure 3. Contents of soluble proteins (a) and soluble carbohydrates (b) in fruits of *Capsicum chinense* Jacq. plants at 134 days after sowing under greenhouse conditions. Plants were grown on substrates containing only soil (control), municipal organic solid waste (MW) (50, 100 or 200% of the fertilizer recommendation) or shrimp waste (SW) (50, 100 or 200% of the fertilizer recommendation). Bars represent the mean values of five replications. Values followed by the same letters are not statistically different according to Tukey's test ($P \leq$ 0.05)

Again, when analyzing the results of the soluble protein and carbohydrate contents, it was observed that MW 200 had the lowest value among the substrates containing MW and exhibited linearity of the results of these solutes in the treatments containing SW.

4. Discussion

4.1 Plant Growth and Number of the Fruits

In the three parameters analyzed (shoot height, stem diameter and number of leaves), a positive effect of the residues on the growth of *C. chinense* Jacq. plants was observed at all concentrations tested compared to plants growing under control conditions. Linearity of the results of the two residues was also observed: a higher concentration of residues in substrates increased plant growth almost linearly.

When applied properly, organic residues have the ability to improve physical, chemical and biological properties of the soil. Some of the effects of these residues in the soil include the following: an increase in concentration of organic matter and capacity of cation exchange; an increase of pH (acid soil correction); an improvement in soil structure, making the soil more stable and favoring better aeration and permeability; and an increase in water retention capacity and higher nutritional quality by making available nutrients such as nitrogen, phosphorus, sulfur and micronutrients (Abreu-Júnior et al., 2005, Aprile & Lorandi, 2012; Bot & Benites, 2005; Cardoso et al., 2013). These benefits mean better cultivation conditions, exerting a positive effect on plant growth and development and providing higher productivity and quality of agricultural products (Abreu-Júnior et al., 2005; Bot & Benites, 2005; Nardi, Pizzeghello, Muscolo, & Vianello, 2002).

The pepper plant growth results found in this study are in agreement with those found by several authors and confirm the benefits of using organic solid waste properly treated in plant cultivation. N. Tzortzakis et al. (2012) studied the performance and suitability of municipal solid waste compost (MSWC) in different content and/or fertigation in pepper (*Capsicum annuum* L.) plants growing under in greenhouse conditions. The addition of MSWC into the soil increased nutritive value (N, K, P, organic matter) resulting in increased EC (electrical conductivity). The results indicated the low content of MSWC added into substrate improved plant growth and maintained fruit fresh weight without affecting plant yield, while fertigation acted beneficially.

F. J. F. Araújo, M. D. Aquino, B. F. Aquino, F. M. L. Bezerra and F. Chagas-Neto (2009) prepared an organic flour from crab waste and used it in different proportions in substrate composition used in the production of cowpea beans (*Vigna unguiculata* [L.] Walp). The substrates analyzed were formed from local soil mixed with different concentrations of chemical fertilizer and/or crab organic flour. Chemical analyses indicate excellent results of this organic product, highlighting the contents of nitrogen, phosphorus, magnesium, calcium and organic matter. The great agronomic potential of the crab waste flour was confirmed by good performance of the plants grown on substrates containing crab fertilizer, providing in some treatments a reduction in the need for chemical fertilizer applications.

A. Chrysargyris, C. Saridakis and N. Tzortzakis (2013) evaluated substrates with different contents of municipal solid waste (MSW) and/or peat in the cultivation of melon (*Citrulus melo* L.) seedlings growing under in greenhouse conditions. The results showed that the high content of MSW affected negatively plant growth, but the low content of MSW may act as an alternative substitute of peat in melon seedling production.

Fachini, Galbiatti and Pavani (2004) studied the effects of substrates with different proportions of pine bark and organic waste compounds in the formation of orange seedlings and found that the mixture of 60% pine bark and 40% municipal waste compound provided better and faster plant development, outside of supplying nutrients, very effective.

High concentrations of nitrogen promote intense plant growth, especially favoring leaf development and, consequently, green mass production. However, in some plants, excess nitrogen can harm fruit production (Keller, 2005; C. M. F. Pinto, P. C. Lima, L. T. Salgado, & F. L. B. Caliman, 2006). This fact was clearly observed in treatments that used shrimp waste. The SW 200 treatment greatly increased plant growth, more than all other treatments. However, the SW 200 treatment did not improve fruit production, possibly due to excess nitrogen.

Considering the results obtained from plant growth and fruit production and also the importance of proper disposal of the largest possible amount of MW and SW, the use of these residues can be indicated for the production of *C. chinense* Jacq. peppers. In this study, the most recommended treatments would be MW 200 and SW 100 or SW 200, depending on the purpose. The SW 200 treatment would be indicated if the greatest interest was green mass production for fodder, and SW 100 would be indicated if the interest mostly involved fruit production.

4.2 Enzymatic Activity

The high activity of antioxidant enzymes can contribute to greater postharvest life. Masia (1998) observed that Golden Delicious apples had a higher activity of superoxide dismutase (SOD) and CAT at harvest compared to Fuji apples, resulting in a longer cold storage period of Golden Delicious apples. Similarly, Lacan and Baccou (1998) verified that Clipper melons showed an increase in the activities of SOD and CAT compared to those of Jerac melons. As a result, Clipper melons had greater postharvest life. Moreover, Y. S. Wang, S. P. Tian and Y. Xu (2005) developed a research with peach fruits (*Amygdalus persica* cv. Okubo) and found that the decrease of SOD and CAT might contribute to the development of chilling injury in the fruits.

Maciel (2012) studied acerola fruits of six clones stored at temperatures of 4 °C and 8 °C (minimum temperature of safety for the acerola) for up to 12 days. The author observed that in the fruits stored at 4 °C, the activities of enzymes were higher than those of fruits stored at 8 °C. This may have been responsible for protection against oxidative stress generated by the cold at 4 °C, maintaining the quality of the fruit.

Several studies demonstrate the induction of an antioxidant enzymatic system as a strategy to increase the tolerance of fruits to cold and thus prolong their shelf lives. Sestari (2010) observed a reduction in the sensitivity of Nanicão bananas when immersed in water at 42 °C for 15 min or at 53 °C for 2 min prior to storage at 6 °C for up to 15 days. According to the author, the low temperature intensified the generation of ROS in fruits, but the heat treatments improved the antioxidant enzymatic system, reducing oxidative stress and resulting in lower rates of injuries in these fruits.

Kluge et al. (2006) found that intermittent warming (6 days at 1 °C + 1 day at 25 °C) in Valência orange, Tahiti lime and Murcott tangor stored at 1 °C reduced injuries caused by cold. Biochemical analyses revealed greater activity or peaks in CAT, APX and glutathione reductase enzymes during the storage period in the fruits of the treatment. These results may be related to increased resistance of the fruits to cold.

In the present study, in general, the treatments with residues provided an increase in CAT activity in pepper fruits; however, the peroxidases (APX and GPX) had their activities reduced. Despite this, it is possible that these results can contribute to greater postharvest life of *C. chinense* Jacq. pepper fruits. CAT is one of the most important enzymes in the process of elimination of hydrogen peroxide, avoiding the formation of the hydroxyl radical and consequent cellular damage (Sies, 1993; Vaidyanathan, Pattathil, Chakrabarty, & Thomas, 2003). Maciel (2012), Sestari (2010) and Kluge et al. (2006) found in their experiments that CAT was the main enzyme involved in the removal of hydrogen peroxide in fruits.

According to Valderrama and Clemente (2004), peroxidase is one of the main enzymes responsible for the deterioration of quality in many fruits. The removal of hydrogen peroxide by this enzyme occurs concomitantly with the oxidation of phenolic compounds to quinones, which rapidly polymerize, forming melanin and pigments of dark color. The enzymatic browning reduces the quality of fruits and their nutritional characteristics (M. I. F. Chitarra & A. B. Chitarra, 2005; Fennema, 2010; Lurie, 2003; Márques, Fleuriet, & Macheix, 1995). The activities of peroxidase are still related to loss of flavor, texture and fruit nutrients (Barrett, Beaulieu, & Shewfelt, 2010; Hui, 2006). Thus, although this enzyme performs an important role in the elimination of hydrogen peroxide, the control of its activities is desirable because it can contribute to the conservation of fruit quality (M. I. F. Chitarra & A. B. Chitarra, 2005; Sousa, 2010). In this study, residues reduced the activity of peroxidase.

In this context, the MW 100 and SW 200 treatments would provide greater shelf life of fruits. However, considering also the growth and production of pepper fruits, the most suitable treatments would be MW 100 and SW 100.

4.3 Organic Solutes

The SW 200 treatment showed the highest results in relation to protein content, elevating the content of this solute in the fruits. The nutritional significance of peppers is due in part to the presence of large amounts of protein in fruits (P. Vaishnava & D. W. Wang, 2003). In plants, proteins are involved in the processes of maturation, germination, cell death, response to oxidative stress processes and others (Bouteau & Bailly, 2008; S. L. T. Lima et al., 2008; Wojtyla, Lechowska, Kubala, & Garnczarska, 2016).

The residues did not increase the soluble carbohydrate content at any concentration, but in MW 100 and SW 200, these solutes were not reduced. The soluble sugars represent a large part of the total soluble solids (TSS), one of the variables that affects directly the quality of the fruit. Plant development conditions, such as fertilization, irrigation and temperature, influence the carbohydrate content in fruits (Razek et al., 2011; Germano, 2000; Hobson & Grierson, 1993; Whiting, 1970; Yahia, 2011). Sacramento, Matos, Souza, Barreto and Faria (2007) related that high contents of TSS are important for the consumption of fresh fruit and for industry because the high contents provide better taste and higher yield in the preparation of the products.

In relation to nutritional quality, the treatments with residues that produced the best results were MW 100 and SW 200. However, in general, the treatments that showed better results considering all parameters analysed were MW 100 and SW 100, although SW 100 provided a reduction in the nutritional quality of fruit.

5. Conclusions

The MW and SW at all concentrations analyzed provided an increase in the parameters of shoot height, stem diameter and number of leaves in *C. chinense* Jacq. plants. The results for both residues were linear such that the greatest plant growth occurred in treatments with higher concentrations of these residues (MW 200 and SW 200).

The numbers of fruits of the treatments were quite variable. However, it can be concluded that the MW 200 and SW 100 treatments provided the highest fruit production.

The best results for the activities of CAT, APX and GPX occurred in the MW 100 and SW 200 treatments. Both provided an increase in CAT and a reduction in APX and GPX, which can contribute to greater postharvest life of the *C. chinense* Jacq. fruits.

The use of MW and SW in nutritional recommendations for cultivation (MW 100 and SW 100) did not contribute to an increase in soluble protein and carbohydrate contents. In the other residue concentrations, the results were quite variable. Thus, it was not possible to establish a precise relationship between application of waste and improvement in the nutritional quality of the fruit.

Thus, considering all parameters analyzed, the most suitable treatments would be MW 100 and SW 100. Therefore, the most suitable treatments indicated are in accordance with the recommendation suggested by the EMBRAPA based on the amount of nitrogen used in the cultivation of pepper. Moreover, the results demonstrate the susceptibility of using substrates with MW and SW in the cultivation of *C. chinense* Jacq. plants.

Acknowledgements

We thank Proof-Reading-Service.com for language corrections.

References

Abreu-Júnior, C. H., Boaretto, A. E., Muraoka, T., & Kiehl, J. C. (2005). Uso agrícola de resíduos orgânicos potencialmente poluentes: propriedades químicas do solo e produção vegetal. *Tópicos Especiais em Ciência do Solo, 4*, 391-470. Retrieved from http://www.scielo.br/scielo.php?script=sci_nlinks&ref=000080&pid=S1415-4366200700010001400001&lng=pt

Amaro, G. B. (2012). *Capsicum chinense. Empresa Brasileira de Pesquisa Agropecuária.* Retrieved from https://www.agencia.cnptia.embrapa.br/gestor/pimenta/arvore/CONT000gn0frh1202wx5ok0liq1mqt5bf5ht.html

Aprile, F., & Lorandi, R. (2012). Evaluation of Cation Exchange Capacity (CEC) in Tropical Soils using four different Analytical Methods. *Journal of Agricultural Science, 4*, 278-289. http://dx.doi.org/10.5539/jas.v4n6p278

Araújo, F. J. F., Aquino, M. D., Aquino, B. F., Bezerra, F. M. L., & Chagas-Neto, F. (2009). Aplicação do composto orgânico produzido a partir de caranguejo uçá *Ucides cordatus cordatus* no cultivo de feijão caupi *Vigna unguiculata* (L.) WALP. *Engenharia Ambiental, 6*, 15-35. Retrieved from http://ferramentas.unipinhal.edu.br/engenhariaambiental/include/getdoc.php?id=952&article=276&mode=pdf.

Barrett, D. M., Beaulieu, J. C., & Shewfelt, R. (2010). Color, flavor, texture, and nutritional quality of fresh-cut fruits and vegetables: desirable levels, instrumental and sensory measurement, and the effects of processing. *Critical Reviews in Food Science and Nutrition, 50*, 369-389. http://dx.doi.org/ 10.1080/10408391003626322

Bot, A., & Benites, J. (2005). *The importance of soil organic matter: key to drought-resistant soil and sustained food production.* Rome: Food and Agriculture Organization of the United Nations.

Bouteau, H. E. M., & Bailly, C. (2008). Oxidative signaling in seed germination and dormancy. *Plant Signaling & Behavior, 3*, 175-182. Retrieved from https://www.ncbi.nlm.nih.gov/pmc/articles/PMC2634111/

Bradford, M. M. (1976). A rapid and sensitive method for the quantification of microgram quantities of protein utilizing the principle of protein-dye binding. *Analytical Biochemistry, 72*, 246-254. http://dx.doi.org/10.1016/0003-2697(76)90527-3

Cardoso, E. J. B. N., Vasconcellos, R. L. F., Bini, D., Miyauchi, M. Y. H., Santos, C. A., Alves, P. R. L., ... Nogueira, M. A. (2013). Soil health: looking for suitable indicators. What should be considered to assess the effects of use and management on soil health? *Scientia Agricola, 70*, 274-289. http://dx.doi.org/10.1590/S0103-90162013000400009

Chitarra, M. I. F., & Chitarra, A. B. (2005). *Pós-colheita de frutas e hortaliças: fisiologia e manuseio.* Lavras: UFLA.

Chrysargyris, A., Saridakis, C., & Tzortzakis, N. (2013). Use of municipal solid waste compost as growing medium component for melon seedlings production. *Journal of Plant Biology & Soil Health, 2*, 1-5. Retrieved from https://www.researchgate.net/publication/288823666_Use_of_municipal_solid_waste_compost_as_growing _medium_component_for_melon_seedlings_production

Dubois, M., Gilles, K. A., Hamilton, J. K., Rebers, P. A., & Smith, F. (1956). Colorimetric method for determination of sugars and related substances. *Analytical Chemistry, 28*, 350-356. http://dx.doi.org/10.1021/ac60111a017

EMBRAPA. (2007). Pimenta (*Capsicum* spp.). *Empresa Brasileira de Pesquisa Agropecuária.* Retrieved from http://sistemasdeproducao.cnptia.embrapa.br/FontesHTML/Pimenta/Pimenta_capsicum_spp/adubacao.html

Fachini, E., Galbiatti, J. A., & Pavani, L. C. (2004). Níveis de irrigação e de composto de lixo orgânico na formação de mudas cítricas em casa de vegetação. *Engenharia Agrícola, 24*, 578-588. http://dx.doi.org/10.1590/S0100-69162004000300010

Fennema, O. R. (2010). *Química de los alimentos* (3rd ed.). Zaragoza: Acribia.

Germano, S. (2000). *Desenvolvimento de bioprocessos para a produção, caracterização e purificação de proteases de Penicillium sp. por fermentação no estado sólido* (Unpublished doctoral dissertation). Universidade Federal do Paraná, Paraná, Brazil.

Havir, E., & McHale, N. A. (1987). Biochemical and developmental characterization of multiple forms of catalases in tobacco leaves. *Plant Physiology, 84*, 450-455. Retrieved from http://www.ncbi.nlm.nih.gov/pubmed/16665461

Hawlliwell, B., & Gutteridge, J. M. C. (2007). *Free radicals in biology and medicine* (4th ed.). Oxford: Oxford University Press.

Hobson, G. E., & Grierson, D. (1993). Tomato. In G. B. Seymour, J. E. Taylor, & G. A. Tucker (Eds.), *Biochemistry of fruits ripening* (pp. 405-442). London: Champman & Hall.

Hui, Y. H. (2006). *Handbook of Food Science, Technology, and Engineering* (Vol. 2). Boca Raton, FL: CRC Press.

Jacobi, P. R., & Besen, G. R. (2011). Solid Waste Management in São Paulo: The challenges of sustainability. *Estudos Avançados, 25*, 135-158. http://dx.doi.org/10.1590/S0103-40142011000100010

Kar, M., & Mishra, D. (1976). Catalase, peroxidase, and polyphenoloxidase activities during rice leaf senescence. *Plant Physiology, 57*, 315-319. Retrieved from http://www.ncbi.nlm.nih.gov/pmc/articles/PMC542015/

Keller, M. (2005). Nitrogen - friend or foe of wine quality? *Practical Winery & Vineyard, 27*, 24-29. Retrieved from http://www.practicalwinery.com/SeptOct05/septoct05p24.htm

Kluge, R. A., Azevedo, R. A., Jomori, M. L. L., Edagi, F. K., Jacomino, A. P., Gaziola, S. A., & Aguila, J. S. (2006). Efeitos de tratamentos térmicos aplicados sobre frutas cítricas armazenadas sob refrigeração. *Ciência Rural, 36*, 1388-1396. Retrieved from http://www.scielo.br/pdf/cr/v36n5/a07v36n5.pdf

Lacan, D., & Baccou, J. C. (1998). High levels of antioxidant enzymes correlate with delayed senescence in nonnetted muskmelon fruits. *Planta*, 204, 377-382. http://dx.doi.org/ 10.1007/s004250050269

Lima, S. L. T., Jesus, M. B., Sousa, R. R. R., Okamoto, A. K., Lima, R., & Fraceto, L. F. (2008). Estudo da Atividade Proteolítica de Enzimas Presentes em Frutos. *Química na Escola, 2*, 47-49. Retrieved from http://qnesc.sbq.org.br/online/qnesc28/11-EEQ-6906.pdf

Lurie, S. (2003). Antioxidants. In D. M. Hodges (Eds.), *Postharvest oxidativestress in horticultural crops* (pp. 138-139). New York: Food Products Press.

Maathuis, F. J., & Diatloff, E. (2013). Roles and functions of plant mineral nutrients. In: F. J. Maathuis (Ed.), *Plant mineral nutrients: methods and protocols* (pp. 1-21). New York: Humana Press.

Maciel, V. T. (2012). *Qualidade e metabolismo antioxidante em frutos de clones de aceroleira armazenados a 4° e 8° C* (Unpublished doctoral dissertation). Universidade Federal do Ceará, Ceará, Brazil.

Márques, L., Fleuriet, A., & Macheix, J. J. (1995). Fruit polyphenol oxidase. New data on and old problem. In C. Y. Lee & J. R. Whitaker (Eds.), *Enzymatic browning and its prevention* (pp. 90-102). Washington: American Chemical Society.

Masia, A. (1998). Superoxide dismutase and catalase activities in apple fruit during ripening and postharvest and with special reference to ethylene. *Physiologia Plantarum, 104*, 668-672. http://dx.doi.org/ 10.1034/j.1399-3054.1998.1040421.x

MMA. (2012). *Plano Nacional de resíduos Sólidos – Versão pós Audiências e Consulta Pública para Conselhos Nacionais.* Brasília. Ministério do Meio Ambiente. Retrieved from http://www.mma.gov.br/port/conama/reuniao/dir1529/PNRS_consultaspublicas.pdf

Møller, I. M., Jensen, P. E., & Hansson, A. (2007). Oxidative modifications to cellular components in plants. *Annual Review of Plant Biology, 58*, 459-481. http://dx.doi.org/ 10.1146/annurev.arplant.58.032806.103946

Nakano, Y., & Asada, K., 1981. Hydrogen peroxide is scavenged by ascorbate-specific peroxidase in spinash chloroplasts. *Plant and Cell Physiology, 22*, 867-880.
Retrieved from http://pcp.oxfordjournals.org/content/22/5/867.short?rss=1&ssource=mfc

Nardi, S., Pizzeghello, D., Muscolo, A., & Vianello, A. (2002). Physiological effects of humic substances on higher plants. *Soil Biology & Biochemistry, 34*, 1527-1536.
http://dx.doi.org/ 10.1016/s0038-0717(02)00174-8

Noctor, G., & Foyer, C. H. (1998). Ascorbate and glutathione: keeping active oxygen under control. *Annual Review of Plant Physiology and Plant Molecular Biology, 49*, 249-279.
http://dx.doi.org/10.1146/annurev.arplant.49.1.249

Pinto, C. M. F., Lima, P. C., Salgado, L. T., & Caliman, F. L. B. (2006). Nutrição Mineral e adubação para pimenta. *Informe Agropecuário, 27*, 50-57.
Retrieved from www.epamig.br/index.php?option=com_docman&task=doc...

Ramakrishna, A., & Ravishankar, G. A. (2011). Influence of abiotic stress signals on secondary metabolites in plants. *Plant Signaling & Behavior, 6*, 1720-1731. http://dx.doi.org/ 10.4161/psb.6.11.17613

Razek, A. E., Treutter, D., Saleh, M. M. S., Shammaa, M. E., Amera, A. F., & Hamid, N. A. (2011). Effect of nitrogen and potassium fertilization on productivity and fruit quality of 'crimson seedless' grape. *Agriculture and Biology Journal of North America, 2*, 330-340. Retrieved from https://www.researchgate.net/publication/215456877_Effect_of_nitrogen_and_potassium_fertilization_on_ productivity_and_fruit_quality_of_'Crimson_seedless'_grape

Rocha, I. P. (2011). Carcinicultura brasileira: processos tecnológicos, impactos socioeconômicos, sustentabilidade ambiental, entraves e oportunidades. *Revista da ABCC*, 13, 36-34. Retrieved from http://abccam.com.br/site/wp-content/uploads/2011/03/carcinicultura%20brasileira%20-%20revista%20abc c%20-%20janeiro%202011.pdf

Sacramento, C. K., Matos, C. B., Souza, C. N., Barretto, W. S., & Faria, J. C. (2007). Características físicas, físico-químicas e químicas de cajás oriundos de diversos municípios da região sul da Bahia. *Magistra, 19*, 283-289. http://dx.doi.org/10.1590/1807-1929/agriambi.v18n08p856–860

Sestari, I. (2010). *Indução de tolerância de frutos às injúrias de frio: aspectos fisiológicos e bioquímicos* (Unpublished doctoral dissertation). Universidade de São Paulo, São Paulo, Brazil.

Sies, H. (1993). Strategies of antioxidante defense. *European journal of biochemistry, 215*, 213-219.
http://dx.doi.org/ 10.1111/j.1432-1033.1993.tb18025.x

Smith, P. G., & Heiser, C. B. (1957). Taxonomy of *Capsicum chinense* Jacq. and the geographic distribution of the cultivated Capsicum species. *Bulletim of the Torrey Botanical Club, 84*, 413-420.
http://dx.doi.org/10.2307/2482971

Sousa, T. P. A. (2010). *Caracterização parcial da peroxidase dos frutos de aceroleira (Malphigia emarginta DC), clones Okinawa e Emepa em três estágios de maturação* (Unpublished master's thesis). Universidade federal da Paraíba, Paraíba, Brazil.

Tzortzakis, N., Gouma, S., Dagianta, E., Saridakis, C., Papamichalaki, M., Goumas, D., & Manios, T. (2012). Use of Fertigation and Municipal Solid Waste Compost for Greenhouse Pepper Cultivation. *The Scientific World Journal*, 1-8. http://dx.doi.org/10.1100/2012/973193

Vaidyanathan, H., Pattathil, S., Chakrabarty, R., & Thomas, G. (2003). Scavenging of reactive oxygen species in NaCl-stressed rice (*Oryza sativa* L.) — differential response in salttolerant and sensitive varieties. *Plant Science, 165*, 1411-1418. http://dx.doi.org/ http://dx.doi.org/10.1016/j.plantsci.2003.08.005

Vaishnava, P., & Wang, D. H. (2003). Capsaicin sensitive-sensory nerves and blood pressure regulation. *Current Medicinal Chemistry-Cardiovascular & Hematological Agents, 1*, 177-188.
http://dx.doi.org/10.2174/1568016033477540.

Valderrama, P., & Clemente, E. (2004). Isolation and thermostability of peroxidase isoenzymes from apple cultivars Gala and Fugi. *Food Chemistry, 87*, 601-606. http://dx.doi.org/ 10.1016/j.foodchem.2004.01.014

Wang, Y. S., Tian, S. P., & Xu, Y. (2005). Effects of high oxygen concentration on pro- and anti-oxidant enzymes in peach fruits during postharvest periods. *Food Chemistry, 91,* 99-104. http://dx.doi.org/10.1016/j.foodchem.2004.05.053

Whiting, G. C. (1970). Sugars. In A. C. Hulme (Eds.), *The biochemistry of fruit and their products* (pp. 1-31). London: Academic Press.

Wojtyla, Ł., Lechowska, K., Kubala, S., & Garnczarska, M. (2016). Different Modes of Hydrogen Peroxide Action during Seed Germination. *Frontiers in Plant Science, 7,* 66. http://dx.doi.org/10.3389/fpls.2016.00066

Worrell, W. A., & Vesilind, P.A. (2012). Solid Waste Engineering (2nd ed.). Stamford: Cengage Learning.

Yahia, E. M. (2011). *Postharvest Biology and Technology of Tropical and Subtropical Fruits: Volume 3 Cocona to Mango.* Cambridge: Woodhead Publishing.

PERMISSIONS

The contributors of this book come from diverse backgrounds, making this book a truly international effort. This book will bring forth new frontiers with its revolutionizing research information and detailed analysis of the nascent developments around the world.

We would like to thank all the contributing authors for lending their expertise to make the book truly unique. They have played a crucial role in the development of this book. Without their invaluable contributions this book wouldn't have been possible. They have made vital efforts to compile up to date information on the varied aspects of this subject to make this book a valuable addition to the collection of many professionals and students.

This book was conceptualized with the vision of imparting up-to-date information and advanced data in this field. To ensure the same, a matchless editorial board was set up. Every individual on the board went through rigorous rounds of assessment to prove their worth. After which they invested a large part of their time researching and compiling the most relevant data for our readers.

The editorial board has been involved in producing this book since its inception. They have spent rigorous hours researching and exploring the diverse topics which have resulted in the successful publishing of this book. They have passed on their knowledge of decades through this book. To expedite this challenging task, the publisher supported the team at every step. A small team of assistant editors was also appointed to further simplify the editing procedure and attain best results for the readers.

Apart from the editorial board, the designing team has also invested a significant amount of their time in understanding the subject and creating the most relevant covers. They scrutinized every image to scout for the most suitable representation of the subject and create an appropriate cover for the book.

The publishing team has been an ardent support to the editorial, designing and production team. Their endless efforts to recruit the best for this project, has resulted in the accomplishment of this book. They are a veteran in the field of academics and their pool of knowledge is as vast as their experience in printing. Their expertise and guidance has proved useful at every step. Their uncompromising quality standards have made this book an exceptional effort. Their encouragement from time to time has been an inspiration for everyone.

The publisher and the editorial board hope that this book will prove to be a valuable piece of knowledge for researchers, students, practitioners and scholars across the globe.

LIST OF CONTRIBUTORS

Majher I. Sarker, Peggy Tomasula and LinShu Liu
Dairy and Functional Foods Research Unit, Eastern Regional Research Center, Agricultural Research Service, US Department of Agriculture, 600 E. Mermaid Lane, Wyndmoor, PA 19038, USA

Kimihisa Itoh, Kohsuke Shimizu and Masahiko Fumuro
The Experimental Farm, Kindai University, Wakayama, Japan

Kazuya Murata, Yuta Nakagaki and Ayaka Shimizu
Faculty of Pharmacy, Kindai University, Osaka, Japan

Yusuke Takata and Shin'ichiro Kajiyama
Faculty of Biology-Oriented Science and Technology, Kindai University, Wakayama, Japan

Tetsuya Matsukawa
The Experimental Farm, Kindai University, Wakayama, Japan
Faculty of Biology-Oriented Science and Technology, Kindai University, Wakayama, Japan

Morio Iijima
The Experimental Farm, Kindai University, Wakayama, Japan
Faculty of Agriculture, Kindai University, Nara, Japan

Hideaki Matsuda
The Experimental Farm, Kindai University, Wakayama, Japan
Faculty of Pharmacy, Kindai University, Osaka, Japan

Shahrokh Khanizadeh, Harvey Voldeng, Xuelian Wang, Allen Xue, Mirko Tabori and Hana Moidu
Ottawa Research and Development Centre, Agriculture and Agri-Food Canada, K.W. Neatby Building, 960 Carling Ave, Ottawa, Ontario, K1A 0C6, Canada

Richard Martin and Allan Cummiskey
Crops and Livestock Research Centre, Agriculture and Agri-Food Canada, 440 University Ave, Charlottetown, Prince Edward Island, C1A 4N6, Canada

Mark Etienne
Dow AgroSciences rental field in Kincardine, Ontario, N0M 1A0, Canada

Ellen Sparry
C&M Seeds, 6180 5th Line, Palmerston, Ontario, N0G 2P0, Canada

Jean Goulet
Centre de Recherche Semican, 1290 Route 116 Ouest, Princeville, Quebec, G6L 4K7, Canada

Vitalis W. Temu, David Johnson and Maru K. Kering
Agricultural Research Station, Virginia State University, Petersburg, Virginia, USA

Victoria Naluyange, Dennis M. W. Ochieno, John Muoma
Department of Biological Sciences, Masinde Muliro University of Science and Technology (MMUST), Kakamega, Kenya

Philip Wandahwa
School of Agriculture, Veterinary Science and Technology, Masinde Muliro University of Science and Technology (MMUST), Kakamega, Kenya

Martins Odendo
Socio-Economics and Statistics Division, Kenya Agricultural and Livestock Research Organization (KALRO), Kakamega, Kenya

John M. Maingi and Omwoyo Ombori
Department of Plant and Microbial Sciences, Kenyatta University, Nairobi, Kenya

Alice Amoding
Department of Soil Science, Makerere University, P. O. Box 7062, Kampala, Uganda

Dative Mukaminega
Faculty of Applied Sciences, Kigali Institute of Science and Technology (KIST), P.O. Box 3900, Kigali, Rwanda

Santhoshkumar K., Prasanthkumar S. and J. G. Ray
Laboratory of Ecology & Ecotechnology, School of Biosciences, Mahatma Gandhi University, Kottayam, Kerala, India

Kimihisa Itoh and Masahiko Fumuro
The Experimental Farm, Kindai University, Wakayama, Japan

Kazuya Murata, Megumi Futamura-Masuda, Takahiro Deguchi, Yuko Ono and Marin Eshita
Faculty of Pharmacy, Kindai University, Osaka, Japan

Morio Iijima
The Experimental Farm, Kindai University, Wakayama, Japan
Faculty of Agriculture, Kindai University, Nara, Japan

Hideaki Matsuda
The Experimental Farm, Kindai University, Wakayama, Japan
Faculty of Pharmacy, Kindai University, Osaka, Japan
3Faculty of Agriculture, Kindai University, Nara, Japan

Biniam M. Ghebreslassie
Department of Horticulture, Hamelmalo Agricultural College, Eritrea
Department of Horticulture, Jomo Kenyatta University of Agriculture and Technology, Nairobi, Kenya

S. M. Githiri
Department of Horticulture, Jomo Kenyatta University of Agriculture and Technology, Nairobi, Kenya

Tadesse M.
Department of Horticulture, Hamelmalo Agricultural College, Eritrea

Remmy W. Kasili
Institute of Biotechnology Research, Jomo Kenyatta University of Agriculture and Technology, Nairobi, Kenya

Ernest Baafi and Joe Manu-Aduening
CSIR-Crops Research Institute, P. O. Box 3785, Kumasi, Ghana

Kwadwo Ofori, Vernon E. Gracen and Essie T. Blay
West Africa Centre for Crop Improvement, University of Ghana, Legon

Edward E. Carey
International Potato Centre (CIP), Ghana

Dhanya Vijayan
Research Scholar, Environment Science Research Lab, St. Berchmans College, Changanacherry, Kerala, India

J. G. Ray
Professor, School of Biosciences, Mahatma Gandhi University, Kottayam, Kerala, India

Daniel Zeru Zelelew
Department of Horticulture, Hamelmalo Agricultural College, Eritrea

Biniam Mesfin Ghebreslassie
Department of Horticulture, Hamelmalo Agricultural College, Eritrea
Deprtment of Horticulture, Jomo Kenyatta University of Agriculture and Technology, Kenya

Gracielle Pereira Bragança and Rosy Mary dos Santos Isaias
Universidade Federal de Minas Gerais, Instituto de Ciências Biológicas, Departamento de Botânica, Laboratório de Anatomia Vegetal. Caixa postal 486.31270-901, Belo Horizonte, Minas Gerais, Brazil

Denis Coelho de Oliveira
Universidade Federal de Uberlândia, Instituto de Biologia, Laboratório de Anatomia e Desenvolvimento Vegetal. Caixa postal 593, Uberlândia, Minas Gerais, Brazil

Laouali Abdou
Université de Diffa, Faculté des Sciences Agronomiques, BP 78, Diffa, Niger

Boubé Morou
Université Dan Dicko Dankoulodo de Maradi, Faculté des Sciences et techniques, Département de Biologie, BP 465, Maradi, Niger

Tougiani Abasse
Institut National de Recherche Agronomique du Niger, BP 429, Niamey, Niger

Ali Mahamane
Université de Diffa, Faculté des Sciences Agronomiques, BP 78, Diffa, Niger
Université Abdou Moumouni, Faculté des Sciences et Techniques, Département de Biologie, Laboratoire Garba Mounkaila, BP 10662, Niamey, Niger

Vitalis W. Temu, Maru K. Kering and Laban K. Rutto
Agricultural Research Station, Virginia State University, Petersburg, Virginia, USA

Soumana Idrissa
Faculté des Sciences Agronomiques, Université de Diffa, BP 78, Diffa, Niger
Institut National de la Recherche Agronomique du Niger (INRAN), BP 429, Niamey, Niger

Fernando Berton Baldo and Adalton Raga
Laboratory of Economic Entomology, Biological Institute Experimental Center, Brazil

Jeferson Luiz de Carvalho Mineiro
Laboratory of Acarology, Biological Institute Experimental Center, Brazil

Jairo Lopez de Castro
Regional APTA, Regional Center of Agribusiness Technological Development of the Southwest of the State of São Paulo, Brazil

Tomoya Oyanagi and Asami Kurita
Graduate School of Environment and Information Sciences, Yokohama National University, Yokohama 240-8501, Japan

Toshihiko Shiraishi
Graduate School of Environment and Information Sciences, Yokohama National University, Yokohama 240-8501, Japan
Faculty of Environment and Information Sciences, Yokohama National University, Yokohama 240-8501, Japan

Hamako Sasamoto
Faculty of Environment and Information Sciences, Yokohama National University, Yokohama 240-8501, Japan
Research Institute for Integrated Science, Kanagawa University, Hiratsuka, Kanagawa 259-1293, Japan

Mei Zhang, Jing-Xiang Meng, Zi-Jie Zhang and Yue Li
National Engineering Laboratory for Forest Tree Breeding, Key Laboratory for Genetics and Breeding of Forest Trees and Ornamental Plants of Ministry of Education, College of Biological Sciences and Technology, Beijing Forestry University, Beijing 100083, China

Song-Lin Zhu
The Forestry Bureau of Xixian, China

Tasnuva Shahrin, Majher I. Sarker, Peggy Tomasula and LinShu Liu
Dairy and Functional Foods Research Unit, Eastern Regional Research Center, Agricultural Research Service, US Department of Agriculture, 600 E Mermaid Lane, Wyndmoor, PA 19038, USA

Yanli Zheng
Laboratory of Food Nutrition and Safety, Tianjin University of Science and Technology, Tianjin, 300457, China

Y. Zubairu, J. A. Oladiran and O. A. Osunde
Department of Crop Production, Federal University of Technology, Minna, Niger State, Nigeria

U. Ismaila
Department of Agricultural Biotechnology, National Biotechnology Development Agency Lugbe Abuja, Nigeria

Francisco Orlando Holanda Costa Filho, Jefania Sousa Braga Amorim, Magnum de Sousa Pereira, Francisca Edineide Lima Barbosa, Rifandreo Monteiro Barbosa, Roberto Albuquerque Pontes Filho and Franklin Aragão Gondim
Instituto Federal de Educação, Ciência e Tecnologia do Ceará (IFCE), Maracanaú Campus, Brazil

Index

www.ingramcontent.com/pod-product-compliance
Lightning Source LLC
Chambersburg PA
CBHW080253230326
41458CB00097B/4439